Implications for Climate Smart Agriculture

Dr. Wajid Hasan, Ph.D. PDF devoted his academic career to Agriculture, particularly in Agriculture Entomology as a Research and Extension. Dr. Hasan is working as a Subject Matter Specialist (Entomology) at KVK Jehanabad, Bihar Agricultural University, Bihar. He was selected for Dr. D.S. Kothari UGC-Post Doctoral Fellow (P.D.F.) in 2010 by the University Grant Commission of India. Dr. Hasan did his doctorate in Agricultural Entomology (2009) from G.B. Pant University of Agri and Tech., Pantnagar. He has made a vibrant link with farmers by advising them plant protection measures when needed for which he has received 11 awards from several societies in the outstanding contribution of the relevant discipline, published 48 research papers, 20 books, 25 book chapters, 38 popular articles, 6 success stories, 2 Practical Manuals, 6 success stories and 47 delivered lecture in different seminar/symposia/conference/workshop and 7 radio talks. He has organized 11 International Conference as Chief-Organizing Secretary, Co-Organizing and Executive Organizing Secretary. Dr. Hasan organized 244 (9896 trainees) Training for Practicing Farmers, Rural Youths, Extension Functionaries, Two Skill training (200hrs), 26 On-Farm Trials (OFT) conducted. Dr. Hasan organized 394 (17896 trainees) Training for Practicing Farmers, Rural Youths, Extension Functionaries, Three Skill training (200hrs), 34 On-Farm Trial (OFT) conducted. Dr. Hasan has a life membership of 5 scientific societies. He is working as Editor-in-Chief in the International Journal of Agricultural and Applied Sciences (IJAAS) and in the Editorial board in 5 scientific International Journals.

Dr. Sachin G. Mundhe, is presently working as a Subject matter Specialist (Agricultural Meteorology) in the Krishi Vigyan Kendra, Durgapur (Badnera), Amravati, Maharashtra. He completed his Ph.D.Degree in Agricultural Meteorology from Bidhan Chandra Krishi Viswavidyalaya, Mohanpur, Nadia, West Bengal, India. He has to qualified the ASRB NET exam in 2018. He has published 7 research papers in journals of national and international repute. He has also presented his research work in 11 national and international conferences; he has received 4 awards from different societies for the outstanding contribution of the relevant disciplines. 5 Book chapters, 34 popular articles, 1 radio talk.

Dr. Abdul Majid Ansari, *Ph.D.*, devoted his academic career to Agriculture, particularly in the field of Horticulture as a Research and Extension worker. Dr. Ansari has been working as a Junior Scientist-cum-Assistant Professor (Horticulture) at ZRS, Chianki, Birsa Agricultural University, Ranchi. He was selected for the best Research Worker in 2021 by BAU, Ranchi. Dr. Ansari did his Doctorate in Vegetable Science (2012) from G.B. Pant University of Agri & Tech., Pantnagar. He has contributed significantly in the field of vegetable breeding and developed a large number of advance lines of eggplant, tomato and vegetable peas and two promising genotypes has been identified for the release in the state of Jharkhand. He has been maintaining a large number of germplasms of different vegetable and fruit crops at the station. He has also made a vibrant link with farmers by advising them to promote scientific management of horticultural crops. He has received 10 awards from several societies in the outstanding contribution of the relevant discipline, published 39 research papers, 3 review articles, 32 oral/poster presentation, 3 lead papers presentation, 5 books, 4 book chapters, 16 popular articles, 4 leaflet/folder, reviewed several research papers of three international journals *viz., Scientia Horticulturae, Fruits Journal* and *Agronomy Journal* and delivered 92 lectures in the different seminar/symposia/conference/workshop. Dr. Ansari has been associated with many externally funded projects in different capacities and Nodal Officer {EAP} in the multi crore World Bank-ICAR funded project named NAHEP-CAAST-IFS, BAU, Ranchi. He has a life membership of 12 scientific societies. He is working as a member of editorial board in the 8 National/International Journal of Agricultural and Allied Sciences and participated in the 12 refresher courses of horticultural importance.

Ms. Shivani Kumari is presently a Research Scholar at the Department of Agronomy at SASRD, Nagaland University, Nagaland. She passed her under graduation degree in Agriculture (Hons.) in 2013-2017 from Lovely Professional University, Phagwara, Punjab, and post-graduation in (Ag.) Agronomy from SHUATS, Prayagraj (U.P.) in the first class with distinction. She is specialized in crop and nutrient management. She has attended a number of workshops and training programs and participated in national/international conferences/seminars *etc.* She has authored and contributed to several book chapters, research papers, review papers, and popular articles to date.

Implications for Climate Smart Agriculture

– Editors –

Dr. Wajid Hasan

Subject Matter Specialist (Entomology),
Krishi Vigyan Kendra (BAU Sabour), Jehanabad, Bihar

Dr. Sachin G. Mundhe

Subject Matter Specialist (Agricultural Meteorology)
Krishi Vigyan Kendra, Durgapur (Badnera), Amravati, Maharashtra

Dr. Abdul Majid Ansari

Junior Scientist-cum-Assistant Professor (Horticulture),
Zonal Research Station (BAU, Ranchi), Chianki, Jharkhand, India

Ms. Shivani Kumari

Research Scholar at Department of Agronomy at SASRD,
Nagaland University, Nagaland

BIOTECH

2022

BIOTECH BOOKS®

ISBN 978-81-7622-501-4

Published by: **BIOTECH BOOKS®**
4762-63/23, Ansari Road, Darya Ganj,
New Delhi - 110 002
Phone: 09899295259, 011-43003222
E-mail: biotechbooks@yahoo.co.in
Website: www.biotechbooks.in

Typeset at:**Classic Computer Services**
Delhi - 110 035

Printed at:**Replika Press Pvt. Ltd.**

PRINTED IN INDIA

Preface

Climate Smart Agriculture (CSA) is an approach to manage agricultural systems to respond effectively to climate change in order to help the farmers. Climate change was a major factor affecting the food and farming system. The CSA approach fulfilled adaptation to change in the climate and reducing greenhouse gas emissions wherever possible. Climate change mitigation encompasses the actions being taken, and those that have been proposed, to limit the magnitude and/or rate of long-term global warming-induced climate change. Natural systems, natural health, human health, and agricultural production have been severely affected by catastrophic changes. According to an estimate of FAO, The population of the world will be increased by one-third of the present population by the year 2050 and the required food for food security would be increased by 60 per cent. The cumulative effect of climate change over the last decades also gave an adverse effect on productivity. Besides this, one important concern is having a high professional risk of being a farmer. Farmer in India considered farming to be very risky and farmers tend to migrate their profession due to the unsecured nature of the weather impact on the farming system. This has a direct impact on the socio-economic process. Climate smart agriculture is a strategy that helps drive efforts to adapt and reorient agricultural systems in order to promote production and food security in a changing climate in an efficient and sustainable way. 'Agriculture' includes the production of crops and cattle, fisheries, and forest management. Agricultural production has to be boosted to feed the world's growing population. Thus, the combined efforts of crop modeling experts, agronomists, field pathologists, breeders, physiologists, and molecular biologists can develop joint stress-tolerant crops that perform well under field conditions.

The idea of climate smart farming (CSA) emerged as a route of transforming and re-orienting farm systems to support food safety in the face of global environmental change. Drastic changes in precipitation and temperature patterns are threatening agricultural production and increasing human vulnerability that depend on Agriculture, including most of the poor world, to their livelihoods. Climate change has an impact on food markets, threatening the provision of food. Threats can be reduced by increasing farmers' adaptive capabilities and by increasing the resilience of agricultural production systems and their efficiency in resource utilization. CSA distinguishes itself from business-to-business approaches by highlighting the ability to implement flexible, context-specific solutions supported by innovative policies and finance. However, little empirical data has been made available to the developers, exacerbating ideological disagreement among the development practitioners. In the decades to come, the challenges of agriculture, food security, and livelihoods will be seriously affected by climate change and its adverse effects. Sustainable agriculture and CSA are highly complementary strategies. The strategies and projects that are included under CSA are Integrated, farmers' knowledge availability, renewable agricultural systems technology, genetic modification, crop relocation, institutional analysis, efficient pesticide management, crop modeling, forecasting, and integration of forecasting and modeling. Sustainable agriculture is going to have a significant impact on food security. In this way, climate change affects different types of crops in different ways. An effective weather forecasting system and early warning systems can prevent climate damage. Applying combined approaches is critical for addressing climate and food security, and there are so many relevant opportunities. Proper agriculture policy will alleviate short-term food insecurity and poverty while it will improve the environment in the long term through agriculture CSA pathways. With per capita income growing and health-conscious population growth rising, demand for horticultural products is on the rise, which is expected to increase further, requiring more production. However, both in terms of quality and price, production has to be competitive. Therefore, the opportunities that exist need to be exploited and gains have to be maintained. Creation of improved varieties with high efficiency, quality characteristics, disease, and pest resistance, and abiotic stress tolerance. Technologies could increase water and nutrient efficiency and variability in yield, quality, and post-harvest crop losses as well. Efforts to ensure the timely availability of quality seeds and planting materials would also be important.

The Publication Implications for Climate Smart Agriculture is expected to benefit researchers, scientists, teachers, academicians, policy planners, progressive farmers, and students and shall act as a good reference base for future advancement in the field. We express our sincere thanks to all contributors for their cooperation and for sharing their valuable experiences and great thoughts in making this book informative.

Wajid Hasan

Sachin G. Mundhe

Abdul Majid Ansari

Shivani Kumari

Contents

Chapter 1

Climate Smart Agriculture and Hi-Tech Farming

***Ambreen Bano*[1]**, *Mariya Ali*[2], *Anmol Gupta*[1], *Neelam Pathak*[3] and *Wajid Hasan*[4]**

[1]IIRC-3, Plant-Microbe Interaction and Immunobiochemistry Lab, Department of Biosciences, [2]Department of Bioengineering, Integral University, Lucknow – 226026, Uttar Pradesh
[3]Department of Biochemistry, Dr.RML Avadh University, Faizabad – 224001, Uttar Pradesh
[4]Krishi Vigyan Kendra, Jeahanabd, BAU Sabour, Bihar

The idea of climate smart farming (CSA) emerged as a route of transforming and re-orienting farm systems to support food safety in the face of global environmental change. Drastic changes in precipitation and temperature patterns are threatening agricultural production and increasing human vulnerability that depend on Agriculture, including most of the poor world, to their livelihoods. Climate change has an impact on food markets, threatening the provision of food. Threats can be reduced by increasing farmers' adaptive capabilities and by increasing the resilience of agricultural production systems and their efficiency in resource utilization. CSA distinguishes itself from business-to-business approaches by highlighting the ability to implement flexible, context-specific solutions supported by innovative policies and finance. However, few empirical data has been made available to the developers, exacerbating ideological disagreement among the development practitioners. Horticulture is an integral part of food safety and nutrition. It is a key component of the economic security of the stakeholder. Hi-tech horticulture is a technology that is modern, less environmentally- dependent and capital intensive, but capable of increasing farmers' productivity and incomes. In the new era of climate change, hi-tech horticulture has become necessary to maintain the productivity and economic stability of Indian farmers. Modern horticulture which is Hi tech farming is beneficial to plant species, the agricultural sector and society in general. The present article examines various aspects of CSA and hi-tech farming that suggest future prospects.

Introduction

Agricultural growth policies have moved from the promotion of one-size-fits-all productivity-enhancing technology to the promotion of better agricultural methods that take into consideration of livelihoods and environmental effects (Sapkota *et al.*, 2019, DeFries *et al.*, 2010, Perfecto *et al.*, 2010, Branca *et al.*, 2010). Lately solution to the Integrated Sustainability Plan is 'Climate Smart Agriculture' (CSA). CSA relates to agricultural systems that improve food protection against Climate Change, enhance the ability of farmers to respond to climate change impacts and reduce climate change where possible (Bogdanski *et al.*, 2010). Today's world, the trend of globalisation and the explosion of technology, hi-tech farming is a key factor that defines the life and income of the farmer. It also helps to enhance the competitive power of agricultural products in the foreign market and to play a role in the socio-economic growth of high-tech agriculture. Then, assessing the reality of the growth of hi-tech agriculture in India

Climate Smart Agriculture (CSA)

It is an approach to the implementation of new agricultural practises to increase yield, Climate change adaptation and mitigation (Lipper *et al.*, 2014), the concept first coined in 2009 and described as "agriculture that sustainably increases productivity, resilience (adaptation), reduces or eliminates greenhouse gas (GHG) (mitigation) and increases accumulation" (FAO 2018). As coined, identified The climate smart agriculture and forestry are an important route towards achieving sustainable development objectives, and presented by the FAO at the Haager Conference on Agriculture, Food Security and Climate Change in 2010. The convergence of sustainable growth in three dimensions (economic, social and environmental) is a synchronic solution to climate challenges and food security. The key priority of CSA is to ensure growth and food protection.

CSA consists of three major pillars:

1. Sustainably growing farm production and revenue
2. Adapting and developing resistance to climate change
3. Greenhouse gas pollution reduction and/or elimination (van Wijk *et al.*, 2020)

These pillars combine economical, social and environment issues into a shared approach to solving challenges to food safety and climate change (Hafezi *et al.*, 2017). CSA is a methodology to implement theoretical concepts and to use financial resources to promote sustainable agricultural production (FAO, 2018). Adaptation is a mechanism that entails managing the risks faced by global change, such as variance (Alam *et al.*, 2017). The analysis of public problems induced by global warming focuses on natural effects, by analysing household or person behaviour while faced with current temperature variations. Variation of global warming is an important problem for experts, programme supervisors and administrators who are mindful that climate change is currently taking shape and is Community and ecological sustainable development are detrimental. Sustainability is a key term for multinational environmental strategy and decision-making of the 21st century (Braimoh and Osaki, 2010). Effective adaptation, which will actually be less than the expense of the consequences that will exist without growth, may be the most important issues of global warming.

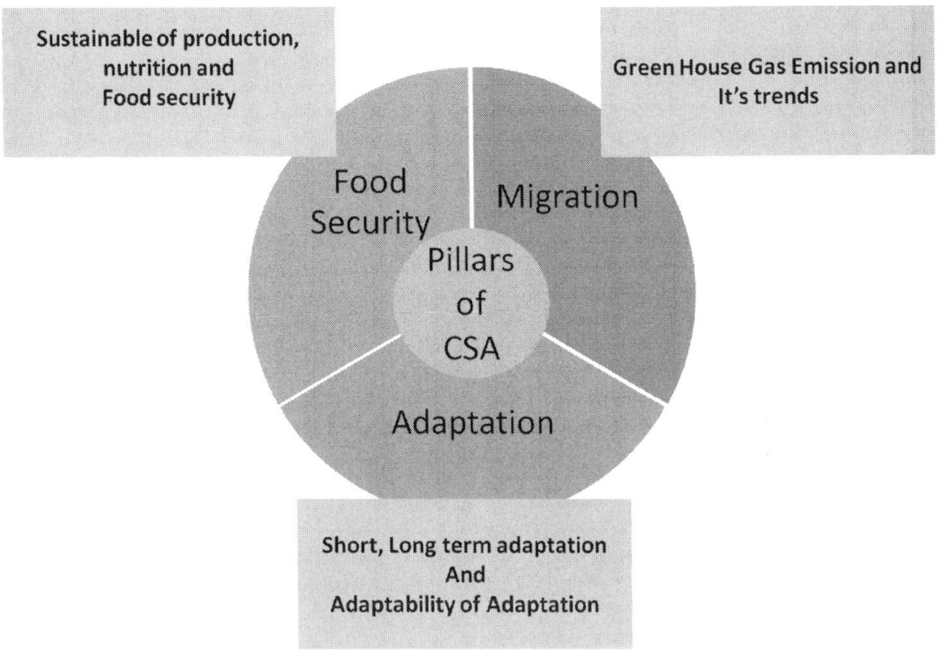

Figure 1.1: Pillars of Climate Smart Agriculture.

CSA is a comparatively recent strategy that helps to improve agricultural productivity and incomes of poor households by maintaining improved farming practises and reducing greenhouse gas emissions. In the evolving environment, CSA is an important solution to improving food productivity and feeding people around the world. The idea that agriculture systems can simultaneously be constructed and applied to boost food safety, support adaptation to climate change and offer additional benefits has been created, even as a paradigm. However, the CSA definition is still emerging and there is minimal study that examines the linkages between CSA and sustainable agriculture and how CSA activities can be encouraged food safety and sustainability goals. Given the ability to achieve the objective of sustainable agricultural production by properly applying the CSA, there is a weak correlation between the components of the CSA at the field level. More specifically, the term is often confused by multiple types of stakeholders. A deeper understanding of the CSA definition and its ties with sustainable agriculture and the mechanisms to help CSA would make for a climate change approach more integrated and a greater implementation of CSA activities to meet food security and sustainable goals for development. Also important to improve and disseminate CSA approaches at the field level in order to create a viable and better future.

CSA is a mixture of practises that have traditionally been utilised in agricultural, Climate change, conservation, agriculture and other sectors. However, the agricultural-climate partnership is weakly known, in particular the industry's dual existence (*i.e.* agriculture systems are a major contributor to global anthropogenic GHG emissions and are simultaneously vulnerable to climate change shocks and stresses. By 2007, statistical evaluations by the Intergovernmental Panel on Climate Change (IPCC. 2012) and policy analyses carried out by development agencies were evident in the dual partnership between climate change and agriculture. The essence and extent of the two-fold effects have become clearer: Agricultural systems are expected to lead to and affect climate instability and transition and have a significant impact on developed countries. The IPCC (2011) in particular concluded with a sense of global urgency that GHGs (CH_4 and N_2O) are growing from agricultural land and that 'interactions can be simultaneously, but in their spatial and geographical characteristics between mitigation and adaptation are occurring in the agricultural sector'. Given the minimum progress made in integrating climate change into agriculture, integrated potential intervention has become imperative: new methods have been required for the shift to climate resistant agricultural production. The global food crisis of 2007–2008 posed a variety of policy concerns concerning the sustainability of agricultural systems in developed countries. For example, food and energy price rises have negatively impacted Consumers with low income and disabilities (Beddington *et al.*, 2012). Primary crops like wheat, maize and soya have grown dramatically, causing social and economic conflicts in poor countries that import foods (Addison *et al.*, 2011). Among the obstacles to exchange and trade, external and internal drivers such as rural development campaignsbiofuel subsidies, poor agriculture policies, land tenure and property rights. The survival of farmers in particular, affected by natural disasters and soil degradation (UNDP, 2013). Development organizations like the Food and Agriculture Organization of the United Nations have already decreased food supply (FAO) and World Bank have raised fears that poverty mitigation initiatives, particularly for rural poor, have been undercut (FAO 2018). The global food crisis of 2007–2008 revealed that food safety remained a precarious problem for the poorest, whereas the sustainability of agricultural development processes in developed countries required a fresh approach to the various interlinked challenges. FAO and World Bank have been led in 2010 (FAO 2018) to formally create the CSA as an approach to transforming trade and subsistence agricultural systems in the developed countries because it needed more sustainable economies, where agriculture was part of the response to climate change – Multilateral donors' core community focus. To recognize the necessity of reconciliation the various changing views of The CSA held The Hague, Netherlands (31 October-5 November 2010), the first Global Policy Conference on that topic, hosted by developing and industrialized countries' governments, together with the FAO and World Bank. At the Global Food Protection and Climate Change Conference, regional organizations and foreign, partners from nations, private sector, Philanthropic, NGOs and science institutions have worked to draw up an action plan (Neufeldt *et al.*, 2013). The Roadmap on Agriculture, Food and Climate Change, with the support of ministers, calls for immediate action to target poor rural people in the world, particularly women farmers (WB, 2010). Very first time, the Roadmap acknowledged agricultural inter linkages, food protection and ministerial Climate Change and recognized the need for integrated policies for CSA. In order to further improve the CSA strategy and collaboration Vietnam (2012) and South Africa conferences have been held (2013). In the Netherlands (2011), USA (2013) and France, successive science seminars were also held (2015). The Global CSA Alliance in September 2014 (GACSA) was introduced at the UN Secretary-Climate General's Summit, with a priority on awareness creation and coordination of global cooperation. To date, the CSA Global Agenda has centred on building a knowledge base on what comprises CSA and how collaborating organizations might foster CSA imperatives.

Agriculture Climate and Climate change may affect the dispersal and output of crops and raise the risks linked to agriculture. Crop yields also had negative effects, emphasising that adaptive measures are necessary (Nelson *et al.*, 2010). Although there can be improved production conditions for some regions, mostly in moderate latitude, the global climate will decrease cereal production by 1 per cent to 7 per cent by 2060 (Parry 2007). The impact of grain, irrigated versus rainfed crops and geographical area [3] may also differ significantly. At least 22 per cent, with up to 56 per 100 of the area of Sub-Saharan Africa' s land, are expected to be harmful to climate change by 2050 (Campbell *et al.*, 2011). Impacts could be comparatively minor by 2050, but are projected to escalate steadily in last half century (Easterling *et al.*, 2007). In addition to shifts in crop yield and output correlated with change in climate, other fields that involve measures for adaptation. Climate induced water shortage due to modifications in time and space rainfall distribution may contribute to escalate competitiveness within and with other sectors of agriculture (Hanjra *et al.*, 2010). In addition, solving these issues will entail the alteration of physical structures, for example irrigation and alteration of the configuration and position of storage facilities (Antle *et al.*, 2010). The elevated danger of flooding and droughts and the changing of fire regimes cause additional risks to farming production (Falloon and Betts 2010, Peterson *et al.*, 2011). Climate regime uncertainty may affect the decision of farmers and if they are investing in the appropriate inputs and services for agricultural fields. In the meantime, Some 30 per cent of the pollution from land use is of world greenhouse gas. An approximate 18 per cent originate from Shift in land use (mainly deforestation) and 10 per cent to 12 per cent more in cultivation (soil erosion and tillage, nitrogen fertilizer, and paddy rice cultivation). Given the total global greenhouse gas emissions and approximately half of agricultural emissions from entry fermentation and field clearing, animal production (including digestion and feed production) contributes about 14.5 per cent and the loss of manures and forest cover. Use of land is the main carbon change possible in several nations. In fact, only sequestrations of carbon on land activities actually provide the prospect elimination of large-scale greenhouse gases (GHGs) from the environment by photosynthesis in soils and perennial plants and carbon sequestration. Carbon accounts for agricultural soil 89 per cent of the technological possibility of sequestration, reflecting projected potential annual CO_2 emissions of between 5.5 and 6 gigatons, approximately equivalent to the overall annual exposure of Global Emission Agriculture. Important quell of the pollution mitigation include better feed and waste management schemes, more effective usage of fertilisers, reduction of degradation and wetland conversion, and regeneration of depleted land (Smith *et al.*, 2007). Changes in land use and land use often can mitigate changes in albedo, evapotranspiration, ground humidity and temperature in the local and regional climate (Desjardins *et al.*, 2007). In addition, certain adaptation steps have major co-benefits of mitigation in agriculture. Increased soil organic matter, for example, increases adaptive ability by growing capacity of soil water and soil productivity, Although carbon is still being sequestered (Falloon and Betts 2010). The links between climate change and farming have been increasingly recognised by the science community, foreign and national public policy initiatives to promote adaptation and mitigation of agricultural products have increased. Since the 15th Conference of the Parties in Copenhagen, aid to increase agricultural importance for climate change negotiations has progressively improved. The first Day of Rural and Agriculture Growth was formally on the agenda for the 2012 conference of the Science and Technical Advice Subsidiary Body (SBSTA) and a related high-level international mechanism to discuss food protection, agriculture and climate change initiated in 2010 in The Hague (Beddington *et al.*, 2012). In addition, the mechanisms have increasingly centred on adaptation, through the establishment of the National Adaptation Program of Action (NAPA) programmes and the more specifically on agriculture in developing countries Adaptation fund. Since 2010, the creation of this theory and the usage of the word itself International organisations were led, in particular the Food and Agriculture Organisation of the United Nations (FAO) and the World Bank (FAO. 2018). The Consultative Group on Foreign Agricultural Research (CGIAR) gave international research community leadership as the concept matured (Moorhead 2009, Vermeulen *et al.*, 2010). Although a novel term for climate change and agricultural production societies, climate smart agriculture incorporates several sustainable land management activities on the field and on the agriculture, already in the literature and commonly practised, such as conservation tillage, agro-forestry, management of residues and others (Pye-Smith 2011). Much of the emphasis of climate smart agriculture has been on the application of these fields and on the forms in which they can be strengthened in the light of climate change. Many others are involved in the debate on farming activities for adaptation and mitigation of climate change, albeit without utilizing climate smart terms (Delgado *et al.*, 2011, Lal *et al.*, 2011).

Pros and Cons of CSA

CSA has several advantages: CSA innovations achieve at least two benefits between efficiency, resilience and prevention, where productivity is a concern for subsistence agriculture in developed countries. Development may be assessed using measures nestled within these large CSA groups compared to a fair baseline. For example, increased efficiency could be calculated as returns, revenues or domestic costs. CSA seeks to exploit and reduce synergies trade-offs through its foundations. CSA is a particular context of both space and time: CSA innovations are socially and culturally suitable for the area in which they are to be implemented. CSA tackles risk: CSA solutions fix environment or weather-related threats when enhancing food protection. Risks discussed can involve catastrophic disasters (such as floods) and risks slowing down (such as delayed onset of seasonal rains).CSA developments can help to boost the effect of these threats both short-term (plant, season, hectare, *etc.*) and long-term output increases (decrease the variability in production over time in spite of climate change).Provided that the biophysical and social dynamics are evolving, whether the CSA technology is or is not complex. In 20 years, what is today CSA at a place will not be CSA at the same place.

The most successful way is to use renewable energy, implement agriculture for recycling, conservation of forests and water, reforestation, *etc.* Some of the solutions to reduce the effects of climate change are the alteration of existing horticultural practices and greater use of greenhouse technologies in order to preserve production. The key strategies for addressing this challenge would be the production of new horticultural crop cultivars tolerant of high temperatures, short term and producing good yield under stress conditions, resistant to pests and diseases, as well as the adoption of hi-tech horticulture and judicious management of natural resources.

Hi-Tech Farming

Hi-tech horticulture is the application of new technology, which is resource intensive, is less reliant on the environment, and has the capacity to grow productivity and production efficiency. On the other hand, Precision Farming requires implementation of technology and concepts to temporal and control spatial uncertainty related to everything facets of horticultural production in order to increase crop efficiency and environmental quality. Hi-tech horticulture includes a wide range of practises, Micro-irrigation, soil-protected/greenhouse culture and fertilizer-based leaf nutrient management, micro-propagation, in-situ moisture-preservation mulchings, germplasm biotechnology, high-density semen, genetically engineered crops, vermiculture, bio-fertilizing use, hi-tech, green-food, soilless crop control and more (Figure 1.2).

Concepts of Farming with High Technology Methods

Hundreds of methods are offered for high-tech agriculture, Just as sustainable and organic agriculture can be carried out in endless ways. High-tech farming can be understood as precision farming (PF). As regards all such terms, it is a technical mark here also with satellite-driven farming, combined with high tech seed and reaping machines, using geographic information systems and other (remote) sensing methods. PF is not just one technology but a set of technologies which can be embedded in a system (Batte and Arnholt 2003) consisting in the majority of cases of existing crop races, which are transgenic features developed by experienced seed companies. New zinc-finger technology that allows transgenes to be injected in higher plants in an accurate location will improve the new breeding. It had been possible under the name "gene targeting," but a much smaller frequency was less manageable (Shukla *et al.*, 2009). Farmers are likely to implement different component technologies in the near future, based on their requirements and additional site-specific features market. PF is a collection of agricultural techniques, introduced mainly in areas which have become more treeless (formerly dominated by prairie vegetation) and fertile North American soils, suitable for modern heavy machinery agriculture and high levels of production, but now quickly expanding worldwide in countries with strong agricultural infrastructure capital. PF can also be used relatively well for remote control limited organised agriculture, such as enhanced discrimination against crops in India (Hyderabad) to differentiate between ca by satellite Remote Sensing (Sai and Rao 2008). A dozen types of plants of agriculture. Many papers of research have nicelyreveal, with vivid images, that satellite imaging is in vogue and can provide ultra-exact imagery in the future (Figure 1.2). However, the operation still focuses on science, and implementations have not yet achieved a large variety, as GM crops (Clive 2009). Remote sensing, combined with intelligent predictive data processing, can promote a significant phase from consistent field management to site-specific management.

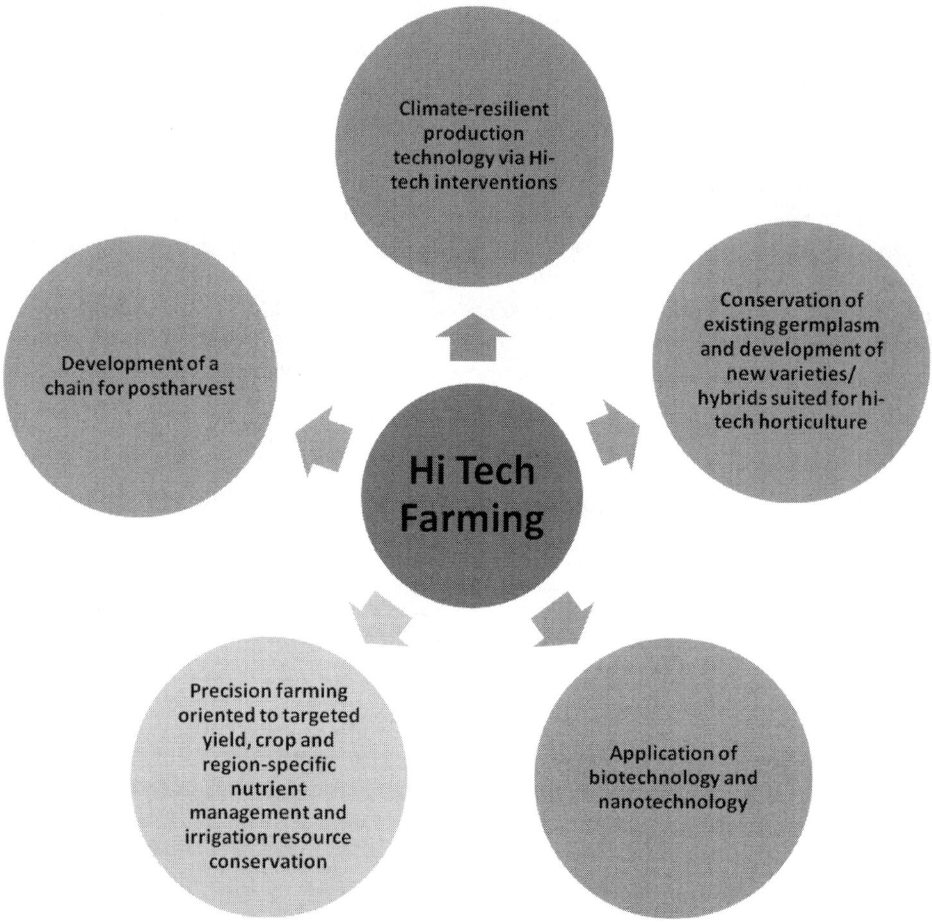

Figure 1.2: Hi-Tech Farming Scopes.

Hi-Tech Farming in India

In the recent period, horticulture is not just a way of diversification, it is an important part of food and nutrition protection and is also a vital element of economic growth. The adoption of horticulture has brought development to several nations, including Karnataka, Maharashtra, Andhra Pradesh and Kerala (Kumar and Kumari 2019, Singh 2014). Considering the burgeoning rise in population, there is a great deal of strain on natural capital, added to global warming and climate change, a reduction in land ownership and a strong demand for premium fresh horticultural goods. The condition demands a move towards new crop production technology in which high-tech horticulture has already taken the lead. Phrase hi-tech horticulture applies to precise processing methods for the effective use of inputs at the required time and quantity to optimise yield and quality in various horticultural crops. It is the introduction of any technology that is new, less environmentally-dependent, capital-intensive and has the potential to increase production and quality of horticultural crops (Chadha, 2001). Hi-tech horticulture has also been described as a new, environmentally sustainable, appropriate, intensive technique to allow farmers to achieve high productivity and high quality products to make more money possible. It is a chain method for growing vegetables, berries, flowers as well as spices with a proper relation between the collection of variety/seeds and the finished product through the phase of modern crop processing techniques through post-harvest management techniques. Hi-tech horticulture has resolved the agro-climate barrier that allows most vegetables and other horticultural items accessible to customers across all calendar months, though at a premium price. Technologies encapsulated in high-tech horticulture involve the usage of genetically engineered (GM) crop varieties originating from organic farming.

Technology and genetic modification, micro-propagation, combined nutrients, water, weeds and pest control, safe agriculture, organic farming, usage of advanced immuno-diagnostic tools for rapid diagnosis of viral diseases, post-harvest innovations, including cold chain (Chadha, 2001).

Hi-tech Cultivars of Farming

Crops Traditional breeding in horticultural crops, particularly fruit trees and planting crops, is difficult and time-consuming. Molecular methods can be used to minimise the time lag and to induce desirable characteristics in the varieties. There is a great deal of space for somoclonal varietal production in horticultural crops. Somoclonal variance offers a modern and alternative method for breeders to achieve genetic diversity reasonably easily and without any advanced technologies in horticultural crops (Krishna *et al.*, 2016). Gene-editing is an emerging field that can be investigated. Traits that need to be targeted include dwarfing, herbicide tolerance, insect resistance, long shelf life and increased nutritional consistency.

Nutrient Management in Hi-tech Farming

Horticulture Hi-tech horticulture depends heavily on the sound application of irrigation and nutrients to horticultural crops. The usage of drip irrigation has resulted in large yields and increased production of fruit (banana, grape, papaya, pomegranate, mandarin, coconut, arecanut and cashew), vegetables (cauliflower, cabbage, gherkin, tomato, baby-corn), cut flowers (carnation, rose and gerbera) and planting crops (coffee, tea and rubber). Plans for fertilisers and leaf nutrient guides have been produced for many horticultural crops (Planning Commission, 2007). However, with advances in biotechnology, planting methods and innovative fertiliser strategies, it is important to review the prescribed doses and schedules. The principles of site-specific, orchard-specific, tree-specific and phenopha-specific application of nutrients are embedded in empirical discourse (Table 1.1).

Table 1.1: Nutrient Removal by different Horticultural Crops

Sl.No.	Crop	Yield ($t\ ha^{-1}$)	Uptake ($kg\ ha^{-1}$)		
			Phosphorus Oxide	Potassium Oxide	Nitrogen
1	Citrus	30	60	350	270
2	Papaya	50	23	130	90
3	Pineapple	50	55	350	185
4	Grapes	20	60	220	170
5	Passion fruit	15	15	75	60
6	Apple	25	45	180	100
7	Banana	40	60	1000	250
8	Mango	15	25	110	100

Hi-tech Propagation

Availability of quality planting content is the essential need to broaden the true horticulture industry. However, the unorganised sector accounts for more than 60 per cent of planting inventory. Furthermore, most nurseries growing horticulture planting materials are engaged in ornamental plant processing and are clustered in the vicinity of towns and cities. As per the Star Rating Scheme of the National Horticulture Board (NHB), of the 689 nurseries ranked none were rated 5-star and only 25 were rated 4-star (www.nhb.gov.in/nursery). This represents the low condition of the acceptance of new technologies by horticultural nurseries. Micro-propagation is an essential method for the efficient development of a large number of plants. It is now being used in crops such as banana (*Musa* spp.), tomato (*Solanum lycopersicum*), chilli (*Capsicum ann*), bell capsicum (*Capsicum ann*) and other ornamental plants. There is considerable scope for micro-propagation of crops, particularly ornamental crops, in India. Protocols also need to be more streamlined in order to increase performance (Chadha, 2001; Sahiijram and Rao, 2017). Plug-plant propagation of horticultural crops is taking root in India, for example, on a wide scale in Chhattisgarh, where it runs. Plug-plant propagation is beneficial for the development of stable, virus-free seedlings, which minimise the pre-harvest time, ensure uniform crop growth and promote harvesting. Vegetables (tomato, chilli, bell capsicum, cole (Brassicaceae), brinjal (*Solanum melongena*), cucurbits (Cucurbitaceae), celery (*Apium graveolens*), parsley (*Petroselinum crispum*), lettuce (*Lactuca sativa*), beet root (*Beta vulgaris*), and onion (*Allium cepa*); fruit strawberry (*Fragaria ananassa*), Papaya (*Carica papaya*), raspberry (*Rubus*) and tomato (*Solanum betaceum*) (Chadha *et al.*, 2017). In India, crops such as guava, mango and grapes have revolutionised the planting and development scenario. Hi-tech methods are effective for greasing. Computerized commercial grafting is possible in Japan for the mass processing of some plants (Chadha, 2001). Such devices can be built natively. The foot-operated graft machines are currently being used in India. Grafting in vegetable crops, particularly watermelon (*Citrullus lanatus*), capsicum and brinjal, is a new hi-tech process. In countries such as Chhattisgarh, large-scale vegetable grafts are made.

Hi-tech Farming Production

Protected cultivation of horticultural crops is the ultimate method of high-tech horticulture in India today. Protected cultivation of horticultural crops is popular in most of India. Among the numerous covered systems, completely automated polyhouses help to be cultivated even under unfavourable climatic conditions. However, they are expensive and commercially feasible only in a small portion of India. Cost-effective systems, such as net houses and low-lying poly-tunnels, are common and increasingly growing in proportion. While the initial costs are substantial, the high value of high-quality output eventually brings a much better income for farmers. These are useful for extracting sufficient yields of higher quality in regular climatic aberrations. Protected cultivation offers space for less land-dependent cultivation. The usage of coco-peat and vermin-compost helps not only improve yield but also minimise soil-borne diseases. The protected cultivation of horticultural crops in India is currently commercially practised for plant breeding and industrial cultivation. Fruits in India, fruit crops are conventionally grown and the output potential of such crops is well below the ability that can be improved by hi-tech approaches. These approaches include high density planting, soil and leaf analysis-based integrated nutrient control, fertilisation, micro-irrigation, proper integrated pest management, weed and disease management, appropriate mechanisation and primary technologies such as pre-harvest luggage and growth regulator usage. In India, hi-tech manufacturing has changed the scenario of fruit crops such as bananas, grapes (*Vitis vinifera*) and pineapples (*Ananas comosus*). One of the most significant innovations for fruit orchards is high density planting (HDP). At the end of the Tenth Five Year Plan, HDP technology was effectively evaluated for pineapple, strawberry, papaya, mango, guava and citrus (Planning Commission, 2007). You may also connect Aonla and sapota to this list (Chadha *et al.*, 2017). High density planting has the ability to raise yields and reduce the expense of growing a variety of horticultural crops. It has already been effectively exploited in orange (*Citrus sinensis*), kinnow, banana and pineapple, to some degree, in mango (*Mangifera indica*) and apple (*Malus pumila*) (Dalvi *et al.*, 2010). There are many ways to accomplish this goal, such as the selection of dwarf cultivars and scissors, the usage of dwarf rootstocks/interstocks, pruning and growth regulators for canopy control (Haldankar *et al.*, 2015). Constraints in HDP are poor availability of dwarf varieties planting material, lack of standardized technology, higher initial establishment cost, professional and scientific competency, and intensive management to avoid negative effects of high density (Goswami, 2014).

Vegetables

It requires not only the safe cultivation of vegetables under greenhouse, net house or polyhouse, but also the usage of precision farming technologies. Integrated

Nutrient Management (INM), Finishing and Mulching. High-tech vegetable production gives the growers distinct advantages of efficiency, sustainability and a desirable selling price. High-quality crops, such as capsicum, cherry tomatoes and tomatoes, are mostly produced either in polyhouses or in net-houses. Net rooms, which not only increase yield and quality of production but also save a significant amount of water, are favoured for summer vegetable cultivation (Chadha, 2001). During the off-season season, vegetable farmers can dramatically increase their income by utilising a range of systems, ranging from basic shade nets to high-tech automated polyhouses (Rai, 2012). Protected cultivation with precision farming is the leading hi-tech solution to vegetable cultivation. At high altitudes, particularly in Ladakh, polyhouses with low-cost technology have been successfully used throughout the year to grow vegetables despite the adverse environment (Mishra *et al.*, 2010). However, the market is for low-cost indigenous products for the building of affordable buildings. The off-season cultivation of cucurbits in low plastic tunnels is one of the most lucrative technology in North India. Walkin-tunnels are ideal and efficient for raising off-season nurseries and increasing vegetables. Insect-proof net houses are used for the virus-free growing of peppers, chilli and capsicum. Polytrenches have been found to be highly useful for growing vegetables under cold desert conditions in the upper reaches of the Indian Himalayas (Singh, 2012). Exotic crop agriculture, hydroponics, aeroponics (soilless cultivation), terrace planting, usage of low tunnel plastics, walk-in tunnels, vertical gardening technology for the development of leafy vegetables in safe structures, aquaponics (fish and vegetable production)

Flowers

Commercial floriculture arose as a hi-tech practise under regulated climatic conditions.

Within the greenhouse, the home and/or the polyhouse. In India, floriculture is seen as a fast growth industry (Mohan, 2016). Commercial floriculture is becoming extremely relevant as it has an immense export potential to gain foreign exchange. India shipped 22,086 MT of floricultural items to the world equivalent to Rs. 548.74 crore (~US$82 million) in 2016-17. (www.apeda.gov.in). Floral goods consist of cut herbs, potted plants, leafy plants, nuts, bulbs, tubers, rooted cuttings and dried flowers or leaves. Significant floral varieties in the international cut flower trade include plum, carnation, chrysanthemum, gerbera, gladiolus, gypsophila, orchids, anthurium, tulips and lilies. Flowers such as roses, gerberas and carnations are cultivated in polyhouses. Maharashtra, Karnataka, Andhra Pradesh, Haryana, Tamil Nadu, Rajasthan and West Bengal have emerged as major centres of floriculture (Mohan, 2016). Indigenous systems for economically and ecologically viable polyhouses tailored to different agroclimatic conditions need to be built. The cut flowers, which are currently being shipped from India, mainly come from high-tech floricultural units. Protected cultivation, although it contributes substantially to overall flower exports in a small region (5 per cent of the total flower crop area). India has a range of advantages and an immense ability to expand the region under intense flower cultivation in order to further increase floricultural exports. This needs more low-cost units in critical positions with solid technical back-up. India is native to a broad variety of ornamental plants. Many of these have an ornamental potential for the usage of loose herbs, cut seeds, potted plants as well as medical and other industrial applications. If policies are formulated to allow optimum use of them by upholding the basic levels of biodiversity protection of these unused plants, they will provide great economic benefits to the farming community. Many of the possible unused animals and their beliefs.

Conclusion

In the decades to come, the challenges of agriculture, food security and livelihoods will be seriously affected by climate change and its adverse effects. The sustainable agriculture and CSA are highly complementary strategies. The CSA concept incorporates three key components: Productivity, Mitigation and Adaptation. CSA's efficacy is ensured through proper management of natural resources, rapid adoption of RCT systems in agricultural sector, and flexible engagement with farmers. The strategies and projects that are included under CSA are Integrated, farmers' knowledge availability, renewable agricultural systems technology, genetic modification, crop relocation, institutional analysis, efficient pesticide management, crop modelling, forecasting, and integration of forecasting and modelling. Sustainable agriculture is going to have a significant impact on food security. In this way, climate change affects different types of crops in different ways. An effective weather forecasting system and early warning systems can prevent climate damage. Climate change was a major factor affecting the food and farming system. Applying combined approaches is critical for addressing climate and food security, and there are so many relevant opportunities. Proper agriculture policy will alleviate short-term food insecurity and poverty while it will improve the environment in the long term through agriculture CSA pathways. With per capita income growing and health-conscious population growth rising, demand for horticultural products is on the rise, which is expected to increase further, requiring more production. However, both in terms of quality and price, production has to be competitive. Therefore, the opportunities that exist need to be exploited and gains have to be maintained. Creation of improved varieties with high efficiency, quality characteristics, disease and pest resistance and abiotic stress tolerance. Technologies could increase water and nutrient efficiency and variability in yield, quality and post-harvest crop losses as well. Efforts to ensure the timely availability of quality seeds and planting materials would also be important. Horticultural production must therefore be seen as an integrated strategy, addressing major gaps in the utilization of potential through focused research with a focus on improving performance. Technology-driven horticulture is therefore expected to solve the problem of complementary and nutritional protection, which eventually contributes to economic growth in health care.

REFERENCES

Addison, T., Arndt, C., and Tarp, F. (2011). The triple crisis and the global aid architecture. African Development Review, 23(4), 461–478. doi: 10.1111/j.1467-8268.2011.00299.x

Alam, G.M.M., Alam, K. and Mushtaq, S. (2017) 'Climate change perceptions and local adaptation strategies of hazard-prone rural households in Bangladesh', Climate Risk Management, Vol. 17, No. 6, pp.52–63.

Antle, J. M., and Capalbo, S. M. (2010). Adaptation of agricultural and food systems to climate change: An economic and policy perspective. *Applied Economic Perspectives and Policy*, *32*(3), 386-416.

Batte, M. T., and Arnholt, M. W. (2003). Precision farming adoption and use in Ohio: case studies of six leading-edge adopters. *Computers and electronics in agriculture*, *38*(2), 125-139.

Beddington, J. R., Asaduzzaman, M., Clark, M. E., Bremauntz, A. F., Guillou, M. D., Howlett, D. J. B. and Wakhungu, J. (2012). What next for agriculture after Durban?. *Science*, *335*(6066), 289-290.

Beddington, J.R., Asaduzzaman, M., Clark, M.E., Fernández Bremauntz, A., Guillou, M.D., Howlett, D.J.B., Wakhungu, J. (2012).What next for agriculture after Durban? Science, 335(6066), 289–290. doi: 10.1126/science.1217941

Bogdanski A, Dubois O, Jamieson C, Krell R. (2010). Making Integrated Systems Work for People and Climate. Rome;: 1–136.

Braimoh, A.K. and Osaki, M. (2010) 'Land-use change and environmental sustainability', Sustainability Science, Vol. 5, No. 1, pp.5–7.

Branca, G., McCarthy, N., Lipper, L. and Jolejole, M.C., 2011. Climate smart agriculture: a synthesis of empirical evidence of food security and mitigation benefits from improved cropland management. *Mitigation of climate change in agriculture series*, *3*, pp.1-42.

Campbell, B., Mann, W., and Meléndez-Ortiz, R. (2011). Agriculture and climate change: A scoping report.

Chadha, K.L. 2001. Hi-tech Horticulture in India. Policy Paper 13. National Academy of Agricultural Sciences, New Delhi.

Chadha, K.L., Singh, S.K., Kalia, P., Dhillon, W.S., Behera, T.K. and Jai Prakash (Editors). 2017. Doubling Farmers Income through Horticulture. The Horticulture Society of India and Daya Publishing House, New Delhi.

Clive, J. (2009). Global status of commercialized biotech/GM crops: 2009. *ISAAA brief*, *41*.

DeFries RS, Rudel T, Uriarte M, Hansen M: (2010). Deforestation driven by urban population growth and agricultural trade in the twenty-first century. Nat Geosci 3: 178–181.

Delgado, J. A., Groffman, P. M., Nearing, M. A., Goddard, T., Reicosky, D., Lal, R. and Salon, P. (2011). Conservation practices to mitigate and adapt to climate change. *Journal of soil and water conservation*, *66*(4), 118A-129A.

Desjardins, R. L., Sivakumar, M. V. K., and De Kimpe, C. (2007). The contribution of agriculture to the state of climate: workshop summary and recommendations. *Agricultural and Forest Meteorology*, *142*(2-4), 314-324.

Easterling W, Aggarwal P, Batima P, Brander K, Erda L, Howden M, Kirilenko A, Morton J, Soussana JF, Schmidhuber S, Tubiello F: Food, fibre and forest products. In Climate Change 2007: Impacts, Adaptation and Vulnerability. Contribution of Working Group II to the Fourth Assessment Report of the Intergovernmental Panel on Climate Change. Edited by Parry ML, Canziani OF, Palutikof OF, van der Linden PJ, Hanson CE. Cambridge, UK: Cambridge University Press; 2007: 273–313.

Falloon, P., and Betts, R. (2010). Climate impacts on European agriculture and water management in the context of adaptation and mitigation—the importance of an integrated approach. *Science of the total environment*, *408*(23), 5667-5687.

FAO (2018). *Upscaling Climate Smart Agriculture. Lessons for Extension and Advisory Services. Occasional Papers on Innovation in Family Farming*. Rome: Food and Agriculture Organization of the United Nations.

FAO IFAD, UNICEF, WFP, and WHO. (2018). *The State of Food Security and Nutrition in the World 2018. Building Climate Resilience for Food Security and Nutrition*. Technical report. Rome: FAO.

Goswami, A. K., Jai, P., and Singh, A. K. (2014). High density planting system in tropical fruits. *HortFlora Research Spectrum*, *3*(3), 298-300.

Hafezi, R., Bahrami, M. and Akhavan, A.N. (2017) 'Sustainability in development: rethinking about old paradigms', World Review of Science, Technology and Sustainable Development, Vol. 13, No. 2, pp.192–193.

Haldankar, P. M., Parulekar, Y. R., Kireeti, A., Kad, M. S., Shinde, S. M., and Lawande, K. E. (2015). Studies on influence of bagging of fruits at marble stage on quality of mango cv. alphonso. *Journal of Plant Studies*, *4*(2), 12.

Hanjra, M. A., and Qureshi, M. E. (2010). Global water crisis and future food security in an era of climate change. *Food policy*, *35*(5), 365-377.

IPCC (2012) Managing the Risks of Extreme Events and Disasters to Advance Climate Change Adaptation, Cambridge University Press, New York, NY 10013-2473, USA.

Krishna, H., Alizadeh, M., Singh, D., Singh, U., Chauhan, N., Eftekhari, M. and Sadh, R.K. 2016. Somaclonal variations and their applications in horticultural crops improvement. Biotechnology 6, 54 pp.

Kumar, A., and Kumari, A. (2019). Precision farming of different horticultural crops by the application of drip irrigation for the Hi-tech horticulture unit at Pusa. *Journal of Pharmacognosy and Phytochemistry*, *8*(4), 1330-1338.

Lal, R., Delgado, J. A., Groffman, P. M., Millar, N., Dell, C., and Rotz, A. (2011). Management to mitigate and adapt to climate change. *Journal of Soil and Water Conservation*, *66*(4), 276-285.

Lipper, L., Thornton, P., Campbell, B.M., Baedeker, T., Braimoh, A., Bwalya, M. and Torquebiau, E.F. (2014) 'Climate smart agriculture for food security', Nature Climate Change, Vol. 4, No. 12, pp.1068–1072.

Mishra, G.P., Singh, N., Kumar, H. and Singh, S.B. 2010. Protected cultivation for food and nutrient security at Ladakh. Defence Science Journal 61, 219-225.

Mohan, V. 2016. Flower Fascination: India set to be floriculture trade leader. The Times of India (Delhi Edition), 14.02.2016.

Moorhead, A. (2009). Climate, agriculture and food security: a strategy for change. *Climate, agriculture and food security: a strategy for change.*

Nelson, G. C., Rosegrant, M. W., Palazzo, A., Gray, I., Ingersoll, C., Robertson, R. and You, L. (2010). *Food security, farming, and climate change to 2050: scenarios, results, policy options* (Vol. 172). Intl Food Policy Res Inst.

Neufeldt, H., Jahn, M., Campbell, B. M., Beddington, J. R., DeClerck, F., De Pinto, A.,...Zougmoré, R. (2013). Beyond climate smart agriculture: Toward safe operating spaces for global food systems. Agriculture and Food Security, 2(12), 1–6.doi: 10.1186/2048-7010-2-12

Parry, M. (2007). The implications of climate change for crop yields, global food supply and risk of hunger.

Perfecto, I. and Vandermeer, J. (2010). The agroecological matrix as alternative to the land-sparing/agriculture intensification model. *Proceedings of the National Academy of Sciences, 107*(13), pp.5786-5791.

Peterson, D. L., Halofsky, J. E., and Johnson, M. C. (2011). Managing and adapting to changing fire regimes in a warmer climate. In *The landscape ecology of fire* (pp. 249-267). Springer, Dordrecht.

Planning Commission. 2007. Report of the Working Group on Horticulture, Plantation Crops and Organic Farming for the XI Five Year Plan (2007-2012). Planning Commission, Govt. of India.

Pye-Smith, C. (2011). Farming's climate smart future: placing agriculture at the heart of climate change policy.

Rai, M. 2012. Protected cultivation of vegetables in Indian plains. In Vegetable Production under Changing Climate Scenario. CAFT Manual, Department of Vegetable Science, Dr. Y.S. Parmar University of Horticulture and Forestry, Solan.

Sahiijram, L. and Rao, M. 2017. Micro-propagation for mass production of horticultural crops. In Doubling Farmers Income through Horticulture (K.L. Chadha, S.K. Singh, P. Kalia, W.S. Dhillon, T.K. Behera, and Jai Prakash, Eds.), pp. 29-33. The Horticulture Society of India and Daya Publishing House, New Delhi.

Sai, M. S., and Rao, P. N. (2008). Utilization of Resourcesat-1 data for improved crop discrimination. *International Journal of Applied Earth Observation and Geoinformation, 10*(2), 206-210.

Sapkota, T. B., Vetter, S. H., Jat, M. L., Sirohi, S., Shirsath, P. B., Singh, R. and Stirling, C. M. (2019). Cost-effective opportunities for climate change mitigation in Indian agriculture. *Science of the Total Environment, 655*, 1342-1354.

Shukla, V. K., Doyon, Y., Miller, J. C., DeKelver, R. C., Moehle, E. A., Worden, S. E. and Urnov, F. D. (2009). Precise genome modification in the crop species Zea mays using zinc-finger nucleases. *Nature, 459*(7245), 437-441.

Singh, B. 2012. High altitude protected vegetable production. In Vegetable Production under Changing Climate Scenario. CAFT Manual, Department of Vegetable Science, Dr. Y.S. Parmar University of Horticulture and Forestry, Solan.

Singh, B. 2014. Protected cultivation of horticultural crops in India - Challenges and opportunities. Agrotechnology 2(4), 1 p.

Smith, P., Martino, D., Cai, Z., Gwary, D., Janzen, H., Kumar, P. and Scholes, B. (2007). O. and Sirotenko, 2007: Agriculture. *Climate change*, 498-540.

UNDP. (2013). Human Development Report 2013. The rise of the south: Human progress in a diverse world. New York, USA: United Nations Development Programme.

van Wijk, M. T., Merbold, L., Hammond, J., and Butterbach-Bahl, K. (2020). Improving Assessments of the three pillars of Climate Smart Agriculture: current achievements and ideas for the future. *Frontiers in Sustainable Food Systems, 4*, 148.

Vermeulen, S. J., Aggarwal, P. K., Ainslie, A., Angelone, C., Campbell, B. M., Challinor, A. J., and Wollenberg, E. K. (2010). Agriculture, food security and climate change: Outlook for knowledge, tools and action.

WB. (2010). Agriculture must play key role in climate solutions. Washington, DC: Author. Retrieved October 10, 2016, from http: //web. worldbank.org/ WBSITE/EXTERNAL/NEWS/0

www.apeda.gov.in accessed on 30.10.2017

www.nhb.gov.in/nursery accessed 30.10.2017

Chapter 2

Why Climate Smart Agriculture (CSA) Needed for Smart Farming

Mallepu. S. Likhitha Reddy[1], Shivani Kumari[2]

[1]Department of Agronomy, UBKV, West Bengal
[2]Department of Agronomy, SASRD, Nagaland University, Medziphema, Nagaland

Climate Smart Agriculture (CSA) is the smart way of farming system which sustainably increasing agriculture productivity, production and incomes, adapting and building resilience to climate change and reducing greenhouse gas emissions. CSA plays a vital role in promoting sustainable agricultural development and ensuring food security and mitigating the negative effect of climate changes on agricultural productivity in India. A number of strategies can be adopted to achieve the objectives of climate smart agriculture. For example, use of integrated renewable energy technologies for farming like wind mills, solar panels, and bio energy-operated water pumps are very crucial for energy-smart food systems. Resource conserving technologies (RCTs) like zero tillage permits the farmers to plant wheat soon after rice or cotton harvest in order to avoid warm weather detrimental for grain filling. The emergence of newly developed varieties which are tolerant to heat, drought and salinity is also a better strategy of CSA. A range of CSA technologies, practices and services have been initiated in climate smart villages as adaptation strategies for coping with climate risks to ensure stability and sustainability in agricultural production. There are exciting opportunities for scaling out and immense potentials of these strategies for enhancing crop yields and farm incomes and reducing greenhouse gas emissions. Strengthening agricultural extension service and agricultural finance to achieve smart farming practices/technologies by linking climate finance to traditional agricultural finance could play a significant role in scaling out the CSA practices and technologies to make agriculture more sustainable and climate resilient and a viable source of livelihood and food security for millions of farmers in the country. Computer aided crop simulation models can guide to find out the possible risk of climate variation on future crop yields, climate smart agriculture development and mitigation procedures. The crop models permit variation of environmental factors such as the water regime and temperature and simulate the crop response through many estimated growth parameters like crop yield.

Keywords: *Climate smart agriculture (CSA), Resource conserving technologies (RCTs), Sustainable agricultural development, Crop simulation models.*

Introduction

Recognizing that climate change is a universal and critical challenge for global food security and the need for improving the existing way of management of agricultural systems and natural resources for effectively achieving food security, FAO (2014) advocates for the necessity of strengthening the adaptive capacity of rural communities to cope with climate changes and increasing climatic variability and building the resilience to climatic shocks. The agricultural sector has to be transformed in order to meet the demand for food of the growing global population. There is a need for changing the prevailing way of management of land, water, soil nutrients and genetic resources to ensure that these are used more efficiently and sustainably to make agriculture more productive and resilient. The agricultural production system requires adaptation to climate change and variability to ensure food and livelihood security to farming communities. The CSA approach contributes towards achieving food and livelihood security and other developmental goals by (a) sustainably increasing agricultural productivity and incomes, (b) adapting and building resilience to climate change and (c) reducing and/or removing GHG emissions, where possible (FAO, 2010, 2013). Broadly, CSA integrates climate change into the planning and implementation of sustainable agricultural strategies and focuses on developing resilient food production systems that can lead to food and livelihood security of farming communities under climate change and variability (Lipper *et al.*, 2014; Vermeulen *et al.*, 2012). It identifies synergies and trade-offs among food security, adaptation and mitigation as a basis for reorienting policy in response to climate change (Lipper *et al.*, 2014). It is designed to identify and operationalise sustainable agricultural development explicitly integrating climate change as a major parameter. Naturally, an integrated approach receptive to specific local conditions is required for CSA to become a reality (FAO, 2014). The CSA approach (a) addresses adaptation and builds resilience to climatic shocks, (b) considers climate change mitigation as a potential co-benefit, (c) is location specific and knowledge-intensive, (d) identifies integrated options that create synergies and reduce trade-offs, (e) identifies barriers to adoption and provides appropriate solutions, (f) strengthens livelihoods by increasing access to services, knowledge and resources and (g) integrates climate financing with traditional sources of agricultural investment (FAO, 2014). Branca *et al.* (2011), Jat *et al.* (2014) and Sapkota *et al.* (2015) argue that agricultural technologies and practices such as minimum tillage, different methods of crop planting, irrigation and nutrient management and incorporation of crop residue can improve crop yields, water and nutrient-use efficiency and reduce

GHG emissions from agricultural activities. Altieri and Nicholls (2013) and (Mittal, 2012) argued that use of improved seeds, ICT-based agro-advisories, crop/livestock insurances and rainwater harvesting can help farmers to reduce the negative impacts of climate change and variability on agricultural activities. Thus, CSA integrates location-specific traditional and innovative technologies, practices and services for adaptation of agriculture to climate change and variability (International Center for Tropical Agriculture (CIAT, 2014). A technology or practice is considered climate smart if it is conducive to achieve at least one of the three objectives of CSA (Khatri-Chhetri *et al.*, 2017).

Why is Climate Smart Agriculture (CSA) Needed for Smart Farming?

- As per FAO estimate, by the year 2050 world population will increase by one –third and food required for food security by 60 per cent.
- Already cumulative impact of climate change since last decade has effect on productivity. Agriculture has become a high risk profession- farmers increasingly prefer to migrate. In India as per National Sample Survey Organization 2005 estimate 60 per cent farmers like to leave farming.
- This has direct impact on socio-economic process. Already countries after countries are facing social turmoil with growing local terrorism and riots for food. In India this is reflected by expansion of Naxalite and MAOs activities in various states. A Movement that encourages young individuals to crimes of extortion and dacoity which has spread over 1/3rd of its total district in which they are present.
- But with available knowledge and experience, it is possible to make agriculture a sustainable livelihood means – but this will require intensive efforts at ground level – local level where agriculture exists and it has to be made climate smart.

CSA uses agriculture as a major tool for mitigation of GHG – CO_2 by laying emphasis on its unique capacity to absorb CO_2 and release Oxygen through photosynthesis process. It envisages achieving this through increased cropping, by reducing rain fed areas through integrated water and river basin management and expansion of agriculture on wasteland, wetland, degraded fallow areas and introducing urban agriculture. CSA is a technique to improve the scientific rule and investment setting to attain sustainable agricultural progress to ensure the food availability under climate change (FAO, 2012). Climate smart agriculture is a way to get sustainable developments as well as green economy goals. It intends towards food availability and takes part to conserve natural assets. It is closely associated with perception of improved growth, as FAO develops it for crop yields (FAO, 2011). Consideration of the impacts of world climate variations on sustainable green growth may help to predict and follow better farming techniques to improve agricultural yields (Fraser, 2008). Adaptation is a process which involves management of risks generated by climate change, including variability. Study of social issues caused by climate change considers the spontaneous impacts, through inspection of household or individual behavior when faced with present climate variability. Farm households experience the risk of output losses because of climate change due to being less adaptable. Adaptation to the climate change is important issue for the researchers, program managers and policy setters who are conscious that climate variations are really happening and are dangerous to challenge community and environmental sustainability. Sustainability is a crucial guiding principle for international environmental policy and decision making in the twenty first century (Braimoh *et al.,* 2010). The worst issues of climate change could be controlled by successful adaptation, which would probably be less than the cost of impacts that would occur without adjustments.

Climate smart Village

The most challenging task for attaining sustainable agricultural development is to adopt appropriate strategies that enhance CSA. The Consultative Group on International Agricultural Research (CGIAR), in its Research Program on Climate Change, Agriculture and Food Security (CCAFS), has been working with rural communities in collaboration with national programmes to develop climate smart villages (CSV) as models of local actions that ensure food security, promote adaptation and build resilience to climatic stresses. CSV is a community approach to sustainable agricultural development where farmers, researchers, local partners and policymakers collaborate to select the most appropriate technological and institutional interventions on the basis of global knowledge and local conditions to increase productivity and incomes, achieve climate resilience and enable climate mitigation. It integrates village developmental and adaptation plans along with local knowledge and institutions into the programme. The greatest strength of the CSV approach is its inclusiveness in bringing together farmers, policymakers, researchers and local organizations to work on a set of climate smart technologies and practices with a view to adapt agriculture to climate change in order to ensure food and livelihood security of farmers in vulnerable regions (Aggarwal *et al.,* 2013).

Building a CSV Involves

(1) selection of the location of CSV based on its climate risk profile, alternate land-use options and the willingness of farmers and local government to participate in the project, (2) working with existing community groups, consisting of farmers, researchers, rural agro-advisory service providers and village officials, (3) conducting a baseline study to capture the current socioeconomic situation, resource availability, average production and income and risk management approaches of village households, (4) stakeholders' discussion to prioritize the climate smart technologies and practices best suited to their local conditions and to indicate the actions they would most willingly carry out, (5) capacity building by organizing regular training sessions for farmers on good agricultural practices related to rain gauges, improved seed varieties, new livestock breeds, tree seedlings, simple machinery such as zero-till machines, subsidies on index-based insurance premiums and discounts on cell phone SIM cards, (6) monitoring and evaluating progress of the chosen farm activities of the participating farmers, and (7) disseminating the message of CSA, videos on success stories and testimonials from the pilot villages,

screening them in nearby villages and publicizing them widely through local, national and international media (Aggarwal *et al.*, 2013).

As a comprehensive approach to sustainable agricultural development, CCAFS' CSVs focus on climate change hotspots in Africa, Asia and Latin America with critical climate smart interventions in key areas to encourage the CSVs to adopt (1) *weather-smart* activities (weather forecasts, ICT-based agro-advisories, index-based insurance, stress-tolerant crops and varieties and climate analogues), (2) *water-smart* practices (resilient water management practices, aquifer recharge, rainwater harvesting, community management of water, laser land levelling, water conservation, drip irrigation, raised bed planting, crop diversification, alternate wetting and drying in rice, direct-seeded rice and on-farm water management), (3) *carbon-smart* practices (agro-forestry, livestock and manure management, conservation tillage, diversified land-use systems and residue management enhancing carbon content in the soil), (4) *nitrogen-smart* practices (leaf-colour charts, hand-held crop sensors and nutrient decision-maker tools for site-specific nutrient management and precision fertilizer application using nutrient expert decision support tools, residue management and legume catch cropping), (5) *energy-smart* technologies and practices (fuel-efficient agricultural machineries, residue management, biogas systems and minimum tillage conserving energy and reducing GHG emissions) and (6) *knowledge-smart* activities (cross-site visits of farmers, farmer-to-farmer learning, capacity enhancement on CSA, seed packets of adapted varieties, community seed and fodder banks and market and off-farm risk management system). These activities and practices would jointly work to build the resilience of a village community to climatic stresses and ensure food and livelihood security to farmers.

Strategies of Climate Smart Agriculture in Farming System

Methane (CH_4) emissions as a result of rice farming can be altered by water management efficiently, getting better management of organic wastes by promoting aerobic decomposition via the process of composting or incorporate soil for the period of the off-term drainage; use of rice varieties with a few infertile tillers, high oxidative roots and higher harvest index; and use of fermented fertilizer as biogas slurry compost instead of unfermented one (Pathak and Wassmann, 2007).

Land use management practices such as reduced tillage, manure and wastes enhance the integration of soil biodiversity, assembly and partial coverage can play a vital role in the accumulation of carbon in the soil (Pathak and Wassmann, 2007). So, the effects of climate change can be solved by climate smart agriculture such as developing new varieties which are tolerant to heat, salinity and resistant to floods and drought, improving water management, following new agricultural technologies such as resource conservation and effective management techniques, crop diversification, improved pest management, crop modeling and better weather forecasting. These strategies are discussed below.

Efficient Resource Management

Resource management is very significant feature of CSA and future climate. Food loses are found through all stages of the food production till food utilization.

Almost one third part of food produced is wasted (Gustavsson *et al.*, 2011). The energy consumed in annually world food losses are almost 38 per cent of the final energy utilized by the total food chain. All the food chains, from agricultural, transport, conservation, processing, cooking and consumption are likely areas for improving energy use efficiency (FAO, 2011). In the continent of Africa, 90 per cent of the removed wood is used for manufacturing household articles and also, for cooking. Better energy saving in cooking stoves can help to decrease deforestation.

Integrated Renewable Energy Technologies for Farming Systems

The suitable energy technologies, tools and different services in farming fields are important to create the stable change to energy smart and proficient food systems. The nature of all these technologies will be governed by natural conditions, transportation and skills available in the labor force. A number of new technologies that can be very important for energy-smart food systems include:

- ✰ Wind mills,
- ✰ Solar panels,
- ✰ Photovoltaic lights,
- ✰ Biogas extraction units,
- ✰ Power generators,
- ✰ Tools for bio-oil mining and purification,
- ✰ Fermentation and distillation processes for ethanol extraction,
- ✰ Hydrothermal conversion tools,
- ✰ Solar-wind electricity production or
- ✰ Bio energy-operated water pumps,
- ✰ Renewable energy-powered vehicles,
- ✰ Equipment for water supply, distribution and purification.

These newly developed technologies increased value to manufacture by the availability of raw resources. These technologies can be interconnected on a farm in terms of integrated food energy systems.

Availability of Technical Knowledge of Farmers

In South Asia farmers mostly belong to poor families and are limited to their recourses, so they are experimenting with climate variability for centuries. They have abundance of knowledge techniques with the number of steps that can be used in the development of technologies for controlling climate vulnerability. This type of knowledge is required to fulfill the modern needs. Conventional environmental knowledge of people improved and developed in the test of time could give different ideas and feasible options for adaptation procedures. Studies have concluded that the observation of traditional farmers was quite evolved in the affairs of earthquakes, landslides and drought and they have developed better techniques to alleviate the effect of natural and climatic variations.

Role of Institutions for CSA Improvement

Those institutions are essential which create and transfer useful information and guide people to interpret the new technologies into understanding and work on it. The institutions such as farmer field schools that guide and facilitate farmers for the implementation of new techniques; Shows of farm radio that share agricultural information which is easily available, useful, useable and weather-related knowledge to local rural people; agricultural plots exhibitions to the community and exchange of ideas between farmers. The profits achieved by using those management techniques which are sustainable for the land usually take time to appear. During this time, the farmers have to tolerate the total expenditure that includes expenditure of labor, land and cash (McCarthy *et al.*, 2011). Since poor farmers have no resources to access credit and markets, and they are unable to adapt to these new techniques for the success of CSA the strong institutions have to maintain agricultural markets and financing mechanisms which are very important. The most efficient technique allows researchers, private sector investors, community members and policy makers to collectively describe troubles that are planned to resolve (Kristjanson *et al.*, 2009).

Resource Conserving Technologies

The resource conserving technologies (RCTs) consists of techniques that enhance efficiency in management of resources or application of inputs and hence use of direct, identifiable and comprehensible economic advantages like decline in production costs; saving fuel, labor and water; and the timely sowing of crops, resulting in improved yields. The RCT such as zero tillage systems (ZT) or no-tillage is a cultivation system in which seeds are directly sowed into uncultivated soil. It can be explained "as the system of cultivating crops into untilled soil by creating thin channels that only have adequate depth and width to attain suitable seed coverage. Remaining soil is left as if no tillage is done with it (Derpsch *et al.*, 2010).ZT permits the farmers to plant wheat soon after the rice harvest or cotton, so that the heads of crops appear and the grains fill before the arrival of warm weather pre-monsoon. As average temperatures in the certain region rise, early planting will be gradually more important for wheat (Pathak *et al.*, 2012).

In the rotation of rice and wheat no tillage is practiced on 5 million hectares in the India, Pakistan, Bangladesh and Nepal (Hobbs *et al.*, 2008). If the disturbance of soil is less, the retention of organic matter can be more in the shape of stored carbon, that's why it does not take part in global warming in the form of CO_2 production. Therefore no-till has much share to greenhouse gas alleviation by improved carbon sequestration in the soil (Zingore *et al.*, 2010). In developing countries only a few long-term studies about soil having organic carbon changes under different tillage regimes have been conducted. The conversion of conventional tillage to no-till is often considered to be an efficient and successful carbon sequestration strategy having a sequestration rate of 367-3667 kg CO_2 ha^{-1} $year^{-1}$ (Tebrugge *et al.*, 2011).

Crops Genetic Modification

Environmental stress mostly affects the decay of organic matter in the soil, availability of nutrients and water to the plant and recycling of water and nutrients.

Nutrient concentration and period of environmental limits conclude level of effect on crop growth cycle and biomass accumulation. Crop yields in Asia are estimated to decrease by 2.5-10 per cent from 2020 onwards and by 5-30 per cent after 2050, with most horrible decrease in South and Central Asia (Cruz *et al.*, 2007). Adjustment techniques can be enhanced by the availability of new crop varieties that are tolerant to heat, drought and salinity and thus reduce the risks of climate situation. Genetic diversity of the seed structure and seed composition has been recognized as a very effective defense against plant disease and pest attack and risks of climate. Similarly, for the compensation of the emerging problems of crop cycle decreases and the other vagaries of the production environment, it is required to develop the varieties. Due to the genetic modification the global yield of vegetables has doubled in the last 25 years and due to this, worldwide trade in vegetables is much higher in volume than that of cereals. Total world's vegetable land is 53.98 million hectares which give total production of 1012.52 million tons (NHB, 2011).

Land-use Management

Changing land use practices, like the location of crops and livestock production, rotation or shifting production between livestock and crops, shifting production out of marginal areas, changing the intensity of the application of fertilizers and pesticides, capital and labor can help minimize the risks from climate change on production of agriculture. Adjustment of sequence of the crop by altering the time of sowing, spraying, and harvesting the crop, in order to take benefits of the altering length of seasons of growth and levels of changing heat and humidity associated is one more option. Changing the time at which the fields are sown can also help the farmers to regulate the length of the growing season for the better adaptation to the altering environment. Adaptations of farmers can also be involved by changing the timing of irrigation or the use of other elements like fertilizers.

Cropping Season Variation

Planting dates can be set to reduce the infertility induced by the increased temperature which may save the flowering period to coincide with the hottest period. Mitigation strategies to reduce the negative effects of increased climate variations as normally experienced in semi-arid tropics and arid regions may consist of changing the sowing or planting dates to take benefit of the wet period and to avoid intense weather events in the growing season. Changed cultivation systems include improving the better cultivars and enhancing the intensity of farming various crops. Farmers will have to manage the changes in different hydrological regimes by adopting changed crop rotations (Pathak *et al.*, 2012).

Crops Relocation

Climate change causes the increase in temperature and CO_2 levels, raises the chances of droughts and floods; all these factors affect crop yields. But the effects will vary between cultures and regions. It is necessary to differentiate regions and crops that are very much prone to climate change/variability so these should be repositioned to more appropriate areas. For example, this is known that as temperature rises it affects the quality of many important crops such as aromatic

Basmati rice and tea. Different other areas that would be more appropriate for these kinds of crops with respect to quality must be identified and evaluated for suitability.

Efficient Pest Management

Variations of temperature and rainfall unpredictably influence pest and disease incidence and extremeness on major crops. It is due to the effect of climate change that will inherently affect the relationship of pest/weed, the host population and pest/weed hosts interactions. Some possible adaptation techniques include:

1. Development of cultivars resistant to diseases and pests;
2. Integrated pest management (IPM) adoption having more dependence on the biological control and change in cultural practices.
3. Adoption of substitute crop production and techniques, as well as places that are resistant to pests and other hazards.

Forecasting of Weather

The weather forecast and early warning systems will very helpful to minimize the threats of climate losses. Information and Communications Technology (ICT) can efficiently help administrators and researchers in planning the contingency programs. In the field of forecasting a new technique Multi Criteria Analysis (MCA) tool has implemented to evaluate climate change policy alert on mitigation and adaptation options (de Bruin *et al.*, 2009). The most appropriate thoughts of the decision makers can implement in their decision making, which is a multi level MCA implemented to label and find possible adaptation techniques. The MCA is developed on the basis of expert decision.

Crop Modeling

Crop modeling is a creative and newly developed tool for dealing risks in Agriculture. Computer aided simulation models play important role to approach the techniques that establish crop management responses and forecast crop yield. Simulation models significantly assist the work of better crop development and recommend valid crop management. Models can guide to find out the possible risk of climate variation on future crop yields, climate smart agriculture development and mitigation procedures. Comparable to new management options and suitable genotypes which are important factors for better yield, crop simulation models are essential tools in field experiments to create new crop management approaches. In this point of view, two crop management systems are APSIM (Agricultural Production System Simulator) and DSSAT (Decision Support System for Agro Technology Transfer) mostly used in the whole world (Ahmad *et al.*, 2014).

Integration of Modeling and Forecasting

Although there remain some uncertainties owing to lack of knowledge, crop simulation models can appreciably estimate and quantify the impact of specific water stress conditions (possible due to climate change) on crop productivity if these are well calibrated and validated in field experiments (Wolf *et al.*, 1996; Grossman-Clarke *et al.*, 2001;). The crop models permit variation of environmental factors such as the water regime and temperature and simulate the crop response via many calculated growth parameters like crop yield. Due to the complexity of the problem, research continues and improvements are constantly being made to models, for example, drought impact assessments. Jamieson *et al.* (1998) compared the models AFRCWHEAT2, CERES (Crop Environment Resource Synthesis)-Wheat, Sirius, SUCROS2 and SWHEAT with measurements from wheat grown under drought. CERES-Wheat has proved to be useful for forecasting drought effects on crops at specific locations (the uncertainty of yield prediction was found to be below 10 per cent). CERES-Wheat is well-referenced and has been successfully tested in a number of studies (Ritchie *et al.,* 1998), especially for climate change effects on growth (Tubiello *et al.,* 1999; Zalud *et al.,* 1999).

GIS Mapping

GIS (Geographical Information System) is used in analysis and mapping which helped in the estimation and computation of the storm course and flooding associated with hot cyclones. The study incorporated population allocation, infrastructure and other under-threat resources. These photographs and images were also used in the investigation of seashore due to hot cyclones and rising sea levels. These regions can be site specific, include provincial or municipal administrative areas and other small national landscapes, like river basins, coastlines and lakes.

Case Studies on CSA

1. Studies based on field-level experiments in various regions of India have demonstrated that zero-tillage farming system has been emerging as one of the important climate smart farming practices in weed management, crop yield enhancement and GHG emission reduction. Zero-tillage practice undertaken in Bihar, Haryana and Uttar Pradesh (UP) yielded higher wheat yields (ranging from 2.8 to 12.4 per cent) relative to the conventional tillage (Table 2.1). This is largely due to saving of time in land preparation, enabling a more timely cultivation of wheat crop after rice in the rice-wheat cropping system. The zero tillage system was found to have advanced crop planting by at least one week and helped reduction of yield losses by 1–1.5 per cent per day after optimum wheat-showing time (Hobbs and Gupta, 2004; Mehla *et al.,* 2000). This system has provided substantial net benefits due to higher yields and reduced cost of cultivation by about 50 per cent, and reduced GHG emissions by allowing alterations in water, tillage and surface residue management practices, contributing significantly to ecosystem sustainability and better environment by reducing the environmental hazards (Grace *et al.,* 2003; Gupta *et al.,* 2004).

   In order to minimize the negative effects of climate change and variability and to maximize economic benefits, small landholders have been implementing a range of CSA practices and technologies such as cropping system improvement (*e.g.* improved crop varieties, diversification, crop rotation and integration of legumes), integrated nutrient management (*e.g.* green manure, compost and site specific nutrient management), resource conservation (*e.g.* minimum/zero tillage), precision water management (*e.g.* planting crops in bed, laser land levelling, mulching with crop residues) and agro-forestry as measures of adaptation to climate change and variability (Khatri-Chhetri *et al.*, 2016). Several studies have reported that the application of these practices/technologies has enhanced crop yields, farm incomes and input-use efficiency (Jat *et al.*, 2009; Kumar *et al.*, 2013; Jat *et al.*, 2014; Sapkota *et al.*, 2014; Aryal *et al.*, 2015).

Table 2.1: Impact of Zero-Tillage Practice as CSA on Wheat Yield in India

Wheat Yield (kg/ha)

Period	*Region*	*Number of Farmers*	*Conventional Tillage System*	*Zero-tillage System*	*Increase in Yield in Zero-Tillage System (Per cent)*
2001-2004	Western UP	27	4,980	5,120	2.81
2000-2003	Eastern UP and Bihar	357	2,980	3,350	12.42
1999-2000	Haryana	124	5,110	5,380	5.30

Source: Gupta and Seth (2007).

Results from a more recent study by Khatri-Chhetri *et al.* (2016) based on a survey conducted with 625 households from the CSVs in Karnal (Haryana) and 641 households from the CSVs in Vaishali (Bihar) districts lend support to these findings. The selected sites with rice-wheat as the dominant cropping system are highly vulnerable to climate change and variability due to rapidly declining groundwater table and increasing soil salinity (Karnal) and occurrence of frequent flood and drought (Vaishali). Farmers in the CSVs are adapting to climate change and variability adopting CSA practices and technologies such as improved crop varieties, varieties suitable for drought/flood, laser land levelling, zero-tillage, residue retention, site-specific nutrient management, legume integration and crop diversification. About 60 per cent of the selected households implemented at least one CSA practice/technology. The CSA adopters prefer to use improved crop varieties (80 per cent), laser land leveling (42 per cent), crop rotations (23 per cent) and zero-tillage (11 per cent).

2. The results reported in Table 2.2 reveal that the farmers have achieved higher output, yield and return in rice-wheat cropping system after implementation of CSA practices and technologies individually or jointly in place of the conventional practices. Among the three CSA interventions, improved seed has the largest impact on production, yield and net return. Laser land leveling has the second highest increase in production and net return, followed by zero-tillage. And zero tillage has the second highest increase in yield, followed by laser land leveling. A combination of improved seeds with laser land leveling/zero-tillage has helped achieve significantly higher yield and net returns than the levels achieved using

laser land leveling or zero-tillage alone. Overall, the results indicate that CSA practices/technologies have helped small farmers in the Indo-Gangetic Plains of India achieve higher productivity and income than the levels they would have without these practices. This suggests that scaling out of such practices and technologies in other regions would provide economic benefits to farmers and help them reduce the adverse impacts of climate change and variability. Farmers' preferences for CSA technologies and their impacts were found to vary according to farmers' socio-economic characteristics (landholding size, total income, access to credit services, *etc.*), location and agro-climatic zones, besides potential benefits and costs of implementation of the technologies. While conducting a primary survey in 16 villages from four dryland districts of Rajasthan.

Table 2.2; Impact of CSA Practices on Output, Yield and Returns in Rice-Wheat System

CSA Practice/ Technology	*Per cent Change in Total Production*	*Per cent Change in Total Input Cost*	*Change in Yield (tones/ ha)*	*Change in Input Cost (Rs/ha)*	*Net Returns (Rs/ha)*
Improved seeds (IS)	19.0	52.0	1.03	1,402	15,712
Laser Land Levelling (LLL)	10.0	9.5	0.33	3,037	8,119
Zero tillage (ZT)	6.0	-41.0	0.36	-1,577	6,951
IS + LLL	17.0	63.0	0.87	1,752	14,194
IS + ZT	16.0	6.0	0.94	234	15,303

Khatri Chhetri *et al.* (2017) found crop insurance, weather-based crop agro advisories, rainwater harvesting, site-specific integrated nutrient management, contingent crop planning and laser land levelling as most preferred technologies, suggesting that the schemes for promoting such technologies should be site specific and most suitable for the local farmers.

Conclusion

FAO estimated that if the present production and consumption rates continue, agricultural production should increase 60 per cent by 2050 to meet the needs of food of world's population. To achieve food security and agricultural development goals with adaptation to climate change will be required to lower emission intensities per output. Improving food protection by moderate climate change, sustainably use the natural resource, use all products more competently, have less inconsistency and greater constancy in their outputs. More fruitful and more flexible agriculture requires a most important change in the way of use of land, water, soil nutrients and genetic resources management by climate smart agriculture techniques.

REFERENCES

Aggarwal, P., Zougmore, R. and Kinyangi, J. (2013). Climate smart villages: A community approach to sustainable agricultural development, CGIAR Research Program on Climate Change, Agriculture and Food Security (CCAFS), Copenhagen, Denmark. Retrieved from http: //www.ccafs.cgiar.org.

Ahmad, M. I., Ali, A., Ali, M. A, Khan, S. R., Hassan, S. W. and Javed, M. M. (2014). Use of Crop Growth Models in Agriculture: A Review, Science International, pp: 26.

Altieri, M. A. and Nicholls, C.I. (2013). The adaptation and mitigation potential of traditional agriculture in a changing climate, Climatic Change, Retrieved from http: //dx.doi.org/10.1007/s10584-013- 0909-y.

Aryal, J. P., Sapkota, T. B., Jat, M. L. and Bishnoi, D. (2015). On-farm economic and environmental impact of zero-tillage wheat: A case of North-West India, Experimental Agriculture, 51(1): 1-16.

Braimoh, A. K., Osaki, M. (2010). Land use change and environmental sustainability, Sustainability Science, 5: 5-8.

Branca, G., McCarthy, N., Lipper, L. and Jolejole, M. C. (2011). Climate smart agriculture: A synthesis of empirical evidence of food security and mitigation benefits from improved cropland management, Mitigation of Climate Change in Agriculture Series 3, FAO, Rome.

Cruz, R. V., Harasawa, H., Lal, M., Wu, S., Anokhin, Y., Punsalmaa, B., Honda, Y., Jafari, M., Li, C. and Huu, N. (2007). Asia Climate Change: Impacts, Adaptation and Vulnerability, pp. 469-506.

Derpsch, R., Friedrich, T., Kassam, A. and Li, H. (2010). Current status of adoption of no-till farming in the world and some of its main benefits, International Journal of Agricultural and Biological Engineering, 3: 1-25.

FAO. (2010). Climate smart agriculture: Policies, practices and financing for food security, adaptation and mitigation. Rome: Food and Agriculture Organization of the United Nations (FAO).

FAO. (2011). Save and grow: a policymaker's guide to the sustainable intensification of small holder crop production, Rome.

FAO. (2012). The Sahel crisis: executive brief. March. Rome.

FAO. (2013). Climate smart agriculture sourcebook. Rome: Food and Agriculture Organization of the United Nations (FAO), Retrieved from http: //www.fao. org/ Publications.

FAO (2014). FAO success stories on climate smart agriculture. Rome: Food and Agriculture Organization of the United Nations (FAO).

Fraser, D. (2008). Understanding animal welfare, Acta Veterinaria Scandinavica, 50(Suppl 1), S1.

Grace, P. R., Harrington, L., Jain, M. C. and Philip Robertson, G. (2003). The long-term sustainability of tropical and sub-tropical rice and wheat systems: An environmental perspective, In Improving the productivity and sustainability of rice–wheat systems: Issues and impacts (pp. 27-43). Madison: ASA Special publication 65.

Grossman-Clarke, S., Pinter, E. J., Kartschall, T., Kimball, B. A., Hunsaker, D. J., Wall, G. W., Garcia, R. L. and LaMorte, R. L. (2001). Modelling a spring wheat crop under elevated CO_2 and drought, New Phytol, 150: 315–335.

Gupta, P. K., Shivraj, S., Nahar, S., Dixit, C. K., Singh, D. P., Sharma, C., *et al.* (2004). Residue burning in rice–wheat cropping system: Causes and implications, Current Science, 87(12): 1713–1717.

Gustavsson, J., Cederberg, C., Sonesson, U., van Otterdijk, R. and Meybeck, A. (2011). Global food losses and food waste: extent, causes and prevention, Rome, FAO.

Hobbs, P. R., Sayre, K. and Gupta, R. (2008). The role of conservation agriculture in sustainable agriculture. Philosophical Transactions of the Royal Society B: Biological Sciences, 363: 543-555.

Jamieson, P. D., Porter, J. R., Goudriaan, J., Ritchie, J. T., Van Keulen, H., Stol, W. (1998). A comparison of the models AFRCWHEAT2, CERES wheat, Sirius, SUCROS2 and SWHEAT with measurements from wheat grown under drought, Field Crops Res, 55: 23–44.

Jat, M. L., Gathala, M. K., Ladha, J. K., Saharawat, Y. S., Jat, A. S., Kumar, V., Sharma, S. K. and Gupta, R. (2009). Evaluation of precision land levelling and double zero till systems in the rice-wheat rotation: Water use, productivity, profitability and soil physical properties, Soil and Tillage Research, 105(1): 112–121.

Jat, R. K., Sapkota, T. B., Singh, R. G., Jat, M. L., Kumar, M. and Gupta, R. K. (2014). Seven years of conservation agriculture in a rice-wheat rotation of Eastern Gangetic Plains of South Asia: Yield trends and economic profitability, Field Crops Research, 164 (August): 199–210.

Khatri-Chhetri, A., Aggarwal, P. K., Joshi, P. K. and Vyas, S. (2017). Farmers' prioritization of climate smart agriculture (CSA) technologies, Agricultural Systems, 151: 184–191.

Khatri-Chhetri, A., Aryal, J. P., Sapkota, T. B. and Khurana, R. (2016). Economic benefits of climate smart agricultural practices to smallholders' farmers in the Indo-Gangetic Plains of India, Current Science, 110 (7): 1251–1256.

Kristjanson, P., Reid, R. S., Dickson, N., Clark, W. C., Romney, D. and Puskur, R. (2009). Linking international agricultural research knowledge with action for sustainable development.

Kumar, V., Saharawat, Y. S., Gathala, M. K., Jat, A. S., Singh, S. K., Chaudhary, N. and Jat, M. L. (2013). Effect of different tillage and seeding methods on energy use efficiency and productivity of wheat in the Indo-Gangetic Plains, Field Crops Research, 142: 1–8.

Lipper, L., Thornton, P., Campbell, B. M., Baedeker, T., Braimoh, A., Bwalya, M., *et al.* (2014). Climate smart agriculture for food security. Nature Climate Change, 4 (December), 1068–1072.

McCarthy, N., Lipper, L. and Branca G. 2011. Climate smart agriculture: smallholder adoption and implications for climate change adaptation and mitigation. FAO, MICCA Series No. 4. Rome, FAO.

Mehla, R. S., Verma, J. K., Gupta, R. K. and Hobbs, P. R. (2000). Stagnation in the productivity of wheat in the Indo-Gangetic plains: Zero-till-seed-cum-fertilizer drill as an integrated solution. New Delhi: Rice–Wheat Systems of the Indo-Gangetic Plains. RWC Paper Series No.8.

Mittal, S. (2012). Modern ICT for agricultural development and risk management in smallholder agriculture in India. Socio-Economic Working Paper 3. Mexico, D.F: CIMMYT.

NHB. (2011). National Horticulture Database, 2011, Ministry of Agriculture, Govt. of India, Gurgaon, pp: 278.

Pathak, H., Aggarwal, P. K. and Singh, S. D. (2012). Climate Change Impact, Adaptation and Mitigation in Agriculture: Methodology for Assessment and Applications, Indian Agricultural Research Institute, New Delhi. Pp: 302.

Pathak, H. and Wassmann, R. (2007). Introducing greenhouse gas mitigation as a development objective in rice-based agriculture: II. Cost-benefit assessment for different technologies, regions and scales, Agril. Syst., 94: 826-840.

Ritchie, J. T., Singh, U., Godwin, D. C. and Bowen, W. T. (1998). Cereal growth, development and yield, In: Tsuji GY, *et al.* (Eds.), Understanding Options for Agricultural data: CO_2 and water interactions. Agron. J. 91: 247–255.

Sapkota, T. B., Jat, M. L., Aryal, J. P., Jat, R. K. and Khatri-Chhetri, A. (2015). Climate change adaptation, greenhouse gas mitigation and economic profitability of conservation agriculture: some examples from cereal systems of Indo-Gangetic Plains, Journal of Integrative Agriculture, 14(8): 1524–1533.

Sapkota, T. B., Majudar, K., Jat, M. L., Kumar, A., Bishnoi, D. K., McDonald, A. J., and Pamplino, M. (2014). Precision nutrient management in conservation agriculture based wheat production of Northwest India: Profitability, nutrient use efficiency and environmental footprint, Field Crops Research, 155: 233–244.

Tubiello, F. N., Rosenzweig, C., Kimball, B. A., Pinter, J. P., Wall, G. W., Hunsaker, D., LaMorte, R. L and Garcia, R. L. (1999). Testing CERES-wheat with free-air carbon dioxide enrichment (FACE) experiment Proceedings of the National Academies of Science 106: 5047–5052.

Vermeulen, S. J., Campbell, B. M. and Ingram, J. S. I. (2012). Climate change and food systems, Annual Review of Environment and Resources, 37: 195–222.

Wolf, J., Evans, L. G., Semenov, M. A., Eckersten, H. and Iglesias, A. (1996). Comparison of wheat simulation models under climate change. I. Model calibration and sensitivity analyses, Clim. Res, 7: 253–270.

Zalud, Z., Dubrovsky, M., Stastna, M. (1999). Modelling climate change impacts on maize and wheat growth and development, In: Proceedings of the International Symposium on Modelling Cropping Systems, Lleida, Spain, June 21–23, 1999, pp. 277–278.

Zingore, S. (2010). SOC sequestration in farming systems in Africa: Potential, opportunities and chal-lenges, In: 2nd Meeting of the Round Table on Organic Agriculture and Climate Change, RTOACC, 2010.

Chapter 3

Engineering of Metabolic Pathways for Abiotic Stress Resistance

V.G. Ishwarya Lakshmi[1], *G. Babithraj Goud*[2] *and K. Divya*[1]

[1]*Department of Genetics and Plant Breeding, PJTSAU, College of Agriculture, Rajendranagar, Hyderabad, Telangana*
[2]*Department of Crop Physiology, PJTSAU, Hyderabad, Telangana*

Drought, salinity and heat are some of the critical abiotic stresses that severely affect the vital production as well as productivity of wide range of crops. A large number of encoding genes for the biosynthesis of these stresses related protecting compounds include glycine betaine, mannitol and heat shock proteins which have been useful for tolerance besides various transcription factors like MAPK, DREB1, WRKY, etc. Advanced technologies like Antisense gene and RNAi have revolutionized the pace of improvement of agri and horti related crops for biotic and abiotic stress tolerance, and for quality enhancement. Besides these, Genetic engineering is a useful mechanism for the deployment of stress tolerance which involves the alteration of genetic structure of an organism by either removing or introducing DNA. In stress metabolism, the process of genetic engineering could also be used to produce metabolites that generally do not occur naturally in the host cell. Genes such as natural and synthetic Cry genes, trypsin inhibitors, cystatin genes and protease inhibitors have made advancements in imparting in stress tolerance in crop plants.

Keywords: *Abiotic stresses, Genetic engineering, Enzymes, Metabolism, Quality, Genes, Transgenic plants.*

Metabolic Pathway Engineering

Normally, plants synthesize vast number of compounds to aid in the development and defense against stress, many of which are produced through species-specific and complex biosynthetic pathways. As of now, 48.6 per cent of anti-cancer compounds and 75 per cent of anti-bacterial compounds for defense against stress are secondary metabolites. Shikimic acid pathway, non-mevalonate (MEP) pathway and mevalonate (MVA) pathway are the pathways involved in secondary metabolites production such as Terpenoids, monoterpene indole alkaloids, iso quinoline alkaloids, flavonoids and anthocyanins. Metabolic engineering involves the manipulation of endogenous pathways to enhance the flux towards desirable molecules involved in it. It is also referred to as the improvement of properties of cells through the change of particular biochemical reactions or the introduction of novel ones for increasing the production of desirable compounds or to reduce the accumulation of undesirable ones.

Why plants for engineering involves the manipulation of plant metabolism helps in attainment of resistance to abiotic and biotic stresses, to production of fuel, food and feed, industrial products, fine chemicals, pharmaceutical products and for novel opportunities in agriculture.

Plant cell metabolism is complex, as they produce over 2,00,000 (including primary and secondary) metabolites with less information available compared to bacteria making it hard to model them.

Organism	Base pairs	Genes	Proteins
E.coli	4.6M	4000-5000	1366 (30 per cent +)
Arabidopsis	135M	30000	3500 (10 per cent +)

Metabolic Pathways

Metabolic pathways are chains of linked chemical reactions catalyzed by enzymes that convert substrates into products through a series of intermediates. The pathways may be cyclical, linear or can be arranged in a three-dimensional grid way. They may also be unidirectional or reversible, with branches and conversion points. There may be enzymatic steps with absolutely specificity or may depend on families of closely related molecules with common group of chemicals. The metabolic pathways products may aggregate in the cell, often within specific compartments, or may be exported. The metabolic pathways with their complexities in plants can be enhanced by shuttling intermediate compounds between the sub-cellular compartments or even with different cell types.

Flow may be controlled by the metabolites themselves, which often impudence the activity of enzymes in the same pathway (or the signaling proteins and transcription factors controlling them) by a process called feedback regulation, thus, making it complex to understand, analyze, reconstruct, and annotate such metabolic pathways. In the past few years significant advances have taken place in metabolic engineering through the imputation of genomics and proteomics to elucidate as well as characterize the pathways in a holistic manner rather than on step by step basis.

Metabolic Engineering for Stress Tolerance

Environmental abiotic stresses such as salinity, drought, and chilling prove to be major obstructions for plant growth hampering crop productivity. Several adaptations to such abiotic stresses have been evolved by certain organisms which are either metabolic or structural. A common adaptation related to metabolism that is followed in certain species is the aggregation of certain organic solutes called as osmoprotectants. These osmoprotectants protect the proteins membranes and protein molecules against damage by high levels of inorganic ions along with protection of metabolic machinery against any oxidative stress damage.

Pathway Engineering for Drought Tolerance

Various genes and genetic material controlling signaling regulatory pathways provide key targets for abiotic stress tolerance Transcription factors (TFs) like DREB1 gene family, Myb gene family play an important role in imparting regulation of the expression of a number of genes involved in controlliong abiotic stress tolerance in plants (Bhatnagar-Mathur *et al.,* 2008). LEA proteins are groups of hydrophillins, which are synthesized as a dessication response in seeds and are often stimulated under various abiotic stress situations like salinity, dehydration, high temperature or even chilling stress in vegetative tissues of various plant plants. Scientists reported transformed normal tomato plants with a DNA cassette containing an Arabidopsis C repeat/dehydration-responsive element binding factor 1 (CBF1) cDNA and a nos terminator, driven by the commonly used cauliflower mosaic virus 35S promoter. These transgenic tomato plants were more tolerant to dehydration stress than the wild-type plants. Enhanced drought and cold stress tolerance in transgenic apple was also reported by with an over-expression of a cold-inducible Osmyb 4 gene extracted from rice crop, encoding a transcription factor. Similarly, over-expression of DREB1b TF gene had also been reported to induce cold tolerance as well as drought tolerance in transgenic grapevine (Jin *et al.,* 2010). Similarly, with a view to explore abiotic stress tolerance in potato, Gangadhar *et al.* (2016) transformed a potato-derived gene StnsLTP1 into potato through an indirect method of gene transfer. The transgenic potato lines produced under stress conditions displayed enhanced cell membrane integrity with a reduction in membrane lipid peroxidation activity and H_2O_2 content along with increased level of antioxidant enzyme activity and aggregation of amount of ascorbates and upregulation of various stress-related genes.

Engineering for Heat Tolerance

Different kinds of reactive oxygen species (ROS) like hydrogen peroxide (H_2O_2) and superoxide are produced inside the plant cells under heat stress conditions, paving a way for enormous kinds of plant physiological disorders affecting growth as well as productivity of crop plants. These reactive oxygen species not only cause enzyme denaturation, but also cause damage to certain cellular components inside cells. Heat tolerance is strongly correlated with the increased capacity of plants to scavenge the reactive oxygen species produced. Thus, it is important to tackle the ROS so as to maintain normal growth and metabolism of plants. Accordingly, plant species have developed numerous techniques to combat ROS with the synthesis of various enzymatic systems like superoxide dismutase (SOD), glutathione reductase (GR) and peroxidase to eradicate peroxide ions. Many works were done for metabolic engineering to heat stress. Improved tolerance to high temperature heat stress in transgenic apple was noticed with the over-expression of cytosolic ascorbate peroxidase gene. Even, transgenic tomato plants with over-expression of the same gene were developed which was very effective by exhibiting high resistance with the exposure to direct sunlight. Similarly, studies on the improved heat stress tolerance was obswerved with the overexpression of Copper or zinc superoxide dismutase gene derived from *Manihot esculenta* under an oxidative stress inducible promoter SWPA2 in potato. On the other hand, production of a variety of chemical compounds including polyamines, carotenoids, *etc.* can directly react with ROS, scavenge them and thus provide protection to the plants against heat stress.

Engineering for Salinity Tolerance

Salt stress or salinity or is one such factor which has an abundance effect on both the quality and quantity aspect of different agriculture and horticulture crops. Tolerance to salinity is a very complex mechanism governed by many genes (Bojorquez-Quintal *et al.*, 2014). Several pathogenesis-related proteins are produced by crop plants that are prone to different abiotic stress conditions in order to overcome the adverse effect of stress conditions. There are several references for the engineering of crops for salinity tolerance. For instance, Subramanyam *et al.* (2011) could successfully enhance the tolerance nature of chilli plants to salinity by the ectopic nature of osmotin gene of tobacco via Agrobacterium tumefaciens-mediated indirect gene transfer technique. The transgenic pepper plants had improved levels of osmoprotectants like proline, glutathione reductase (GR), glycine betaine, superoxide dismutase (SOD), ascorbate peroxidase (APX), and relative water content (RWC) in as revealed in the biochemical analysis. Similarly, transgenic tomato plants expressing tolerance to chilling and salt stress were developed by incorporation of cytosolic ascorbate peroxidase (cAPX) gene, that quenches hydrogen peroxide in plant cells was derived from pea. Thus, protection against oxidative injury could be observed. The transgenic plants also exhibited better seed germination rate than the wild type, proving to be an added advantage to the plant. On the whole, abiotic stresses including salinity, chilling and oxidative stresses are the critical factors limiting the cultivation and productivity crop plants for which engineering can provide a solution.

Metabolic engineering usually aims to enhance the aggregation of desirable target products as it is naturally produced in less quantity. This may have any direct intervention to improve the aggregation of target product by enhancing flux through the pathway or by diminishing flux through catabolic pathways that eliminate intermediates or the target products itself. This engineering may also be useful in producing metabolites that are not usually synthesized naturally in the host cell as the corresponding metabolic steps might be completely missing.

This necessitates the import of either partial or complete heterologous pathway to improve the metabolic capabilities of the host.

Practically, there are several interventions that are to be crossed, like nurturing enough knowledge about the important endogenous pathways in order to completely know the best intervention ways, identifying as well as developing the most suitable metabolically active genes, expression of those genes to synthesize a functional enzyme with a heterologous background, and, finally, achieving the accumulation of target compounds without harming the host plant.

Approaches for metabolic engineering include

1. Single gene intervention
2. Multiple gene intervention

1. Single Gene Intervention

Metabolic engineering through single gene intervention includes strategies like

1. Accumulation of End Product
2. Accumulation of Intermediate Product
3. Controlling the Flux via Branch Point
4. Searching alternative Routes to a Specific Product

Enhancement

Through a target pathway, flux can be increased by enhancing enzyme expression in upstream pathways for ensuring sufficient presence of precursors. The starting enzyme expression can be enhanced in the target compound pathway so that flux is cautiously directed towards the metabolite that is targeted. Improvisation of rate-limiting enzymes activities that create bottlenecks is useful in causing the making of inhibitory intermediates.

Suppression

Includes suppressing the competing pathways that are downstream of branching points for avoiding intermediates being diverted along with the suppressing catabolic steps so as to allow the target metabolite not to be completely degraded or converted into other molecules.

Sequestration

Includes creation of sink compartments that actually store the target metabolites, all of which shift the chemical equilibrium towards the increased production side.

Limitations of Single Gene Intervention

- Inability to solve multiple bottlenecks
- Increased activity of multimeric enzymes containing heterologous subunits
- Inability to target multiple metabolites

2. Multiple Gene Intervention

Metabolic engineering through multigene intervention includes strategies like

Gene Stacking

Two or more genes of interest into a single plant

Re transformation

Once a stable transformation procedure has been established, a metabolic engineering strategy must be identified. Metabolic engineering technology can be used for enhancing natural product yield using two primary strategies leading to increased ability of a plant for adapting towards the environmental conditions (*e.g.*, temperature, pests, drought, UV). Improving the overall nutritional value of a food product including antioxidant capacity, fatty acid production and vitamin production allows for increased yield of a single product for even medicinal and nutraceutical applications.

Increasing accumulation of several compounds simultaneously

I. Accumulating an End Product

This is one of the simplest and easiest metabolic engineering ways as it can usually be obtained by enhancing the available upstream precursors and/ or maximizing flux by boosting the activity of committed pathway enzymes, particularly those representing rate-limiting steps. The final end products can cause inhibition of previous-acting enzymes, as a result of which the strategies to be developed must be enough to prevent feedback inhibition. The engineering strategies described also took the potential for feedback regulation into account.

II. Accumulating an Intermediate Product

An intermediate product accumulation can be more challenging than that of an end product because intermediates tend to be converted into end products. Pathway intermediates targeting strategies may involve

a) Combination of boosting upstream flux
b) Blocking downstream flux
c) Creating metabolic sinks for removing intermediates and prevention of further conversion

a. Combination of Boosting Upstream Flux

Metabolic engineering in the carotenoid pathway must satisfy for several phenomena, including the availability of downstream enzymes with intermediates that can exert both positive and negative feedback reactions.

b. Blocking Downstream Flux

Using antisense RNA in potato tubers, the β-carotene hydroxylase genes CHY1 and CHY2 were silenced resulting in a 38-fold increase in β-carotene levels and a 4.5-fold increase in total carotenoids. Similarly, by co-suppression or antisense

RNA, silencing the zeaxanthinepoxidase (ZEP) gene prevented the conversion of zeaxanthin to violaxanthin, thus increasing zeaxanthin levels by up to 130-fold.

c. Creating Metabolic Sinks to Remove Intermediates and Prevent Further Conversion

Diverting the intermediate into a sub-cellular compartment that lacks the enzymes necessary for further conversion, thus creating a sink metabolically that shifts the equilibrium towards total production in the cell. Instead of completely preventing downstream steps, it may be intellectual to divert the intermediate into a sub-cellular compartment that lacks the useful enzymes which are very needful for further conversion, making a metabolic sink that could shift the equilibrium in the cell towards production side.

III. Controlling Flux through a Branch Point

IV. Alternative Routes to a Specific Product

Although many metabolic products are synthesized by specific converging or linear pathways, others can be synthesized through alternative routes, thus providing a range of strategies for the technique of metabolic engineering. The major approaches used to boost ascorbate accumulation are to increase the flux through one or more of the de novo synthesis pathways, specifically targeting rate-limiting steps if possible, to increase the efficiency of recycling by over-expressing enzymes that recycle oxidation products and to inhibit enzymes that promote oxidation. Interventions have been successful to changeable degrees including the over-expression of GDP-L galactosephosphorylase (GGP), which enhanced the ascorbate levels by two-to sixfold in tomato, potato, and strawberry. The main enzyme responsible for recycling ascorbate in plants is dehydroascorbatereductase (DHAR), and a six fold increase in ascorbate levels was achieved by expressing the rice DHAR gene in maize under the control of an endosperm-specific promoter. Two DHAR iso-forms are present in leaves of potato, one expressed constitutively in the cytosol and the other expressed specifically in green tissue and targeted to the plastids. Transgenic potatoes accumulated higher ascorbate levels in green tissues and tubers which were expressing cytosolic DHAR, but those expressing plastidial DHAR accumulated higher levels of ascorbate only in the leaves. Thus, it seems that this ascorbate metabolic engineering is very sensitive to the source of the heterologous enzyme and its spatio-temporal expression profile in the transgenic plants, is likely to reflect the complex regulatory systems that control the accumulation of ascorbate in different plant tissues.

V. Increasing Accumulation of Several Compound Simultaneously

a) Products from the Same Metabolic Pathway

Because several carotenoids have specific nutritional roles some are essential nutrients and most are antioxidants that reduce the risk of chronic diseases, it is appropriate to devise metabolic engineering strategies that lead to the accumulation of more than one product in the pathway. When the desirable outcomes include a large number of intermediates and end products, the most appropriate solution is to funnel flux in such a way that it becomes restricted progressively, *i.e.*, by eliminating early rate-limiting steps but reducing the efficiency of subsequent reactions progressively, allowing the accumulation of particular intermediate compounds.

b) Products from Different Pathways

Metabolic engineering focuses on either single compounds or pathways as the target is usually to enhance the abundance of a particular molecule that is not present naturally in adequate quantities. Main draw back in Nutritional engineering of plants is that no single intervention is good enough only nutritionally complete staple crops will address the challenge of malnutrition satisfactorily. The only way to achieve nutritional completeness is the simultaneous metabolic engineering of different pathways.

Two further considerations are important when engineering different pathways simultaneously:

1. It is necessary to consider the potential for crosstalk and the role of shared resources. If multiple pathways drain the same group of precursors or interlink in other ways, then it would be more effective to strive for moderate enhancement of these pathways rather than maximizing the flux, because the drain on resources could result in antagonism between the pathways and have a negative overall impact on the plant.
2. It is necessary to consider potential interactions between end products. Not only is the co-accumulation of ascorbate and folate nutritionally beneficial, but the ascorbate acts as an antioxidant and increases the stability of folate, which is sensitive to oxidative degradation.

Current Challenges

This concept of metabolic engineering can be summarized as a ladder that needs to be climbed, with each run representing a novel challenge, but the path to success is hurdled with many potential pitfalls. An understanding of at least the key enzymes and intermediates in a metabolic pathway is necessary before metabolic engineering can be attempted, and this generally begins with the isolation of corresponding genes. Even in plants with complete genome sequences, such as rice, potato, maize the absence of functional annotations makes it difficult to identify the components of complete metabolic pathways. The inefficiency of data is could be addressed using the large-scale sequencing methodologies along with proteomics projects targeting the metabolically related tissues, which could produce a range of important secondary metabolite products. RNA-seq is used to produce large numbers of sequences that can be studied to identify the specific candidate enzymes, which can then be tested either in vitro conditions or in bacteria for determining their precise roles in catalysis. Even having no precise information of the complete pathway, progress can be achieved by supplementing the supply of precursors and analyzing the impact on the expression of gene, based on the fact that pathway intermediates often regulate genes involved in the same pathway.

Once a candidate gene has been selected, the next phase is to attain the adequate expression of a functional enzyme in an appropriate background. Promoter choice is important to ensure the gene is expressed robustly, but even when a strong promoter is used, the level of active enzyme accumulating in the cell may be low because of feedback regulation and/or epigenetic phenomena that lead to transgene silencing either immediately or in subsequent generations. In some cases, such endogenous regulatory effects can be avoided by using a heterologous transgene because the regulatory mechanisms are not conserved between species—*e.g.*, bacterial prephenate dehydrogenase (TYRA) is resistant to feedback inhibition by downstream products and is therefore preferred for the engineering of tocochromanols. Feedback regulation can be regulated by wantingly mutating the transgene either to remove or modify protein parts which are responsible for the response. The enzyme expression in storage tissues is a straightforward approach to prevent different inhibition forms as the seeds, fruits and tubers tend to be more forgiving when metabolism is disturbed, mainly because they do not affect the primary metabolic processes ongoing in vegetative tissues but also because they are naturally adapted to store large amounts of particular compounds.

Conclusion

The Metabolic engineering is a simple process of improving the capacity of existing pathways, along with controlling the distribution of flux with bolting on additional functionalities so that new novel compounds can be produced. The revolution of metabolic engineering from single simple steps to multistep approaches, along with the combination of high-throughput methods aligned for gene discovery and functional analysis will pave a way for enhancing the predictive precision of early development stages, thus refining engineering strategies and reducing the need for trial-and-error method of testing. The important nutritional and pharmacological properties of many vital secondary metabolites in combination with their low levels of production show active research into novel strategies for engineering will become increasingly important.

These challenges could be addressed by utilizing the most advanced molecular biology techniques like high-throughput screening of RNA-seq and DNA-seq data for identifying missing steps in the metabolic pathways, including an eye on on high-throughput assay development in order to identify related genes even in incomprehensible plant species. It is tricky to target the best inter- vention points and accurately predict the outcome. So we need to go for Knowledge-based metabolic engineering that begins with vast datasets including genomics, proteomics, transcriptomics as well as metabolomics for allowing the rapid identification of genes, regulatory hierarchies with feedback loops. The concentration on metabolic pathways paving for industrial and nutritional products also suggests the implications of these new technological approaches for a rapid transition to the commercial environment. Also, High-throughput genomic technologies for abiotic stress can provide vast number of technological tools to refine by accurately identifying genes, their pathways and regulatory controls.

REFERENCES

Bojorquez-Quintal, E., Velarde-Buendýa, A., Ku-Gonzalez, A., Carillo-Pech, M., Ortega- Camacho, D., Echevarrýa-Machado, I., Pottosin, I., and Martínez-Estévez, M. (2014). Mechanisms of salt tolerance in habanero pepper plants (*Capsicum chinense* Jacq.): proline accumulation, ions dynamics and sodium root-shoot partition and compartmentation. *Frontiers in Plant Science*, 5, 10–3389.

Jin, W.M., Dong, J., Yan, A.L., Wang, Y. Cheng., M.X., and Han, Z.H. (2010). Studies on adventitious shoots regeneration and transformation system of grape in vitro. *Acta Horticulturae Sinica*, 35, 27–32.https: //www.scimagojr.com/journalsearch.php?q=21100790326 and tip=sid and clean=0

Gangadhar, B.H., Sajeesh, K., Venkatesh, J., Baskar, V., Abhinandan, K., and Moon, S. H. (2016). Identification and characterization of genes associated with thermo-tolerance using virus induced gene silencing in *Nicotiana benthamiana*. *Plant Growth Regulation*. 1–12.

Subramanyam, K., Sailaja, K.V, Subramanyam, K., Rao, D.M, and Lakshmidevi K. (2011). Ectopic expression of an osmotingene leads to enhanced salt tolerance in transgenic chilli pepper (*Capsicum annum* L.). *Plant Cell, Tissue and Organ Culture*, 105(2), 181-192.

Chapter 4

Breeding for Abiotic Stress Tolerance in Crop Plants to Mitigate Climate Change Impacts

Rani Jadhav and J.R. Katore

AICRP on Linseed, College of Agriculture, Nagpur
Dr. Panjabrao Deshmukh Krishi Vidyapeeth (PDKV), Akola, Maharashtra

Human manipulation of earth's resources gives rise to new problems day by day and hence different methods are developed to overcome the same. World population is likely to increase by 1.8 billion as of 2030 and by 2.5 billion as of 2050. By 2030 food demand is probable to increase by 50 per cent because of continued population growth and higher incomes (www.popcouncil.org). If worldwide population reaches 9.1 billion by 2050, the FAO says that world food production will need to rise by 70 percent, and food production in the developing world will need to double. Instead food productivity is declining because of the effect of various abiotic stresses; therefore reducing these losses is a key area of concern for all nations to deal with the increasing food requirements. However, abiotic stresses and climate change are becoming increasingly serious threats to crop production worldwide at a time when food staple supply will need to be significantly higher to meet the demand of the growing human population. Drought is the main abiotic factor followed by soil salinity. At present, agriculture has new enormous challenges due to population growth, the pressure on agriculture liability on the environmental conservation, and climate change. To deal with these new challenges, many plant breeding programs have changed their breeding scope to stress tolerance in the last years. One such example is breeding for abiotic stress, exploiting non arable land too, for feeding the ever growing population. Research is the key to meet the challenges of modern agriculture in a sustainable and positive fashion. So, in this chapter, emphasis has been given on importance of abiotic stresses specially on drought and salinity, conventional plant breeding methods such as selection and introduction, pedigree method, shuttle breeding, mutation breeding, back cross method and recurrent selection to develop improved cultivars tolerant to most sorts of abiotic stresses.

Introduction

Plants are exposed to a wide range of environmental stresses which lower downs and restricts the productivity of agricultural crops. Two types of environmental stresses are come across to plants which are classified as (1) Abiotic stress and (2) Biotic stress. The abiotic stress causes the loss of major crop plants universal and it includes radiation, salinity, floods, drought (water stress), extremes in temperature (cold, frost and heat), heavy metals, submergence and nutrient deficiencies and mineral toxicity *etc.* They adversely impact growth, development, yield and seed quality of crop plants. While, biotic stresses comprise attacks by several pathogens such as bacteria, fungi, nematodes, oomycetes and herbivores. These stresses reduce the crop production. In future it is expected that fresh water insufficiency will increase and in the end intensity of abiotic stresses will increase. Therefore there is a purpose to develop crop varieties that are tough to abiotic stresses to ensure food security in coming years.

Nowadays climate has changed all around the globe by continuously increase in temperature and atmospheric CO_2 levels. The distribution of rainfall is uneven due to the change in climate which acts as an important stress as drought. After drought is enacted on crop plants stop in growth is the first response subjected to plants. Plants cut their growth of shoots under drought conditions and lessen their metabolic demands. After that defensive compounds are being synthesized by plants under drought by mobilizing metabolites required for their osmotic adjustment.

Need to Study Abiotic Stress Tolerance in Plants?

1. Abiotic stresses, for instance high or low temperature, drought and salinity are most important factors that limit the geographical distributions of plants on earth.
2. Abiotic stresses generally cause severe loss of crop yields.
3. Abiotic stresses affect worldwide crop production system and threaten food security of human being.

Importance of Abiotic Stress

According to world estimates, an average of 50 per cent yield losses in agricultural crops is caused by abiotic factors (Wang *et al.*, 2007). These comprise mostly of high temperature (40 percent), salinity (20 percent), drought (17 percent), low temperature (15 percent) and other forms of stresses (Ashraf *et al.*, 2008). Only 9 per cent of the world area is conducive for crop production, while 91 per cent is afflicted by various stresses. As per the current estimates (ICAR, 2010), 120.8 million ha constituting 36.5 per cent of geographical area in India are degraded due to soil erosion, salinity/alkalinity, soil acidity, water logging, and other edaphic problems. In the context of global warming and climate change, further severe productivity loss due to escalating adverse effects of abiotic stresses can be expected. Among environmental factors, as given in Table 4.1 abiotic stress are the major constraints limiting the productivity (Buchanan *et al.*, 2000).

Table 4.1: Record Yield, Average Yields and Yield Losses Due to Biotic and Abiotic Stress of Eight Major Crops

Crop	*Record Yield (kg/ha)*	*Average Yield (kg/ha)*	*Biotic Stresses*	*Abiotic Stresses*	*Yield Loss (Per cent of Record Yield)*	
					Abiotic Stress	*Biotic Stress*
Corn (*Zea mays* L.)	19300	4600	1952	12700	65.8	10.1
Wheat (*Triticum aestivum* L.)	14500	1880	726	11900	82.1	5.0
Soybean (*Glycine max* L.)	7390	1610	666	5120	69.3	9.0
Sorghum (*Sorghum bicolor* L.)	20100	2830	1051	16200	80.6	5.2
Oat (*Avena sativa* L.)	10600	1720	924	7690	75.1	8.7
Barley (*Hordium vulgare* L.)	11400	2050	765	8590	75.4	6.7
Potato (*Solanum tuberosum* L.)	94100	28300	17775	50900	54.1	18.9
Sugar beet (*Beta vulgaris* L.)	121000	42600	17100	61300	50.7	14.1

Source: Buchanan *et al.*, 2000.

Stress-Definition

"When some factors of the environment interferes with the complete expression of genotypic potential, it is called as stress". In an ideal environment there is no intervention by any ecological factor with the complete expression of the genotypic potential of an individual or line; such an environment is called stress-free environment.

Stress during the reproductive phase of a crop causes far more economic loss than comparable stress during earlier phases. In some cases, adverse effects of a stress during an earlier phase may be almost entirely compensated, for example, moisture stress during the vegetative phase of sorghum. Stress in plants refers to 'outside conditions that badly affect growth, development or yield of plants'. A plant stress usually reflects some quick changes in environmental condition. Abiotic stress is caused by environment which may either physical or chemical, whereas biotic stress is due to biological units like diseases, insects, *etc.* If the stress is mild plants can be improved from injuries while if it is severe then it leads to death of crop plants by preventing flowering, seed formation and induce senescence.

Classification of Stresses

Stresses are classified as biotic (pathogens, pests, weeds, *etc.*) and abiotic depending on their biological or abiological nature. In this chapter the emphasis is given on abiotic stresses in plants.

Abiotic Stresses

Abiotic stresses are the primary sources of yield losses, which account for 71 per cent of total reductions in crop yields. Management of abiotic stress is one of the most important challenges facing agriculture. Abiotic stress makes to limit choice of crops and agricultural production over large areas and extreme events can lead to total crop failures. They adversely affect the livelihoods of individual farmers and their families so that national economies and food security. It include potentially adverse effects of Salinity, Drought, Flooding, Metal toxicity, Nutrient deficiency, High temperature and Low temperature.

Drought and Rainfall Pattern

Drought is one of the major abiotic stresses of agricultural plants restraining crops return globally (Tuberosa *et al.*, 2003). It does affect the crops growth and the quality of yield.

Generally effects of drought stress on plants are;

- ✰ Decrease in seed germination and development
- ✰ Poor growth in plants
- ✰ Reduced reproductive growth
- ✰ Reduction in leaf mass
- ✰ Reduced photosynthesis
- ✰ Reduced stomatal conductance and
- ✰ A substantial reduction in the total dry matter

28 per cent of the world's land is too dry to support vegetation (Kramer and Boyer, 1995). The world's production of grain has, in some years, been reduced by five per cent or more as a result of severe droughts in key countries. It has been estimated that drought causes an average 17 per cent yield loss annually in the tropics (Edmeades *et al.*, 1992), but losses can be much more severe and total crop failures are not unknown.

Temperature

A continuous rise in the temperature due to climate change over time will result in reduced rainfall, wind speed and snow cover which will also lead to a lessened length of growing season for crop plants. It will affect negatively the quality of the crops and yield. Variations in the environmental temperature occur more rapidly than changes in stress factors. The increase in temperature is the consequence of the global warming which varies regionally. One of the fresh study on special effects of frost and extreme temperatures on wheat production (*Triticum aestivum* L.) resulted that frost caused sterility and abortion of formed grains while extreme heat lead to reduction in the number of grains formed in the filling period of the grain (Barlow *et al.*, 2015)

Waterlogging/Flooding

Among the environmental factors those affecting plant growth and development, water stress is one of the most important factor. A water stress may arise either from an insufficient water or drought stress or from excessive water activity or water logging (Azhar and McNeilly, 2004).

Climate change has transformed the hydrological cycle processes because of which there is reduction in crop growth in so many areas around the world. Waterlogging occur as result of leakage from irrigation canals or pure surface drainage but predominantly caused by heavy rainfall in such area (Ashraf and Mehmood, 1990). The available oxygen (O_2) for plants cells is abridged because in ponding water the diffusion process of O_2 is slow in (Alexandre, 2019). Due to the partial supply in O_2, anaerobic bacteria discharges toxic amount of iron ion, manganese ion and sulphide.

Effects of Flooding Stress on Plants

1. Decay and death of leaves
2. Wilting
3. Abscission
4. Epinasty
5. Lenticels formation

Salinity

The utmost common forms of land degradation are results from soil salinization. Almost all of the continents have saline soils. Salinity is common to arid and semi-arid regions of the world where evapotranspiration surpasses rainfall, and as a result leads to insufficient rain to filter away the soluble salts from the root zone (Vijayalakshmi, 1999). Lands with salinity stress problem covered at least 7 per cent of the world land area four decades ago (Vijayalakshmi, 1999) and has been seen to double every two decades. It shows that the salinity stress problem is speedily increasing every day because of climate change. The ability of plants to take up water and other nutrients from the soil reduces drastically because of salinity stress which leads to stunted growth; salt deposits find their way into the transpiration stream and damage the cells in leaves causing leaf burn; it also affects the enzyme activity within the plant. The salinity of a soil is measured in terms of electrical conductivity (EC), and these soils have a mixture of salts of Sulphate, Sodium, Magnesium, Chlorine and Calcium.

Salt stress has become an constantly increasing danger to food production. It is a major factor limiting the crop productivity. Increased salinization of arable lands is likely to have a disturbing global effect, resulting in 30 per cent land losses in the next 25 years and up to 50 per cent by the year 2050 (Zheng *et al.*, 2004). Soluble salts can cause harm to plant, if they are in high concentration in water or soils.

Soil salinity poses a global threat to world agriculture by reducing the yield of crops and ultimately the crop productivity in the salt affected areas. Salt stress reduces growth of crops and yield in many ways. There are mainly two primary effects which are forced on crop plants by salt stress *i.e.* osmotic stress and ion toxicity. The osmotic pressure in the soil solution surpasses the osmotic pressure in plant cells under salinity stress, due to the presence of more salt which limits the ability of plants to start uptake of water and minerals like K^+ and Ca_2^+. So above primary effects of salinity stress leads to certain secondary effects alike assimilate production, reduced cell expansion and membrane function along with reduced cytosolic metabolism.

Heat

Increase in temperature all over the world has become a great worry, which not only affect the development of plants but their productivity particularly in agricultural crops. When plants come across heat stress the ratio of seed germination, photosynthetic efficiency and yield declines. Under heat stress, during the reproductive growth period, the function of tapetal cells is vanished, and the anther is no viable.

Toxin

The increased dependence of agriculture on chemical fertilizers and sewage waste water irrigation and rapid industrial development has further added toxic metals to agriculture soils causing harmful effects on soil-plant environment system.

Table 4.2: Important Abiotic Stresses Limiting Productivity of Major Pulse Crops In India

Crop	*Season*	*Abiotic Stress*
Chickpea	Timely sown	Low temperature, nutrient stress
	Early	Terminal drought, salt stress
	Late	Terminal drought, cold, nutrient stress
Pigeonpea	Kharif-early	Waterlogging, nutrient stress
	Medium late	Cold, terminal drought, waterlogging
	Pre-rabi	Cold, terminal drought
Mungbean	Kharif	Pre-harvest sprouting, terminal drought
	Zaid	Pre-harvest sprouting, temperature, drought stress
	Rabi	Terminal drought
Urdbean	Kharif	Terminal drought
	Zaid	Pre-harvest sprouting, temperature, drought
	Rabi/rice fallow	Terminal drought
Lentil		Moisture, temperature
Clusterbean		Moisture and nutrient stress

Source: Reddy, 2006.

Mechanisms for Abiotic Stresses

There are three types of stress resistance mechanisms occurs in plants *viz.*

1. Avoidance mechanisms: These prevent exposure to stress.
2. Tolerance mechanisms: These permit the plant to withstand stress.
3. Acclimation: Plants alter their physiology in response stress

Breeding Methodology

Type of reproduction of the species should be considered when adopting any breeding method for any crops whether it is self-pollinating, cross-pollinating or asexual. The cultivar type associated to the genetic control of the traits. Because individual plants react in a different way to similar abiotic stress factors, it can be difficult to breed a species for more than one resilient trait at a time, but that is precisely what plant breeders are looking to do. The plant breeding methods for drought resistance are similar for yield and other economic characters. In common, pedigree and bulk method can be used for self-pollinated crops and recurrent selection for cross-pollinated crops. However, if transfer of few traits relating to drought resistance to a high-yielding genotype is the aim, then back cross is the appropriate methodology. On the other hand, biparental mating (half-sib and full-sib) maintains the broad genetic base and provides the scope to develop the desired genotype of drought resistance (Yunus and Paroda, 1982). Drought tolerant crop varieties can be developed through introduction, selection, hybridization and mutations breeding approaches. Salinity tolerant variety were developed through pedigree, modified bulk pedigree and anther culture approach. Complex traits of abiotic stress phenomena in plants make genetic modification for efficient stress tolerance difficult to achieve (Wang *et al.*, 2007). Breeding methodology for drought resistance has been discussed by Hurd (1969), Rao *et al.* (1971) and Chang *et al.* (1975). Abiotic stress breeding is essential ways to fight yield drop. Developing such crop varieties which have salt, drought and heat tolerance are known to be the most promising, less resource consuming, economically viable and socially acceptable approach.

Conventional Approaches

The choice of suitable breeding programme for the development of tolerant cultivars to a defined abiotic stress depends upon a number of factors: screening techniques, sources and mechanisms of tolerance *etc.*

Selection and Introduction

Breeding for any trait start with the assembly of genetic variation through the collection and evaluation of the available germplasm. If the desirable variability is not available within a locality or species, introduction of the exotic germplasm can be resorted to. Some varieties with salinity tolerance have been developed universally by using selection and introduction approaches. Mahsuri is one example: it was introduced to India and other countries from Malaysia and become popular with farmers in eastern India because of its moderate resistance to lodging and good grain quality (Rao Balakrishna and Biswas, 1979).

Pedigree Method

Pedigree method consists two lines in crossing, each providing desirable genes which are to be incorporated into the resulting lines. The F_2 segregating generation is grown and selection is made on selfed plants. This is the classical method in which the lineage of the plant selection in the segregating generation is maintained until it is stabilized in the F_7 or F_8 generation. Rice varieties from Bihar, Rajendra Dhan, Panidhan-2, Radha, Shakuntala, and Mahsuri have been released, of which Mahsuri is popular among the farmers. The modified pedigree method (single panicle selection) was adopted by breeders at Narendra Deva University of Agriculture and Technology (NDUAT), Faizabad, Uttar Pradesh (Mishra *et al.*, 1996). Pedigree selection method was fund effective to produce new lines tolerant to drought stress (Tammam *et al.*, 2004) CSR0-10, CSR-13, CSR-23, CSR-27, CSR-30, CSR-36, CR Dhan-402, CR Dhan-403, TRY-1, TRY-2, TRY3, White ponni, CO-43 *etc.* were developed through recombinant varieties (Reddy *et al.*, 2014).

Shuttle Breeding

A collaborative Shuttle Breeding Program began in the *kharif* season of 1992 to improve the yield potential of the rainfed lowland rice varieties of eastern India (Assam, Bihar, Chhattisgarh, Eastern Uttar Pradesh, Odisha and West Bengal) with an ICAR-IRRI (Reddy *et al.*, 2013) for improvement of rice. In this approach, pre breeding material or advanced breeding materials are evaluated at different locations for their adaptability and the best adapted materials are again crossed and evaluated at different target site in replicated trials. In the next step functions multiplication testing for advanced breeding materials is done. CSR-23 and CSR-27 the salt tolerance rice varieties were developed through this approach under an ICAR-IRRI collaborative project (Mishra, 1994).

Mutation Breeding

This method is used in exhausted gene pool condition. Chemical mutagenes EMS provide broad spectrum genetic changes with lesser sterility effects, as compared to other mutagens. Mutation is used when the desired genes are not available in the germplasm. This method was followed for the development of the first salt tolerant rice variety, CSR-10. With the use of mutation breeding, two broadly accepted rainfed lowland rice cultivars, Biraj and Jagannath, were developed from OC-1393 and T141, respectively. These varieties are suitable for use in 5-30 cm standing water (Rao Balakrishna and Biswas, 1979).

Back Cross Method

It is a form of recurrent hybridization in which a gene for a superior trait may be added to a desirable variety. To transfer of traits for drought tolerance to a high yielding cultivar, backcross method of breeding is suitable (Edmeades *et al.*, 1992). Large number of introgression lines with enhanced abiotic stress tolerance has been developed in a massive backcross breeding program using three recurrent parents and 203 donor lines with tolerance to several abiotic stresses in rice (Ali and Abo-El- Wafa, 2006).

Table 4.3: Drought Tolerance Genotypes of different Crop Species in India

Crop	*Genotype*
Rice (*Oryza sativa*)	ARC-10327, DJ-129, DV-110, DZ-41, DNJ-60
	Khush and Coffman, 1977 MTU-17, CH-45, PTB-29, B-76, TKM-1, N-22, Sudha
	Sahbhagi Dhan (IR74371-70-1-1)
Wheat (*Triticum aestivum*)	Triticum sphaerococcum, T. vavilovii, T. aestivum cv. C-306
Maize (*Zea mays*)	ZPBL-1304
	HI-209, HI-295, HI-536, HI-1040
Cowpea (*Vigna radiata*)	TVu-11979, TVu-14914
	IT-90K-59-2, Kanannado, Dan Ila
Chickpea (*Cicer arietinum*)	BG-365, BG-364, Pusa-362, Pusa-3351, Pusa-256 and IPC-94-32
Mustard/rape (*Brassica* spp.)	Oscar, range, Tarnab-2

Source: Meena *et al.*, 2017.

Abiotic Stress Management

Managing abiotic stresses in plants is a very challenging job because of its complex nature. As abiotic stress tolerance is multigenic in nature, the modern trend is moving in the direction of genetic transformation of multiple genes or transcription factors. Generally, abiotic stresses can be managed by genetically improving the genes and transcription factors, or by using cultural practices which contains modification and change of planting time and crop density in the field so as to reduce the effect of these abiotic stresses. Another method is to apply plant hormones, signalling and trace elements and also by using osmoprotectants.

Table 4: Drought Tolerant Cultivars/Lines of different Crops Developed through Conventional Breeding at different Centers/Institutions

Crop	*Cultivar/Line*	*How Developed*	*Centers/Institutions Involved*
Peanut (*Arachis hypogaea* L.)	ICGV 87354	Derived from a cross between Argentine and PI259747 and developed through nine generations of bulk selection	Plant Materials Identification Committee of the International Crops Research Institute for the Semi-Arid Tropics (ICRISAT), India
Safflower (*Carthamus tinctorius* L.)	Morlin	Derived from a single plant selection from F11 population	Eastern Agricultural Research Center and Montana Agricultural Experiment Station, Sidney
Chickpea (*Cicer arietinum* L.)	FLIP 87-59C	Developed by crossing ILC3843 with FLIP87	International Center for Agricultural Research in the Dry Areas (ICARDA)
Wheat (*Triticum aestivum* L.)	Willow Creek	Through breeding in single replication observation (SROB) nurseries	Montana Agricultural Experiment Station, Sydney

Crop	*Cultivar/Line*	*How Developed*	*Centers/Institutions Involved*
Soybean (*Glycine max* (L.) Merr.)	R01-416F and R01-581F	Both lines were originated after selection from F9 population. developed from a cross between Jackson and KS4895	Arkansas Agricultural Experiment Station, USA
Barley (*Hordeum vulgare* L.)	Lenetah	Developed using a pedigree selection procedure with all early generation population and selected from the cross, 94Ab12981 × 91Ab3148	Agricultural Research Service, Aberdeen, ID, in cooperation with the Idaho Agricultural Experimental Station
	Giza 2000	The pedigree breeding method was used for development and it was originated from the cross between the Egyptian local cultivar Giza 121 and the line 366/13/1 (Giza 117/Bahteem 52/ Giza 118/FAO 86)	Barley Research Department, Agricultural Research Center at Giza, Egypt
Maize (*Zea mays* L.)	Oh605 22	Selected full-sib progenies from the AAE population were intermated with 30 full-sib progenies obtained from OhS3267LAN plants Ohio State University (OSU), Ohio	Agricultural Research and Development Center, USA
	ND2005 and ND2006	Developed through an integration of recurrent selection and pedigree breeding methods	North Dakota Agricultural Experiment Station, USA

Source: Ashraf, 2010.

Various non-monetary inputs or techniques, *viz*. selection of crops, selection of drought-resistant/tolerant/early-maturing/short-duration varieties, healthy and disease-free seeds, seed treatment need to be adopted for mitigating abiotic and biotic stress for maximizing productivity and profitability of the dryland farmers. Better crop management practices like nutrient, water and weed management, insect-pest and disease management are essential to withstand these stresses. Climate resilient crop varieties along with other suitable adaptation and mitigation strategies will help overcome the adverse impact of climate change by lowering the yield losses under stress condition (Maheswari *et al*., 2015).

Conclusion

It is predicted that earth's temperature will rise by 3–5°C in the coming 50–100 years. As there is constant increase in temperature and irregular rainfall the changes of flood and drought is always in consideration. The anthropogenic actions such as extreme use of fertilizers, unsuitable irrigation and exploitation of metal resources can lead to salt stress to a large amount. Under these situations, plants will perhaps come across more frequently, parallel to both biotic and abiotic stresses. It is the responsibility of plant breeders to develop stress tolerant cultivars in order to safe food security and to ensure security to the farmers. Molecular work is to be done at the genetic level to develop mechanisms in plants so as to prevent them from various types of stress conditions. Unless and until the responsive mechanisms are not developed against biotic and abiotic stresses, the plants will continuously exposed to such stresses and finally will prove a great danger to world agriculture. It can be seen that climate change rigorously strengthens the effect of these abiotic stresses on plants. If the stresses are not properly managed it will create impossible to meet global food demand. The tolerance of plants to abiotic stress is as a result of events of diverse stress-responsive genes that relate with other elements of stress-transducing pathways. A very good knowledge of the physiological processes in response to different abiotic stress will help plan methods and tools to transform in order to improve tolerance of different abiotic stress in plants. Thus, from crop plants, model plants and microorganisms some stress-tolerant gene resources, need to be identified and developed by taking use of suitable genetic transformation technologies. Recent experimentation with transgenic plants has led to increased salinity tolerance, with emphasis on the areas of ion homeostasis, osmotic regulation and antioxidant protection. As different abiotic stresses are inter-related (*e.g.* salinity and osmotic stress), our ability to improve crop performance may well be determined by combining different, apparently unrelated approaches for introducing several stress tolerance mechanisms into specific crop plants. Abiotic stress may trigger a series of responses in plants that include changes in gene expression and cellular metabolism. The duration, severity, and frequency with which a stress is imposed, and the affected organs and tissues, developmental stage, and genotype also influence plant responses to stress.

REFERENCES

Ali, M.A. and Abo-El-Wafa, A.M., 2006. Inheritance and selection for earliness in spring sunder heat stress. *Assiut J. Agric. Sci.*, **37**: 77-94.

Alexandre, D. O. 2019. Abiotic and Biotic Stress in Plants. *Intech open* 77845.

Ashraf, M. and Mehmood, S. 1990. Effects of waterlogging on growth and some physiological parameters of four Brassica. *Plant and Soil.***121**: 203-209.

Ashraf, M. 2010. Inducing drought tolerance in plants: Recent advances. *Biotech. Advances*, **28**: 169–183.

Ashraf, M., Athar, H.R., Harris, P.J.C. and Kwon, T.R. 2008. Some prospective strategies for improving crop salt tolerance. *Adv. Agron.*, **97**: 45–110.

Azhar, F. M. and McNeilly, T. 2004. The genetic basis for salt tolerance in *Sorghum bicolor* (L) Moench seedlings. *Plant Breeding*. **101**: 114-121.

Barlow, K., Christy, B., O'leary, G., Riffkin, P., Nuttall, J. 2015. Simulating the impact of extreme heat and frost events on wheat crop production: A review. *Field Crops Research*. **171**: 109-119.

Buchanan, B.B., Gruissem, W. and Jones, R.L., 2000. Biochemistry and molecular biology of plants. American Soc. Plant Physiologists. Rockville. Maryland. USA FAO. (1985). *Production year book*, **39**. Roam: FAO.

Chang, T.T., De Datta, S.K. and Coffman, W.R., 1975. Breeding methods for upland rice. In: Major research in upland rice. IRRI, Los Banos, Philippines. 143-157.

Edmeades, G.O. J. Bolaoos and H. R. Lafitte. 1992. Progress in selecting for drought tolerance in maize. In D. Wilkinson (ed.), Proc. 47th Annual Corn and Sorghum Research Conference, Chicago, December 9-10, 1992. ASTA, Washington. Pp. 93-111.

Hurd, E.A., 1969. A method of breeding for yield of wheat in semi-arid climates. *Euphytica.* **18**: 217-226.

Kramer., P. J and J. S. Boyer. 1995. Water relations of plants and soils. Academic Press., San Diego.

Maheswari, M., Sarkar, B., Vanaja, M., Srinivasarao, M., Srinivasarao, Ch., Venkateswarlu, B. and Sikka, A.K. 2015. Climate resilient crop varieties for sustainable food production under aberrant weather conditions. Technical Bulletin, ICAR-Central Research Institute for Dryland Agriculture, Hyderabad, p. 47.

Meena H. P., Bainsla, N. K. and Yadav D. K.2017. Breeding for abiotic stress tolerance in crop plants. *Recent advances in plant stress physiology*. 329-378.

Mishra, B. Akbar, M. Seshu, D.V. and Senadhira, D., 1996. Genetics of salinity tolerance and ionic uptake in rice. *Int. Rice Res. Newsl.* **21**: 38-39.

Mishra, B. 1994. Breeding for salt tolerance in crops. In: Salinity management for Sustainable agriculture: 25 years research at CSSRI. Central Soil Salinity Research Institute, Karnal, India, pp. 226-259.

Rao Balakrishna, M.J. and Biswas, S., 1979. Rainfed lowland rice in India. In: Rainfed lowland rice: selected papers from the 1978 International Rice Research Conference. Los Banos. International Rice Research Institute, P.O. Box 933, Manila, Philippines, 87-94.

Rao, M.J.B., Murth, K.S., Choudhury, D., Srinivasulu, K. and Gangacharan, G., 1971. Breeding for drought and upland conditions in rice. *Oryza*, **8**(2): 75-84.

Reddy, A.A. 2006. Impact assessment of pulses production technology, Research Report No 3, Indian Institute of Pulses Research, Kanpur, Uttar Pradesh, India.

Reddy, M.A., Francies, R.M., Rasool, Sk. N. and Reddy, V.R. P., 2014. Breeding for tolerance to stress triggered by salinity in rice. *Internat. J. Appl. Biol. and Pharmaceutical Tech.* **5**(1): 167-176.

Reddy, J. N., Patnaik, S.S.C., Sakar, R.K., Das, S.R. Singh, V.N. Indrani Dana, Singh, N.K., Sharma, R.N., Ahmed, T., Sharma, K.K., and Verulkar, S., Collard, B.C.Y., Pamplona, A.M., Singh, U.S., Sarkarung, S., Mackill, D.J., and Ismail, A.M. 2013. Overview of the eastern india rainfed lowland shuttle breeding network (eirlsbn). *SABRAO Journal of Breeding and Genetics*. **45** (1): 57-66.

Tammam, A.M., El-Ashmoony, M.S.F., El-Sherbeny, A.A. and Amin, A.L., 2004. Selection responses for drought tolerance in two bread wheat crosses. *Egypt J. Agric. Res.* **82**: 1213-1226.

Tuberosa, R., Grillo, S. and Ellis, R. P. 2003. Unravelling the genetic basis of drought tolerance in crops. In: Toppi LS, Pawlik Skowronska B, editors. Abiotic Plant Stresses. Dordrecht: Springer.

Vijayalakshmi, D. 1999. Abiotic stresses and its management In agriculture. In: Stresses in Plants. India: Springer;

Wang, W., Vinocur, B. and Altman, A. 2003. Plant responses to drought, salinity and extreme temperatures: towards genetic engineering for stress tolerance. *Planta*. **218**: 1-14.

Wang, J.P., Raman, H., Zhou, M.X., Ryan, P.R., Delhaize, E., Hebb, D.M., Coombes, N. and Mendham, N. 2007. High-resolution mapping of the Alp locus and identification of a candidate gene HvMATE controlling aluminium tolerance in barley (*Hordeum vulgare* L.). *Theoretical and Applied Genetics*. **115**: 265–276.

Yunus, M. and Paroda, R.S., 1982. Impact of biparental mating on correlation coefficients in Bread wheat. *Theor. Appl. Genet*. **62**: 337-343.

Zheng, X. Y, Wang, Y. J, Song, S. H, Li, L. and Klocke E. 2004. Detecting molecular markers associated with heat tolerance of Chinese cabbage. *Acta Horticulturae*. **637**: 317-323.

Chapter 5

Role of Phytoalexins in Stress Management of Oilseed and Pulse Crops

Rashad Khan[1], Sameena Begum[2]* and R. Divya Bandhavi[1]

[1]Odisha University of Agriculture and Technology, Odisha
2[1]Professor Jayashankar Telangana State Agricultural University, Hyderabad, Telangana

Oilseed and pulse crops are often attacked by a great variety of pathogens that causes large economic losses. Plants defend themselves against pathogens by producing antimicrobial compounds, which include phytoalexins. In this chapter, we summarize the current knowledge on phytoalexins synthesize by oilseeds and pulse crops such as groundnut (resveratrol), soybean (glyceollins), sunflower (scopoletin and ayapin), mustard (spirobrassinin, cyclobrassinin and rutalexin), pigeon pea (cajanin, cajanol and cajaninstil bene acid) and chickpea (medicarpin and maackiain). The speediness of phytoalexin accumulation is related to resistance in plants to stress caused by abiotic and biotic factors, although the genetic information for phytoalexin synthesis is found in susceptible and resistant plants. This chapter deliberates the contribution of phytoalexins to biotic and abiotic resistance in these crops.

Introduction

Crop loss due to pest and pathogen attack is a serious problem throughout the world. Plants are frequently attacked by many pathogens and respond by the activation of defense genes, the formation of reactive oxygen species (ROS), the synthesis of pathogenesis-related (PR) proteins, localized cell-wall reinforcement and the production of antimicrobial compounds. In plants, stress conditions induce the production of secondary metabolites with antimicrobial activity that plays a key role in the plant defense mechanism; these metabolites are named as phytoalexins. Phytoalexins are a heterogeneous group of metabolites that exhibit biological activity towards a variety of pathogens and are considered as molecular markers of disease resistance. In this chapter, some of the progresses made in phytoalexin studies were presented, with a focus on the major oilseed and pulse crops. However, we provided a broader viewpoint on the research of phytoalexins improvement after diverse biotic and abiotic stresses and their role in mitigating these stresses.

Oilseeds

Oilseeds serve as an inimitable source of income for resource-poor farmers, with a substantial manifold utilization in many households in India. Principal oilseed crops such as soybean and groundnut supply more than 35 per cent of the world's processed vegetable oil. Both biotic and abiotic constraints account for lower yields of oilseeds. Biotic constraints related to diseases are key among the factors working against profitable oilseed production. The induced synthesis and accumulation of secondary metabolites such as phytoalexins is one of the most widely studied chemical plant defense responses to pathogen infections (Hammerschmidt, 1999).

Groundnut

The peanut plant infected with a microbial pathogen produces a distinct set of stilbene-derived phytoalexins. These stilbenoids are major contributing factors to the plant's resistance against diseases. Previously, it has been reported that six phytoalexins were identified from peanut kernel slices including trans-resveratrol (Figure 5.1), trans-arachidin-1, trans-arachidin-2, trans-arachidin-3, trans-3-isopentadienyl-4, 30, 50-trihydroxystilbene and SB-1 (Sobolev *et al.*, 2009). In one ounce of commonly edible peanuts, the average amount of resveratrol was found to be 73 lg (Sanders *et al.*, 2000). Resveratrol were reported at extremely low concentration in uninfected tissues of peanut plants (Yu *et al.*, 2005). In recent times, significantly higher amount of trans-resveratrol contents (0.11 lg/g) were found in peanut cultivar "Runner"(Chukwumah *et al.*, 2012). A rapid increase in trans-resveratrol ranging from 22.5–88.2 lg/g was found in sprouting peanuts (Pae *et al.*, 2011).

Biotic and Abiotic Stress Resistance

Peanut produces stilbene-derivative phytoalexins in response to fungal infection. The presence of resveratrol in peanut was first identified in the hypocotyls infected by *Helminthosporium carbonum* and implicated as a phytoalexin (Ingham, 1976). Maize and groundnut are highly susceptible crops and major source of aflatoxins contamination to humans.

Aflatoxin B_1 produced by *A. flavus* and *A. parasiticus* in the groundnut is considered to be the major cause of liver cancer along with the chronic infection with hepatitis B and C viruses and acute aflatoxicosis. Phytoalexins especially resveratrol synthase/stilbene synthase were the most abundant enzymes that

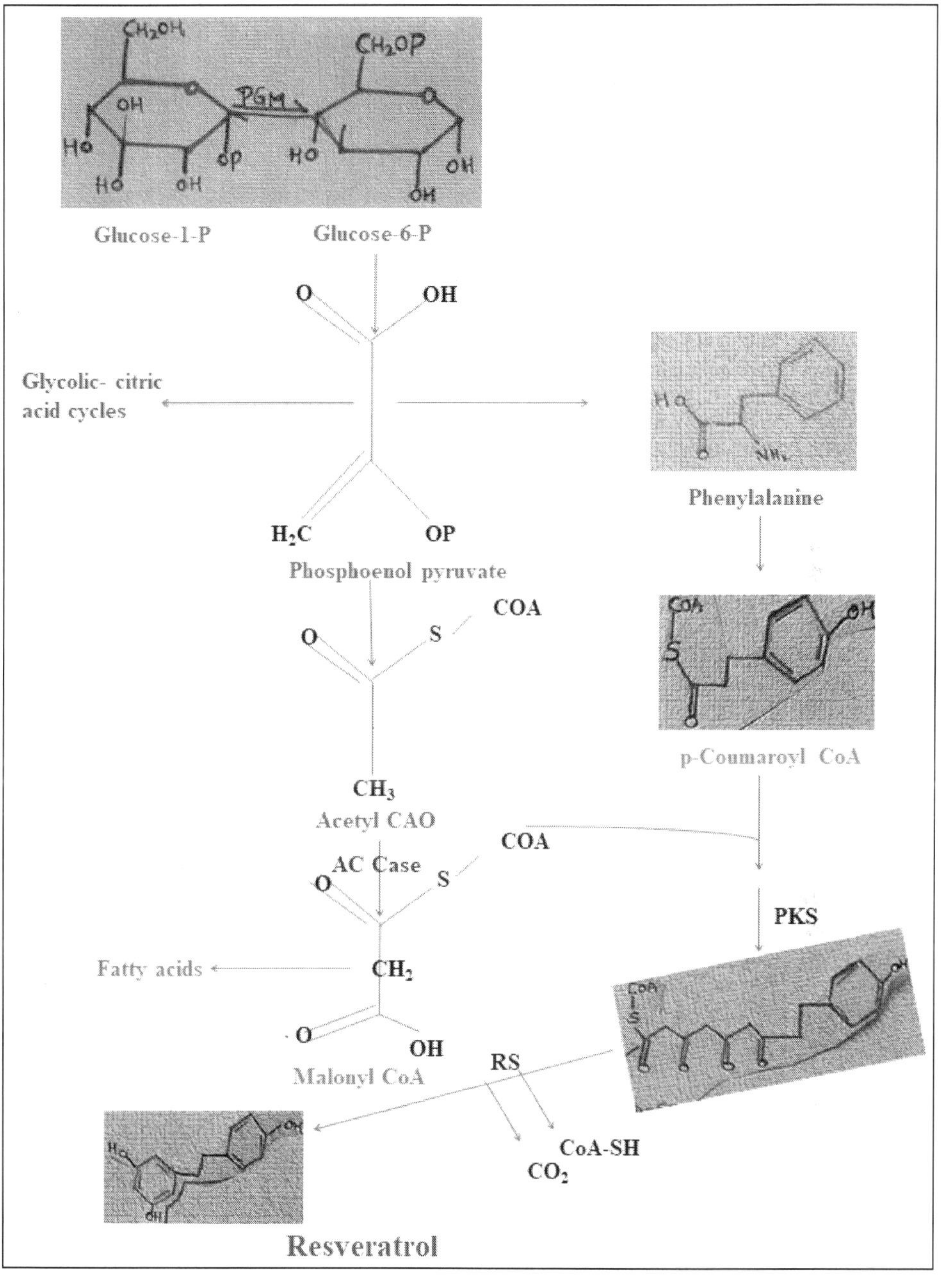

Figure 5.1: Biosynthesis Pathway of Resveratrol in Groundnut.

were found to produce in response to *Aspergillus* infection during the early stages of infection. The level of expression of these genes was higher upon the trigger from pathogen attack (Spurthi N Nayak *et al.,* 2017). The infection by *Aspergillus caelatus* in groundnut induces production of resveratrol, arachidin and arahypin phytoalexins (Sobolev and Cole, 1999). Recently a report on the effect of resveratrol on *A. flavus* revealed that it reduces growth and aflatoxin production in fungus (Wang, 2015). Germinated peanuts have been shown to produce phytoalexins, such as resveratrol, arachidinsan diso pentadie-nyl-3, 5, 40-tri hydroxyl stilbene and up to 45 stilbenoid phytoalexin derivatives after inoculation with the food grade fungus *Rhizopus oligosporus* (Wu, 2011).

Also, the damage caused by insects and other stresses may explain the increase of trans-resveratrol in peanut leaves. Previously, the induction of resveratrol by yeast extract was shown in the cultures of peanut cell suspension (Lanz *et al.*, 1990). However, all kinds of microorganisms do not elicit resveratrol in peanuts. *Pseudochrobactruma saccharolyticum*, a gram-negative bacterium isolated from the peanut callus failed to elicit resveratrol (Yang *et al.*, 2010).

In peanut seeds, abiotic resveratrol elicitors were studied using post-harvest techniques such as UV light, size reduction method and ultrasound (Rudolf and Resurreccion 2005). The peanut seeds were reduced in size by various methods and then subjected to one of the abiotic stresses. It was found that these postharvest abiotic stresses caused an increase in trans-resveratrol concentrations over peanut seeds not treated.

A noticeable increase in the concentration of resveratrol in peanut leaves, by more than 200-folds, was observed after peanut leaves illuminated with an ultraviolet lamp at 1.35 mE/m^2 s for 2h and 12h incubation (Chung *et al.*, 2003). The UV treatment increased the resveratrol concentration from 0.25 to 11.97 lg/g in callus, while it remained constant in other tissues throughout the experiment (Kuo-Lung *et al.*, 2005). Ultrasound exposure caused a significant increase in resveratrol concentration in all peanut samples after 24 h of incubation in dark. It was found that ultrasound treatment causes a swift increase in the enzymatic activity of PAL, which is the main enzyme responsible for the deamination of phenylalanine for the biosynthesis of resveratrol precursor coumaryl CoA (Wu and Lin 2002).

The highest amount of resveratrol contents (3.96 lg/g) was observed in sliced peanut kernels treated with an ultrasound treatment at 39.2 mW/cm^3 for 4 min following 36 h incubation. Also, over 20-folds increase of resveratrol contents was detected in the peanut leaves treated with paraquat, whereas, two-fold increase was observed by wounding and H_2O_2. On the other hand, imbibed cotyledons of peanuts accumulated resveratrol in response to wounding induced by slicing (Arora and Strange 1991). .

Soybean

Glyceollins are a group of phytoalexins that are produced by the soybean plant (*Glycine max*) under the conditions of stress (Graham). Glyceollins are novel phytoalexins derived from soybeans that exist as three isomers (glyceollin I, II and III) (Figure 5.2). This group of compounds is generated as a plant defense mechanism by an interaction between the soybean plant and many microorganisms. Moreover, glyceollins function as antiestrogenic agents and may be used for the prevention or treatment of prostate, breast and ovarian carcinomas. The glyceollins were found

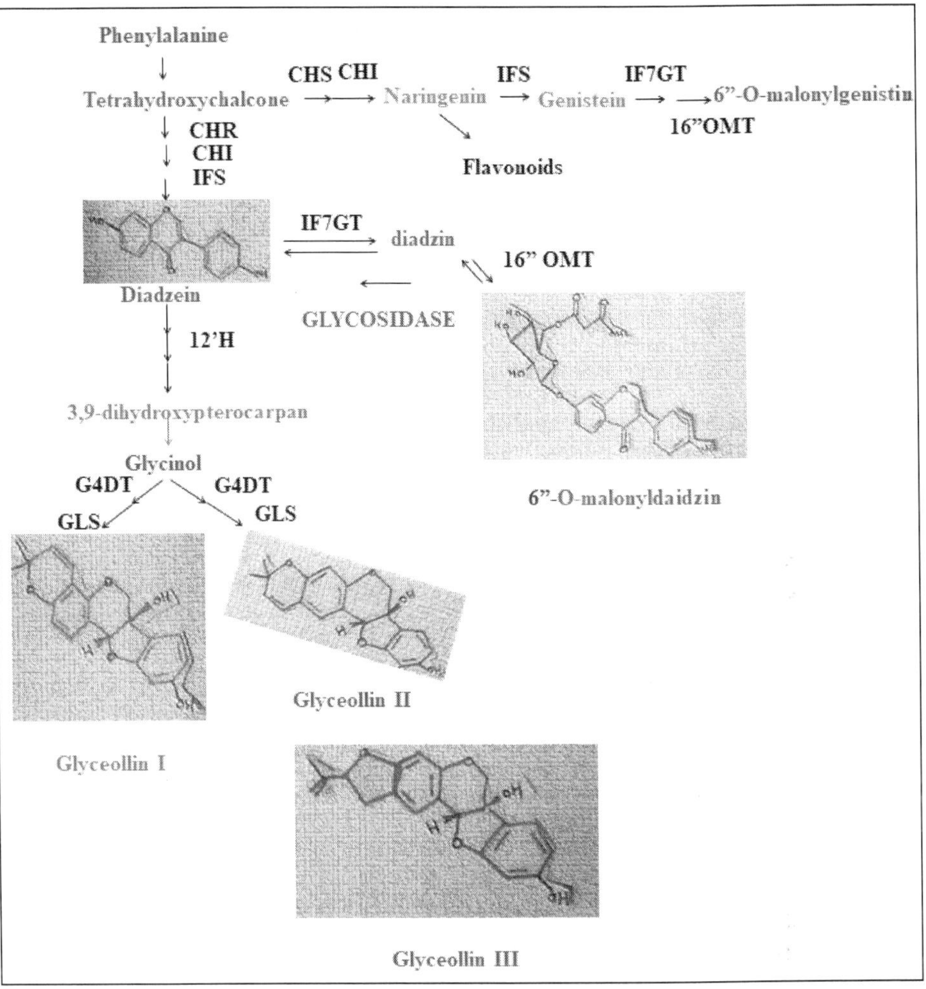

Figure 5.2: Glyceollin Biosynthetic Pathway.

to be involved in biological functions such as anticontractile activity in vascular smooth muscle through the RhoA/Rho kinase signaling pathway, insulin sensitivity and anti-melanin synthesis activity by inhibiting cAMP in vitro (Lee *et al.*, 2010). Consequently, glyceollins are toxic to nematodes, fungi, insects and bacteria for soybean self-protection (Bhattacharya *et al.*, 1992).

The soybean phytoalexins are pterocarpans derived from the phenylpropanoid pathway and appear as an isomer series (I-IV). Another anti-microbial isoflavone, isoflavone daidzein which is the immediate precursor of the glyceollins and genistein (tri hydroxylated isoflavone) gets accumulated in soybean tissues against fungal elicitors (Graham and Graham 1991). The soybean seedling tissues release both daidzein and genistein as large pools of pre-formed conjugates by hydrolysis in response to incompatible interaction with pathogens or when treated with fungal or plant elicitors (Graham, 1995).

Biotic and Abiotic Stress Resistance

Live and autoclaved spore suspensions of *Mucorramosis simus* were shown to be potent inducers of phytoalexin production in soybean cotyledons (Pelicice *et al.*, 2000). A branched hepta-β-glucoside from the phytopathogen *Phytophthora sojae* capable of eliciting phytoalexin accumulation in soybean cotyledons was found at concentrations of 10-8 to 10-9 M (Sharp *et al.*, 1984). Chitin and chitosan oligomers with DP from 4-6 have also been described as potent inducers of the phenylpropanoid biosynthetic pathway in soybean leaves.

Studies have shown that exposing soybean plants to infective strains of *Brady rhizobium japonicum* led to the exudation of glyceollin, a major phytoalexin of soybean (Schmidt *et al.*, 1992). However, the amount of glyceollin released in the presence of *Brady rhizobium japonicum* was less compared to levels showed in the presence of *Phytophthora* pathogen (Schmidt *et al.*, 1992). With *Sin orhizobium fredii* (another soybean symbiont), however, the level of glyceollin exudation was comparable to that caused by pathogenic *Phytophthora* (Schmidt *et al.*, 1992). During the early stages of interaction between host and microsymbiont, rhizobia act very much like plant pathogens and higher levels of glyceollin were noticed in the rhizosphere of *G. max* after 20 h of infection with *B. japonicum* (Schlaman *et al.*, 1989). Higher concentrations of glyceollin were found in mycorrhiza-infected roots of *G. max* than in axenic roots. Previous reports (Rudolf and Resurreccion 2005)showed that both pathogens and rhizobial symbionts can stimulate exudation of phytoalexin in N_2-fixing legumes.

The less complex ptero carpan glycinol, a biosynthetic precursor of glyceollin, has been reported to be metabolized by *E. carotovora*. Also, it was noted that the fungi *Aspergillus niger, Aspergillus oryzae* and *Aspergillus flavus* were all able to induce glyceollin in soybean (Boue and Raina 2003). Glyceollin showed a significant antimicrobial effect against *Phytophthora capsici* and *Sclerotinia scrotiorum* (Kim *et al.*, 2010). The abiotic stresses such as the presence of heavy metals or UV radiation strongly induce glyceollin phytoalexin in soybean (Ebel and Grisebach, 1988).

Sunflower

Sunflower (*Helianthus annus* L.) is one of the most important oilseed crops in the world, characterized by its short growing period, high yield potential and high content of edible vegetable oil. Unfortunately, it is susceptible to a variety of fungal pathogens, which cause great losses in its yield and oil quality. Sunflower contains various secondary products including terpenes, flavonoids and simple phenolic compounds (Tal and Robeson, 1986) which have been involved in the resistance of sunflower to both insect pests and fungal pathogens (Tal and Robeson, 1986). The majority of these compounds are present as constitutive metabolites in healthy plants. However, the coumarins scopoletin and ayapin (Figure 5.3) have been observed to accumulate in sunflower tissues following infection with pathogenic and non-pathogenic fungi (Tal and Robeson, 1986).

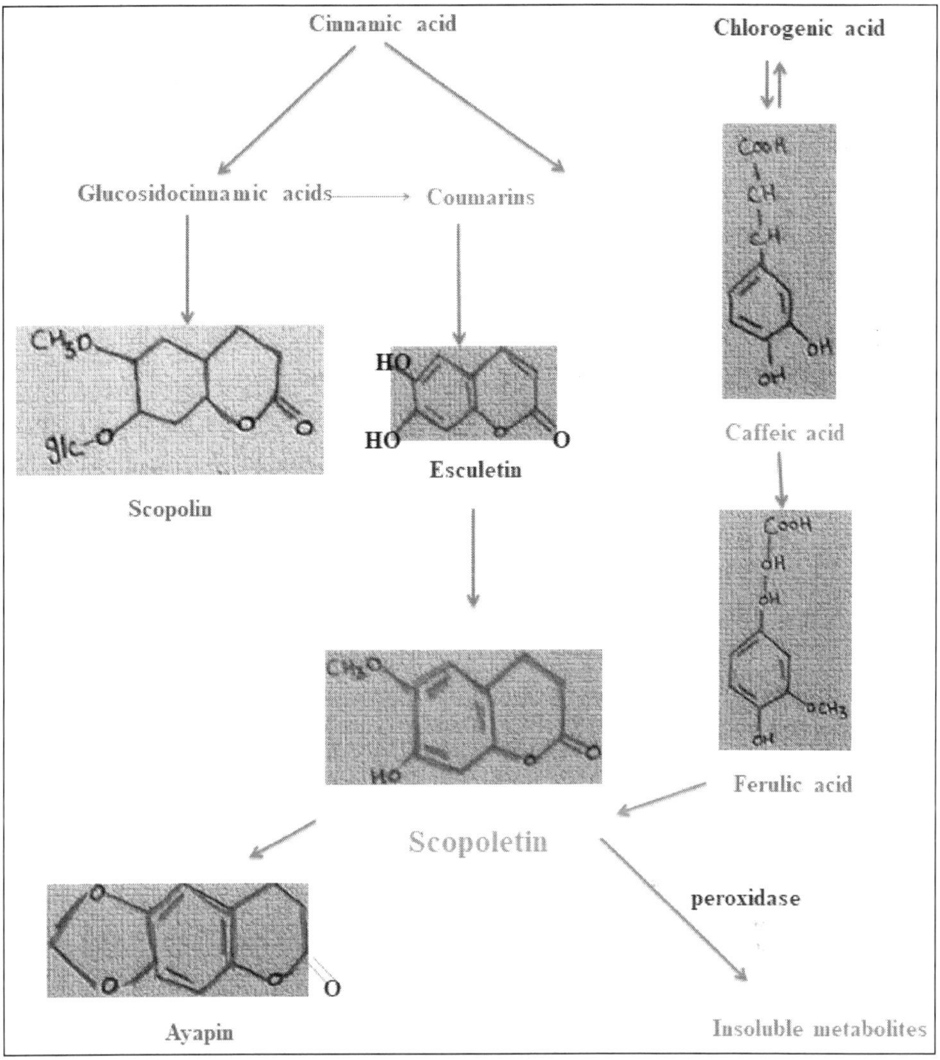

Figure 5.3: Potential Routes of Biosynthesis and Metabolism of Scopoletin in Sunflower.

Biotic and Abiotic Stress Resistance

In sunflower, coumarins such as scopolin, scopoletin and ayapin have been described as phytoalexins, since their synthesis is only induced in response to abiotic elicitors and fungal infection. Furthermore, the accumulation of chlorogenic, ferulic and caffeic acids was reported along with scopoletin, after the infection of sunflower with downy mildew (Saftiæ-Pankoviæ *et al.*, 2006). Similarly, the increase in preformed phenolics, as well as the production of scopoletin and ayapin in the leaves of sunflower plants in defense against *M. phaseolina* was confirmed

(Al-Wakeel *et al.*, 2013). Scopoletin-dependent total peroxidase (POD) activity was low in sunflower cultivars susceptible to *Sclerotinia sclerotiorum*, both constitutively and after the infection (Prats *et al.*, 2006). When stems were inoculated with fungal conidia of *Helminthosporium carbonum* or *Alteraria helianthi* scopoletin was rapidly induced being the predominant phytoalexin with ayapin accumulating more slowly (Tal and Robeson 1986). Studies with fungal pathogens of sunflower, such as *Phomamac donaldii*, have revealed that ayapin is more resistant to degradation by the fungi than scopoletin and can selectively accumulate in pathogen-infected stems. The coumarins scopoletin and its glucosyl derivative, scopolin, and ayapin may also be part of the sunflower defense strategy against parasitic *Orobanche cernua*, by acting as allele chemicals preventing *O. cernua* seed germination. The phytoalexins accumulation blocks the radicle penetration through the host root which prevents the connection of parasite with the host vascular tissues leading to its death (Khadija Serghini *et al.*, 2001). The $CuC1_2$ treatment selectively increases the metabolism of scopoletin to form ayapin or this elicitor stimulates the selective metabolism and degradation of scopoletin.

Mustard

Significant amounts of antimicrobial compounds called glucosinolates were reported in brassica species. Glucosinolates constitute a group of sulphur-containing glucosides that following tissue damage, are converted by the endogenous thioglucosidase myrosinase into breakdown products such as isothiocyanates (ITCs) and nitriles which exhibit toxicity towards bacteria, fungi and insects (Kliebenstein 2004). Species of *Alternaria* are accountable for significant losses in economically important oilseed crops and vegetable species (Thomma 2003). Earlier studies have established that the major glucosinolate-breakdown products (ITCs) and some crucifer phytoalexins have both *in vitro* and in planta inhibitory effects towards fungal pathogens (Tierens *et al.*, 2001). Continuous production of phytoalexin (spirobrassinin, cyclobrassinin, and rutalexin) and phytoanticipin (indolyl-3-acetonitrile, arvelexin, caulilexin C, and 4-methoxyglucobrassicin) was noticed in canola and rapeseed in response to the elicitation of biotic and abiotic stresses.

Crucifers produce high levels of indolic phytoalexins and glucosinolate-breakdown products against the necrotrophic fungi *Alternaria* spp. that cause black spot diseas. Brassica *napus* and *Brassica carinata* both have some mechanism that inhibits spore germination. In *Brassica rapa* the phytoalexins cyclo brassinin and brassilexin were found to be more inhibitory to *A. candida* than rutalexin. The pathogen may apparently redirect the phytoalexin pathway towards rutalexin, avoiding cyclobrassinin and brassilexin accumulation. On the other hand, *A. candida* may detoxify both cyclobrassinin and brassilexin, like necrotrophic plant pathogens. Phytoalexin production in infected or $CuCl_2$ stressed leaves suggested that *A. candida* escape the plant defense mechanisms by re-directing the biosynthetic pathway of phytoalexin (Pedras *et al.*, 2008).

A phytoalexin with a molecular formula $C_9H_6N_2S$ was isolated from the leaves of *Brassica juncea* cv. Aurea inoculated with *Leptosphaeria maculans*. Then phytoalexin was purified and quantified using reversed-phase and silica HPLC.

It inhibited spore germination and hyphal growth of *L. maculans* at 12 µg ml^{-1} in water. This phytoalexin was also accumulated with sprays of $AgNO_3$ or $CuCl_2$ solutions. Phytoalexin accumulation kinetics were compared in *B. juncea* cv. Aurea (hypersensitive to *L. maculans*) and *B. napus* cv. Brutor (susceptible to *L. maculans*) with $CuCl_2$ as a non-specific elicitor. The phytoalexin was detected in *B. juncea* 6h after challenge and in *B. napus* 18 h after challenge. Moreover, higher phytoalexin accumulation was found in *B. juncea* (4 to 10 times more) than *B. napus after* 48 h of elicitation (Thierry Rouxel *et al.*, 1989).

Pulse Crops

Pulse crops are important from a nutrition and production point of view but biotic and abiotic diseases are major challenges to control. Infection of this disease can significantly reduce yield (by up to 25 per cent) by decreasing the photosynthetic leaf area. The resistance of plants to different pathogens depends on the synthesis and level of various defense enzymes like hydrolases; peroxidases and antimicrobial compounds like phytoalexins.

Pigeon Pea

Pigeon pea, (*Cajanus cajan*), is a perennial legume (subfamily Papilionoideae) grown in the semi-arid tropic and subtropic regions of many developing countries. A total of 27 flavonoids consisting of six flavones, eight isoflavones, four flavonols, two anthocyanins, three flavanones, three isoflavanones and a single chalcone were identified (structure shown in Figure 5.4–5.8). Pigeon pea leaves are the richest flavonoid containing organ in the plant. The flavanone, cajaflavanone was reported from the roots (Dahiya 1991).

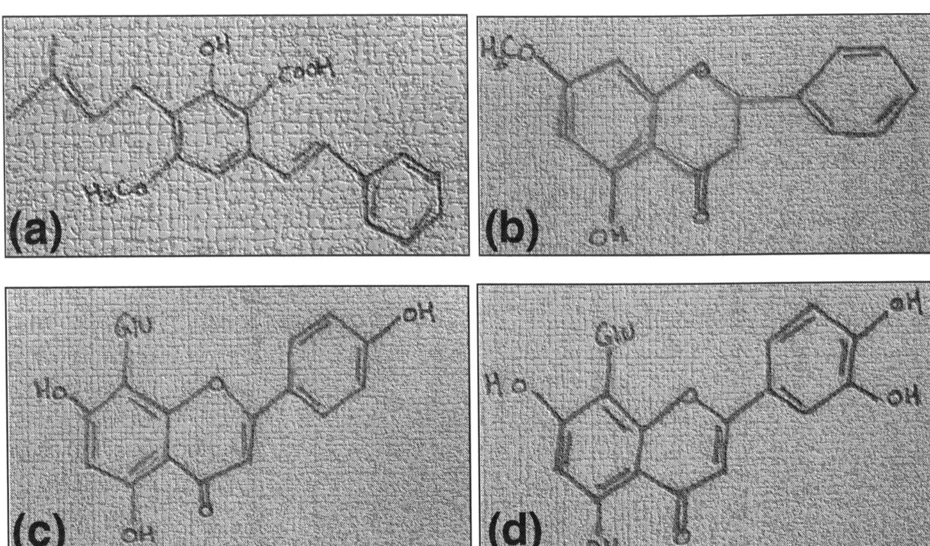

Figure 5.4: Chemical Structures of (a) Cajaninstilbene acid, (b) Pinostrobin, (c) Vitexin and (d) Orientin. Glu: glucose.

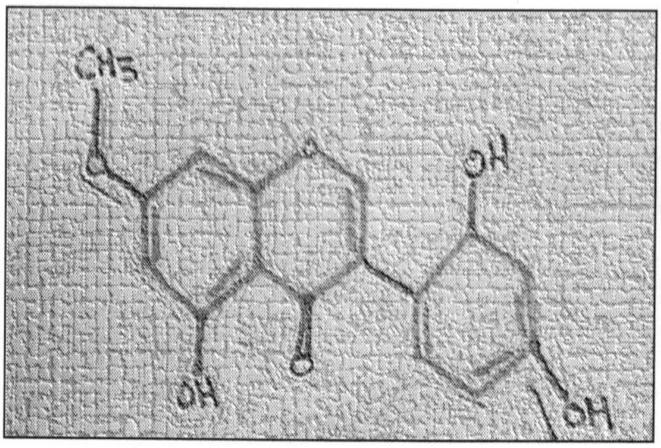

Figure 5.5: Structure of Cajanin.

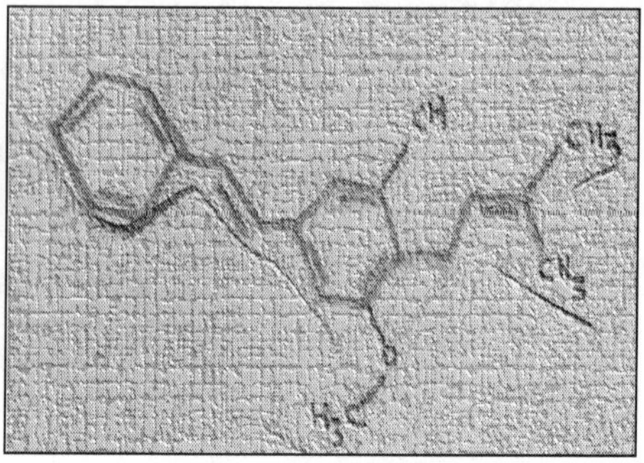

Figure 5.6: Structure of Longistylin A.

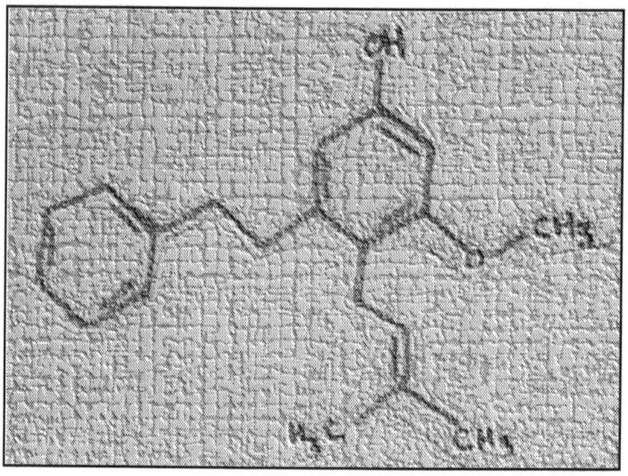

Figure 5.7: Structure of Longistylin C.

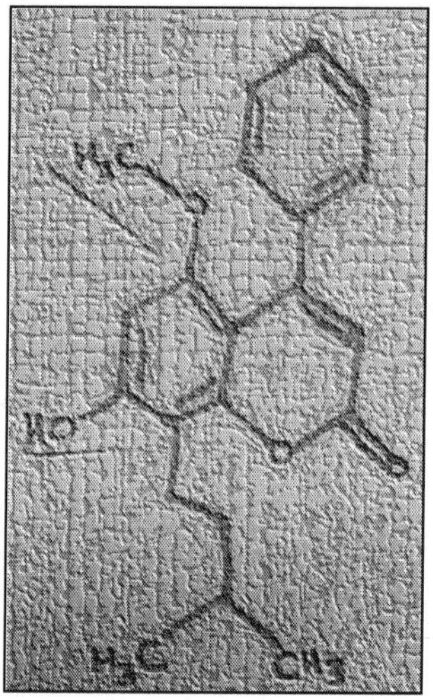

Figure 5.8: Structure of Cajanuslactone.

Biotic and Abiotic Stress Resistance

Pinostro binchalcone is the only chalcone described from pigeon pea and its occurrence is reported from the leaves which were challenged with the necrotic fungus *Botrytis cinerea* (Cooksey *et al.*, 1982). It is also assumed to play a role in photo-protection. Etiolated stems of pigeon pea inoculated with the fungus *H. carbonum* showed the accumulation of iso flavones formononentin, genistein, 22 -hydroxygenistein and cajanin, and the cajanol while being limited to trace levels in uninfected plants (Ingham 1976). The iso flavanone cajanone also showed an inhibition effect on the growth of the pigeon pea wilt pathogen *Fusarium oxysporum* (Preston 1977). Behavioral studies of larvae of the agricultural pest, *Helicoverpa armigera* confirmed quercetin 3-methyl ether increased their feeding behavior while cajanin stilbene acid (CSA) was found to be a feeding deterrent (Green *et al.*, 2003). Certain stimuli such as UV light, NO, H_2O_2, jasmonic acid and fungal infections induce the accumulation of stilbenes and they take part in both constitutive and inducible defense mechanisms in plants (Chong *et al.*, 2009).The effects of UV application on post-harvest pigeon pea leaves established that orientin, luteolin, apigenin and apigenin-6,8-di-C-α-é-arabinopyranoside all increased in concentration in response to variable levels of UV exposure (Wei *et al.*, 2013).

Chickpea

Chickpea (*Cicer arietinum* L.) is the third most important pulse crop in the world, but a major constraint in its world production is a destructive fungal disease known as ascochyta blight caused by *Ascochyta rabiei* (Pass.) L. (Coram and Pang 2007). Plants activate certain responses against pathogen infection that works as a disease resistance mechanism. These responses include the hypersensitive reaction, the *de novo* synthesis of phytoalexins, proteinase inhibitors, hydrolytic enzymes and increased production and deposition of lignin into the plant cell wall (Lawton and Lamb 1987). The best-characterized phenyl propanoid derived phytoalexins are the pterocarpans, isoflavans and isoflavanones of legumes including chickpea, bean, alfalfa, pea and soybean (Dixon *et al.*, 2002). Pterocarpans make up a group of natural isoflavonoids, which are usually involved as phytoalexins in the defense against pathogens of leguminous plants. Medicarpin and maackiain are two main phytoalexins in chickpea (Figure 5.9). Two common fungal pathogens of chickpea and other legumes *viz.*, *Nectriahaemato cocca* and *Ascochytarabiei* can degrade these phytoalexins by converting them to less toxic isoflavan, isoflavanone, or 6a-hydroxyptercocarpan through NADPH-dependent reductase (Vanetten *et al.*, 1989). The bean pathogen, *Fusarium solani f. sp. phaseoli*, can also detoxify four major bean phytoalexins, namely kievitone, phaseollin, phaseollidin, and phaseollinn isoflavan. Kievitone detoxification occurs through kievitonehydratase-catalyzed hydration of the isopentenyl side chain to produce the less toxic kievitone hydrate (Vanetten *et al.*, 1989). There is an indication that *Rhizobium* and *Brady rhizobium* strains transform various flavonoid molecules within their vicinity (Rao *et al.*, 1991), signifying that bacterial and fungal symbionts probably overcome iso flavonoid phytoalexin toxicity in tissues and soil by biotransforming these compounds.

A study of a root rot disease of chickpea caused by *N. haematococca* MP VI suggested that metabolism of the chickpea phytoalexins (-)-maackiain and (-)-medicarpin (8, 9) could be important for pathogenesis virulence. It was reported that during time course studies in peas, application of a new class of insect elicitor 'Bruchin B' increased the expression of isoflavone synthase gene within 8h of elicitor treatment followed by an increase in the level of iso flavones phytoalexin 'pisatin' that reached the highest level at 32- 64 h after elicitation (Cooper *et al.*, 2005). In another study, the elicitor from fungus *Ascochyta rabiei* induced a large amount of ptero carpan phytoalexins medicarpin and maackiain in cell suspension cultures of chickpea within 8 h in resistant varieties, whereas, in susceptible cultivar, only a small amount of phytoalexins were accumulated 12 h after elicitor application (Kessman *et al.*, 1988). More recently, studies concluded that chickpea phytoalexins (the pterocarpans maackiain and medicarpin) are important components of the resistance mechanism of this chickpea to fusarium wilt (Stevenson *et al.*, 1997).

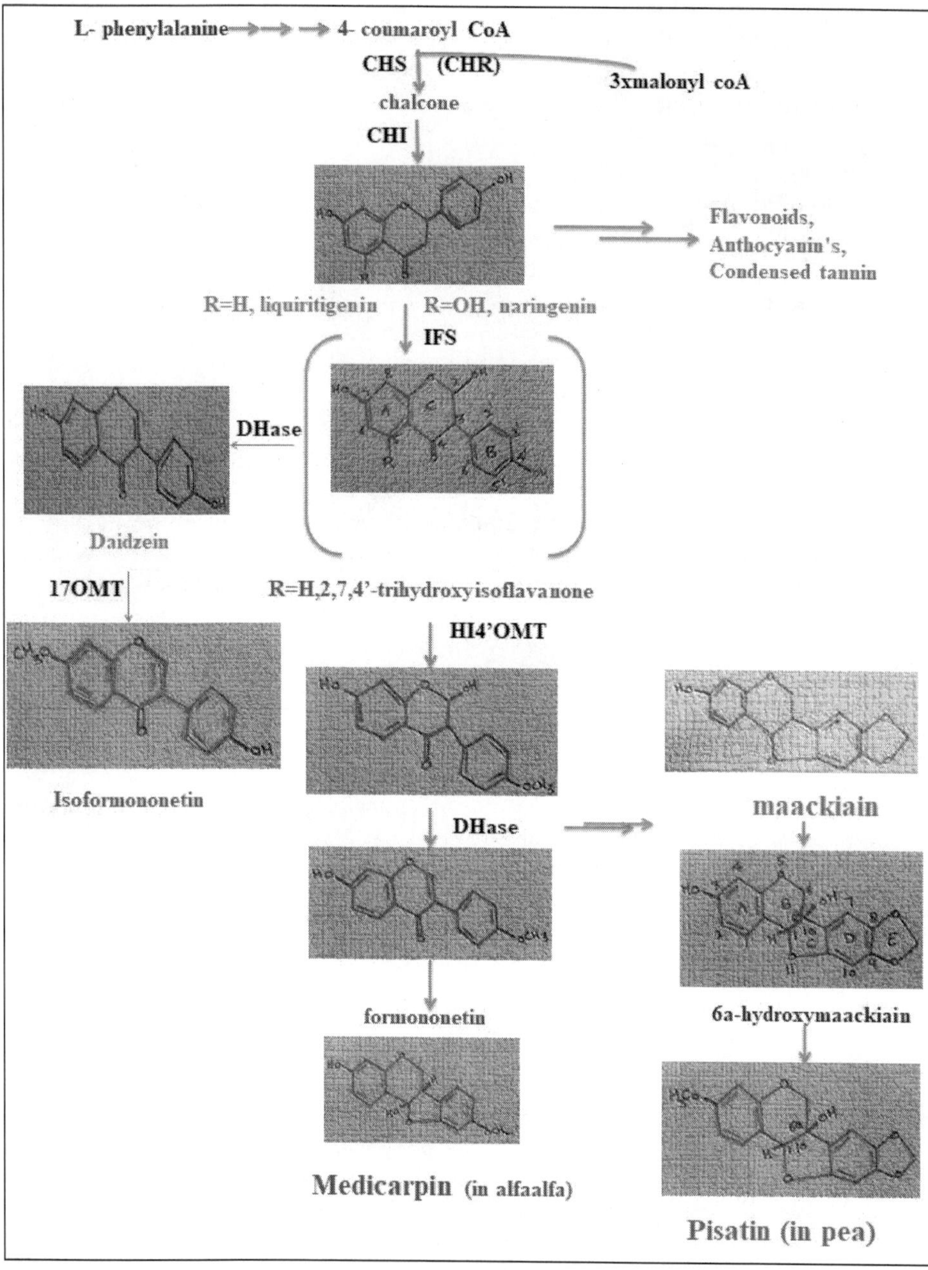

Figure 5.9: Biosynthesis of Maackiain, Medicarpin and Pisatin.

Conclusion

Though phytoalexins are deliberated as important for plant resistance against pathogens, the phytoalexins of most species and cultivars have yet to be characterized. Recent approaches, such as genome-wide analyses, could open the door to study regulatory networks controlling the metabolism of phytoalexins and provide a better insight of the phytoalexin's role in defense against pathogens. Better knowledge of the mode of action of phytoalexins and the mechanisms used by pathogens to bypass this line of defense could bring to light new opportunities for the directed control of phytoalexin production in specific tissues and at particular developmental stages.

REFERENCES

Al-Wakeel SM, Gabr M M A, Abu-el-soud WM and Saleh A M. (2013). Coumarin and salicylic acid activate resistance to *Macrophominaphaseolina* in *Helianthus annuus*. ActaAgronomicaHungarica, 61(1), 23–35.

Arora MK and Strange RN. (1991). Phytoalexin accumulation in groundnuts in response to wounding. *Plant Sci.*, 78, 157–163.

Bhattacharya D, Siddiqui K A I and Ali E. (1992). Phytotoxic metabolites of *Macrophomina phaseolina*. *Indian Journal of Mycology and Plant Pathology*, 22, 54–57.

Boue SM and Raina AK. (2003). Eûects of plant flavonoids on fecundity, survival, and feeding of the Formosan subterranean termite. *J. Chem. Ecol.*, 29, 2575–2584.

Chong JL, Poutaraud A and Hugueney P. (2009). Metabolism and roles of stilbenes in plants. *Plant Sci.*, 177, 143–155.

Chukwumah Y, Walker L, Vogler B and Verghese M. (2012). Profiling of bioactive compounds in cultivars of Runner and Valencia peanut market-types using liquid chromatography/APCI mass spectrometry. *Food Chem.*, 132, 525–531.

Chung IM, Park MR, Chun JC and Yun SJ. (2003). Resveratrol accumulation and resveratrol synthase gene expression in response to abiotic stresses and hormones in peanut plants. *Plant Sci.*, 164, 103–209.

Cooksey CJ, Dahiya JS, Garratt PJ and Strange RN. (1982). Two novel stilbene-2-carboxylic acid phytoalexins from *Cajanuscajan.Phytochemistry*, 21(12),2935–2938.

Cooper LD, Doss RP, Price R, Peterson K and Oliver JE. (2005). Application of Bruchin B to pea pods results in the up regulation of CYP93C18, a putative isoflavone synthase gene, and an increase in the level of pisatin, an isoflavone phytoalexin. *J. Exp. Bot.* 56, 1229-1237.

Coram TE and Pang ECK. (2007). Transcriptional profiling of chickpea genes differentially regulated by salicylic acid, methyl jasmonate and amino cyclopropan carboxylic acid to reveal pathways of defense-related gene regulation. *Functional Plant Bio.*, 34, 52-64.

Dahiya JS. (1991). Cajaflavanone and cajanone released from Cajanuscajan (L. Millsp.) roots induce nod genes of *Bradyrhizobium* sp. *Plant Soil*, 134(2), 297–304.

Dixon RA, Achnine L, Kota P, Liu CJ, Reddy MSS and Lang LJ. (2002). The phenylpropanoid pathway and the plant defense: A genomic perspective. Mol. *Plant Pathol.*, 3, 371-390.

Ebel J and Grisebach, V. (1988). Defense strategies of soybean against the fungus *Phytophthora megasperma* f. sp. glycine: a molecular analysis. *Trends Biochem. Sci.*, 13, 23-27.

Graham TL and Graham MY. (1991). Glyceollin elicitors induce major but distinctly different shifts in isoflavonoid metabolism in proximal and distal soybean cell populations. *Mol Plant Microbe Interact*, 4, 60–68.

Graham TL. (1995). Cellular biochemistry of phenylpropanoid responses of soybean to infection by *Phytophthora sojae*. In Handbook of Phytoalexin Metabolism and Action. (M. Daniel and R.P. Purkayastha, eds.). Marcel Dekker, New York, 85-116.

Green PWC, Stevenson PC, Simmonds MSJ and Sharma HC. (2003). Phenolic compounds on the pod-surface of pigeon pea, *Cajanus cajan*, mediate feeding behavior of *Helicoverpa armigera* larvae. *J Chem Ecol.*, 29(4), 811–821.

Hammerschmidt R. (1999). Phytoalexins: what have learned after 60 years? *Annu. Rev. Phytopathology*, 37, 285–3062.

Ingham JL. (1976). Induced isoflavonoids from fungus-infected stems of pigeon pea (*Cajanus cajan*). Naturforsch(C). 31(9–10), 504–508.

Ingham JL. (1976). Trihydroxystilbene as a phytoalexin from groundnuts (*A. hypogaea*). *Phytochem.*, 15,1791–1793.

Kessman H, Daniel S and Barz W. (1988). Elicitation of pterocarpon phytoalexins in cell suspension cultures of different chickpea (*Cicer arietinum* L.) cultivars by an elicitor from the fungus *Ascochytarabiei*. J. Biosci., 43, 529-535.

Khadija Serghini, Alejandro Pérez de Luque, Mercedes Castejón Muñoz, Luis García Torres and Jesús V. Jorrín. (2001). Sunflower (*Helianthus annuus* L.) response to broomrape (*Orobanche cernua* Loefl.) parasitism: induced synthesis and excretion of 7 hydroxylated simple coumarins. *Journal of Experimental Botany*, 52, 2227–2234.

Kim HJ, Suh HJ, Kim JH, Park S, Joo YC and Kim JS. (2010). Antioxidant activity of glyceollins derived from soybean elicited with *Aspergillus sojae*. *Agric. Food Chem.*, 58, 11633–11638.

Kliebenstein DJ. (2004). Secondary metabolites and plant/environment interactions: a view through *Arabidopsis thaliana* tinged glasses. *Plant, Cell and Environment*, 27: 675–84.

Kuo-Lung K, Chang PS, Cheng YC and Lien CY. (2005). Production of stilbenoids from the callus of *Arachishypogaea*: a novel source of the anticancer compound piceatannol. *J Agric Food Chem.*, 53, 3877–388.

Lanz T, Schroder G and Schroder J. (1990). Differential regulation of genes for resveratrol synthase in cell cultures of *Arachis hypogaea* L. *Planta*, 181, 169–175.

Lawton MA and Lamb CJ. (1987). Transcriptional activation of plant defense genes by fungal elicitor, wounding and infection. *Mol. Cell. Biol.*, 1, 335-341.

Lee Y, Kim H, Lee K, Jeon HW, Cui S and Lee YM. (2010). Inhibitory effect of glyceollin isolated from soybean against melanogenesis in B16 melanoma cells. *BMB Rep.*, 43, 461–467.

Morandi D, Bailey J A and Gianinazzi-Pearson V. (1984). Isoflavonoid accumulation in soybean roots infected with vesicular-arbuscular mycorrhizal fungi. *Physiol. Plant Pathol.*, 24, 357-364.

Pae SB, Ha TJ, Lee MH, Hwang CD, Shim KB and Park CH. (2011). Evaluation of characteristics of peanut sprout using Korean cultivars. *Korean J Crop Sci.*, 56(4), 394–399.

Pedras MS, Zheng QA, Gadagi RS and Rimmer SR. (2008). Phytoalexins and polar metabolites from the oilseeds canola and rapeseed: differential metabolic responses to the biotrophAlbugo candida and to abiotic stress. *Phytochemistry*, 69(4), 894-910.

Pelicice FM, Dietrich SMC and Braga MR. (2000). Phytoalexin response of fifteen Brazilian soybean cultivars. *Journal of physiology*, 12, 45-53.

Prats E, Bazzalo ME, Leon A and Jorrin JV. (2006). Fungitoxic effect of scopolin and related coumarins on *Sclerotiniasclerotiorum*. A way to overcome sunflower head rot. *Euphytica*, 147, 451-460.

Preston NW. (1977). Cajanone: an antifungal isoflavanone from *Cajanus cajan*. *Phytochemistry*. 16(1), 143–144.

Rao JR, Sharma ND, Hamilton JTG, Boyd DR and Cooper IE. (1991). Biotransformation of the penta hydroxy flavone quercetin by Rhizobium loti and *Bradyrhizobium* strains (lotus). *Appl Environ Microbiol.*, 57, 1563–1565.

Rudolf JR and Resurreccion AVA. (2005). Elicitation of resveratrol in peanut kernels by application of abiotic stresses. *J Agric Food Chem.*, 53, 10186–10192.

Saftiæ-Pankoviæ D, Veljoviæ-Jovanoviæ S, Pucareviæ M, Radovanoviæ N and Mijiæ A. (2006) Phenolic compounds and peroxidases in sunflower near-isogenic lines after downy mildew infection. *Helia*, 29, 33–42.

Sanders TH, McMichael RW and Hendrix KW. (2000). Occurrence of resveratrol in edible peanuts. *J Agric Food Chem.*, 48, 1243–1246.

Schlaman HRM, Spaink RJH, Okker BJJ and Lugtenberg. (1989). Subcellular localization of the nodD gene product in *Rhizobium leguminosarum*. *J. Bacteriol.*, 171, 4686-4693.

Schmidt PE, Parniske M and Werner D. (1992). Production of the phytoalexin glyceollin I by soybean roots in response to symbiotic and pathogenic infection. *Bot Acta.*, 105, 18–25.

Sharp JK, Valent B and Albersheim P. (1984). Purification and partial characterization of a β-glucan fragment that elicits phytoalexin accumulation in soybean. *Journal of Biological Chemistry*, 259, 11312-11320.

Sobolev VS and Cole RJ. (1999). Trans-resveratrol content in commercial peanuts and peanut products. *J Agric Food Chem.*, 47,1435–1439.

Sobolev VS, Neff SA and Gloer JB. (2009). New stilbenoids from peanut (*Arachis hypogaea*) seeds challenged by an *Aspergillus caelatus* Strain. Journal of Agricultural and Food Chemistry, 57, 62–68.

Spurthi N Nayak, Gaurav Agarwal, Manish K Pandey, Hari K Sudini, Ashwin S Jayale and ShilpPurohit. (2017). *Aspergillus flavus* infection triggered immune responses and host-pathogen cross-talks in groundnut during in-vitro seed colonization. *Scientific Reports.*, 7, 9659.

Stevenson PC, Turner HC and Haware MP. (1997). Phytoalexin accumulation in the roots of chickpea (*Cicer arietinum* L.) seedlings associated with resistance to *fusarium* wilt (*Fusariumoxysporum* f. sp. *ciceris*). *Physiological and Molecular Plant Pathology*, 50, 167– 78.

Tal B and Robeson DJ. (1986). The metabolism of sunflower phytoalexins ayapin and scopoletin. Plant-fungus interaction. *Plant Physiol*, 82, 167-172.

Tal B and Robeson DJ. (1986). The induction by fungal inoculation of ayapin and scopoletin biosynthesis in *Helianthus annuus*. *Phytochemistry*, 25, 77-79.

Thierry Rouxel, Alain Sarniguet, Albert Kollmann and Jean-Francois Bousquet. (1989). Accumulation of a phytoalexin in *Brassica* spp in relation to a hypersensitive reaction to *Leptosphaeria maculans*. *Physiological and Molecular Plant Pathology*, 34, 507-517.

Tierens KF, Thomma BP, Brouwer M, Schmidt J, Kistner K, Porzel A, Mauch-Mani B, Cammue BP and Broekaert WF. (2001). Study of the role of antimicrobial glucosinolate-derived isothiocyanates in resistance of *Arabidopsis* to microbial pathogens. *Plant Physiology*, 125, 1688–99.

Vanetten HD, Mathews DE and Mathews PS. (1989). Phytoalexin detoxification: importance for pathogenicity and practical implications. *Ann Rev Phytopathol.*, 27, 143–164.

Wang, H. (2015). Deep sequencing analysis of transcriptomes in Aspergillusflavus in response to resveratrol. *BMC Microbiology*, 15, 182.

Wei ZF, Luo M, Zhao CJ, Li CY, Gu CB, Wang W, Zu YG, Efferth T and Fu YJ. (2013). UV-induced changes of active components and antioxidant activity in postharvest pigeon pea [*Cajanus cajan* (L.) Millsp.] leaves. *J Agric Food Chem.*, 61(6),1165–1171.

Wu J and Lin L. (2002). Ultrasound-induced stress responses of Panax ginseng cells: enzymatic browning and phenolic production. *Biotechnol Prog.*, 18, 862–866.

Wu, Z. (2011). Food grade fungal stress on germinating peanut seeds induced phytoalexins and enhanced polyphenolic antioxidants. *J. Agric. Food Chem.*, 59, 5993– 6003.

Yang MH, Kuo CH, Hsieh WC and Ku KL. (2010). Investigation of microbial elicitation of trans-resveratrol and trans-piceatannol in peanut callus led to the application of chitin as a potential elicitor. *J Agric Food Chem.*, 58, 9537–9541.

Yu J, Ahmed M and Goktepe I. (2005). Effects of processing methods and extraction solvents on concentration and antioxidant activity of peanut skin phenolics. *Food Chem.*, 90 (1–2), 199–206.

Chapter 6

Herbicide Resistance Crops

Varsha Nakala[1], N. Lavanya[2] and Rajap Shiva Kumar[2]

[1]Department of Agronomy, Junagadh Agricultural University, Junagadh – 362001, Gujarat
[2]Department of Agronomy, Professor Jayashankar Telanagana State Agricultural University, Rajendranagar – 500030, Telangana

With the discovery of the resistance in weeds lead to the development of herbicide resistant crops (HRC's) by scientists. Herbicides prove to be the better option for the weed control in critical weed competition period. But the use of non selective herbicides like glyphosate and glufosinate ammonium in the crop is restricted due to the damage caused to the crop by them even though they give abroad spectrum and efficient weed control. In this situation, HRC's come to help which allow the use these herbicides helping the farmers to control all types of herbicides with a single herbicide and reducing the cost of cultivation. These HRC's have both pros and cons. The pros are multifold include the lesser use of herbicide, environmentally friendly, economical viable, while the cons include the geneflow and development of super weeds, build up resistance in weeds to these herbicides and also many theories stating that they may be harmful to the biodiversity. The usage of these herbicides in India is restricted and they are undergoing field trails to check their effect on the ecological balance.

***Keywords**: Resistance, Glyphosate, Glufosinate, Weed control, Transgenic, Herbicide.*

Introduction

In 1967, the Weed Science Society of America defined weed as "a plant growing where it is not desired" (Buchholtz, 1967). In 1989 the Society's definition was changed to define a weed as "any plant that is objectionable or interferes with the activities or welfare of man" (Humburg, 1989; Vencill, 2002).

Methods of Weed Control

Weeds in crop production can be managed in various methods which may be classified into one or more of five main categories. The 5 general methods of weed control are: Preventative, Cultural, Mechanical, Biological and Chemical.

Preventative Weed Control

Preventative weed control refers to any control method that targets to prevent weeds from being germinated or established in cultivated crop. Preventative weed control practices include using certified weed free seed, feeding animals with the fodder that is weed free, making sure farm equipment is cleaned before moving from one location to another and complete decomposition of compost or FYM.

Cultural Weed Control

Cultural weed control refers to methods which help to maintain field so that weeds are either established in minimal quantities or increased in count at slower rate. Cultural weed control consists of crop rotation, stale seed bed, avoiding overgrazing of pastures or rangeland, use of fast growing species or competitive varieties and maintaining good soil fertility.

Mechanical Weed Control

Mechanical weed control refers to methods that require the use of farm equipment to control weeds. Manual weeding, hoeing, mowing *etc.*, come under mechanical weed control.

Biological Weed Control

The usage of natural enemies of weeds to control the germination of weed seeds or suppress the growth of those weeds is called as biological method of weed control. *Cassia tora* and *Hyptis saveolans* inhibit the growth of *Parthenium*, *Neochetina bruchii* can help to control *Eichhornia crassipes* in aquatic systems *etc.*,

Chemical Weed Control

Application of a chemicals *i.e.* herbicides onto foliage of weeds or soil to control the germination or growth of the weed species is chemical weed control. Few herbicides used to control weeds are 2,4-DB, pendimethalin, quizalophos and paraquat.

Herbicides are a class of pesticides that are used in the management of undesired plants or weeds. Extensive and continuous dependence on herbicides for weed control in various cropping systems across the world has resulted in the evolution of resistance to herbicides.on an interesting note, this issue of weed resistance to herbicides has fetched the visionary scientists with a crucial idea to develop crops resistant to previously non-selective herbicides (Reddy and Nandula, 2012).

Herbicide resistance is defined by WSSA, as the acquired ability of a weed population to survive an herbicide application that previously was known to control the population. Herbicide tolerance is the inherent ability of a species to survive and reproduce after herbicide treatment. There has been no selection acting on the tolerant weed species, and there has been no change in the weed species lack of response to the herbicide over time.

The first case of herbicide resistance was reported in the United States in the 1950's. Field bindweed resistant to 2,4-D was discovered in Kansas in 1964, and common groundsel resistant to triazine herbicides was registered in Washington in 1970. Currently, around 514 unique cases (species x site of action) of herbicide resistant weeds are reported globally in 262 species which include 152 dicots and 110 monocots. Weeds are known to be evolved resistance to 23 of the 26 herbicide sites of action and to 167 different herbicides. Herbicide resistant weeds have been reported in 93 crops in 70 countries(www.weedscience.org). The discovery of herbicide-resistant weeds triggered the curiosity among the scientists for use this mechanism in crop breeding which lead to the development of Herbicide resistant crops (HRC's). HRCs are grown commercially since 1984, with the first triazine-resistant oilseed rape cultivar (OAC Triton), developed by methods of traditional breeding and introduced on the Canadian market Triazine resistance had been backcrossed from *Brassica rapa* L. into a commercial variety of oilseed rape (Hall *et al.*, 1996). However, the transgenic herbicide-resistant (HR) crop varieties were first introduced in 1996 and offered the promise of more effective weed control. More than 79 million hectares worldwide were planted to HR varieties of soybean, maize, canola, cotton, alfalfa, and sugar beets by 2008 (James, 2009).

Table 6.1: Few Non-transgenic Herbicide Crops which are Developed by Traditional Breeding/Selection Techniques

Selection Method	*Herbicide Ffamily*	*Crop*	*Year of Disclosure*
Whole plant	Triazine	Canola	1984
Microspore selection	Imidazolinone	Canola*	1989
Mutagenesis	Sulfonylurea	Soyabean*	1987
	Imidazolinone	Wheat*	1991
	Imidazolinone	Rice	1998
	Triazine	Wheat	2006
Pollen mutagenesis	Imidazolinone	Corn*	1992
Tissue culture	Accase inhibitor	Corn	1992
	Imidazolinone	Corn*	1991
	Triazine	Soyabean	1996
	Sulfonylurea	Canola	2002
Cell selection	Imidazolinone	Sugarbeet	1998
Transfer from weedy relative	ALS inhibitor	Sunflower*	2000
	ALS inhibitor	Sorghum	2008
	Accase inhibitor	Sorghum	2008

Adapted from Duke (2005); Green and castle (2010).

Accase: Acetyl-coA carboxylase; ALS: Acetolactate synthase.

Crops indicated with* are commercially available with resistance to correspondence herbicide family (in the same row).

About 60 per cent (87 of 144) of all transgenic/biotechnological events reported involved herbicide resistance traits (CERA, 2010). An unparalleled e87 fold increase has been observed in transgenic crop hectarage, from 1.7 million ha in 1996 to 148 million ha in 2010. while the number of countries cultivating transgenic crops increased to 29 in 2010 among which 19 countries are developing countries and rest of the 10 are developed countries. Transgenic crops accounted for 10 per cent 1500 million ha of cropland across the world (James, 2010). Among the HRCs, the most dominant transgenic crop in 2010 was soybean, occupying 73.3 million ha or 50 per cent of global area planted to transgenic crops. In 2010, herbicide resistance crops: soybean, maize, canola, cotton, sugar beet, and alfalfa accounted for 61 per cent (or 89.3 million ha) of the global transgenic area (148 million ha) (Reddy and Nondual, 2012).

Table 6.2: Present Transgenic Herbicide Resistant Crops and their Associate Trait Genes

Crop	*Resistant Trait*	*Trait Gene*	*First Sales*
Alfalfa	Glyphosate	Cp4 epsps	2005
Canola	Glyphosate	Cp4 epsps and goxv 247	1996
	Glufosinate	Pat	1995
	Bromoxynil	Bxn	2000
Cotton	Bromoxynil	bxn	1995
	Glyphosate	Cp4 epsps	1996
		Two ep4 epsps	2006
		Zm-2mepsps	2009
Corn	Glufosinate	Bar	2004
	Glyphosate	3 modified ep4 epsps	1998
		Two ep4 cpsps	2001
	Glufosinate	Pat	1997
	Glyphosate+glufosinate	Double stack	Not clear
Soyabean	Glyphosate	Ep4 epsps	1996

Benefits of Herbicide Resistant Crops

Control of Weeds

Broader Spectrum of Weeds Controlled

With the advent of the HRC's the scope for utilisation of non-selective herbicides such as glyphosate and glufosinate has widened. The main benefit of HR crops is the elimination of any injury from previously non-selective herbicide which provide the broad spectrum of weed control.

Reduced Crop Injury

The use of various herbicides more specifically the post emergence cause damage to the crop there by leading to the yield loss. This reaction is more common with the increased dosage application of the herbicides and the symptoms are more severe when the crop is under stress or unfavourable environmental conditions occur. HRC's help to overcome this problem with their increased selectivity towards the herbicides.

Flexibility and Simplicity

This technology being simple to use neither requires special skills nor training is flexible to use without much restrictions, which is probably one of the reasons for such wide adoption by farmers across the globe. They allow the use of wider window of application of herbicides which enables the effective weed control even during poor weather conditions like incessant rainy days.

Environment

Less Herbicide Usage

HRC's reduce the use of herbicides owing to use of herbicides with broad spectrum control. the two or more post emergence herbicides for control of grasses, sedges and broad-leaved weeds are replaced by single broad-spectrum herbicide.

Less Herbicide Carry-Over

Generally, the non selective herbicides are less persistent and known for non residual nature. Thus, the sowing of the HRC's enables the use of this non residual, non selective herbicides having lesser carry over effect.

Use of Herbicides which are Environmentally Friendly

Non selective herbicides like glyphosate and glufosinate have lower toxicity to humans and animals compared to some other herbicides because they are absorbed by the organic particles in the soil and decompose rapidly.

Economics

Higher and more Profitable Yields

Weeds which are major pests of the crop causing huge losses and those which compete for water, nutrients and sunlight can be managed easily with the adopting of herbicide resistant crops from the initial seedling stage itself. This gives a competitive advantage to the crop and finally leads to increased higher yields along with better quality of the product.

Reduction in Cost of Cultivation

By the introduction of HRC results in a price reduction for conventional herbicides. With the increasing resistance of weeds towards selective herbicides making it difficult to manage. The HRC's help to replace the selective herbicides and also reduce the cost for two or more herbicides to control the wide spectrum of herbicides in the fields.

Other Benefits

With the discovery and report of resistance almost 40 years ago, weed resistance to herbicides has been a limiting factor for the usage of the herbicides. For example, biotypes that are known to be resistant to triazine herbicides include 40 dicot and 15 monocot species. Around 262 species were found resistant to 167 different herbicides which is becoming a major drawback for the usage of herbicides. Therefore, HRCs in an IWM programme can provide a new mode of action when used as an aid in resistance management. Since single or multiple herbicide resistance in weeds is a serious problem in certain parts of the world, the use of HRCs can provide solutions.

HRC's have a wider scope in conservation agriculture which demand the limited tillage and fuel consumption. Inclusion of them in CA helps to control weed population after the harvest of the previous crop with non selective herbicides like glyphosate.

Demerits of Herbicide Resistant Crops

Yield Lag and Yield Drag

Yield drag and yield lag are two major concerns identified by researchers. yield reduction due to addition of foreign genes is a yield drag. The potential yield depression due to the age of the variety in which the gene is inserted is the Yield lag.

Single Selection Pressure and Weed Resistance

Repeated application of same herbicide creates selection pressure on the weed population which may finally lead to development of weed resistance to the non selective herbicides which are used.

Weed Shifts

Weed shift are common in the common cultivated crops. Weedy and invasive species take advantage of the available niches by easily adapting to changes in production practices in. Species that don't adapt to management changes become "less frequent" compared to those who do adapt. Weed shifts will be occurred in each inside a population of an exact species (*e.g.* living mutants), or inside a plant community (*e.g.* certain species). Therefore, though glyphosate and glufosinate can control many weed species, they do not control all plant species.

Gene Escape

The "escape" of genes implies herbicide resistance through pollen from HRCs to other plant species is major concern, especially to closely-related wild relatives of weed species. Gene escapes from HRCs are not a new phenomenon. An herbicide resistance gene was naturally transferred through pollen from herbicide-tolerant IMI-wheat to jointed goat grass (*Aegilops cylindrica*) in north-western US.

Gene Flow and Contamination of Organic Crops

Contamination of non-GMO crops occurs by pollen from GMO crops through the Gene flow. It is a major concern for organically-grown crops. For example, over usage of glyphosate in resistant soybean can cause problems for the production of organic soybeans due to contamination by glyphosate-resistant genes via cross-pollination from adjacent fields with HRC soybean. This also applies for organic corn production. As the tests can detect very minor quantities of cross-contamination, organic farmers are concerned that such cross-contamination may limit their ability to market organic crops which must contain no GMO seeds.

Herbicide Drift

A general concern when using any herbicide is drift and non-target movement. The concern becomes more with the use of non-selective herbicides such as glyphosate and glufosinate. Improper application and misidentification of fields planted with non-HRCs can occur if proper care is not taken to identify such fields and to avoid drift onto nearby fields with crops that are sensitive to the herbicide applied.

World Markets and Food Labelling

Current anti-biotechnology sentiment in Europe and Japan have caused a reduction in grain imports from the US to these countries. There is already an estimated 30 per cent reduction in US exports of various products related to glyphosate-resistant soybeans. This is the result of opposition to biologically engineered crops by consumers in these countries. In addition, many countries are considering, or have implemented, labelling regulations for GMO crops and grain products.

Method of Herbicide Resistance

The three principal requirements for manipulating herbicide resistance in plants through biotechnology are: 1. Identifying a gene that can provide resistance to an herbicide 2. Modification of this gene so that it can be expressed in the target crop plant and 3. Transfer of the modified gene into the crop plant.

Identifying a Gene that can Provide Resistance to Herbicide

Identification of Genes for Herbicide Resistance a number of mechanisms are to be used by the plants to tolerate herbicides such as reduced uptake of the herbicide, increase in the metabolism, and a target site which is not affected by the herbicide. Genes can be identified that would alter any of these steps to provide resistance to a particular herbicide. In general, the two mechanisms that are favoured for genetic engineering are modification of the target site and metabolism of the herbicide. The revolutionary aspect of this new technology is that genes for such traits need not come from the traditional germplasm resources used by plant breeders. Herbicide resistance genes may be found in any organism, including microbes. After the identification of those genes can be isolated, modified and transferred to the desired crop. The techniques and methods used to identify genes for herbicide resistance depend on the nature of the herbicide and its mode of action. Identification of microbes that either survive in the presence of the herbicide or use the herbicide as a food source has proved to effective approach. There are a number of advantages to screen the herbicide resistance genes in bacteria: a. through simple procedures large numbers of bacteria can be tested, such as screening for bacteria that grow on agar medium added with the herbicide, b. large collections of bacteria can be obtained simply from soil samples or other sources, c. Genes that are responsible for herbicide resistance can be isolated quickly because bacterial genomes are small and molecular genetic techniques facilitate this process.

Modification of Genes for Expression in Plants

Two essential pieces of information will be carried by the genes. Primarily, they will carry information to produce proteins. The cell will decode the sequence of bases in the DNA to direct the synthesis of a specific protein. Then the sequence of bases in DNA will be transcribed by the machinery of the cell into a similar sequence of bases in RNA. Now it is translated into a sequence of amino acids in the protein. From bacteria to plants and animals, the DNA code for proteins will be generally the same in all organisms. Consequently, if the gene is expressed after being transferred from a bacterium to a plant will encode the same protein. In addition to this, specific protein, each gene will also carry instructions about where, when, and to what level that gene should be expressed- that is, activated or switched on. For the function of all organisms these regulatory features are critical. In the sequence of bases in the DNA the regulatory information is also carried and it is referred to a promoter. Where and when that gene is expressed and whether the protein will be made in high amounts or at a low level is determined by a promoter of a gene. However, regulatory instructions are not the same from one organism to another, as if like the codes for proteins. As a result, unless the promoters are tailored to work in the host into which this gene is being moved genes transferred between organisms will not be expressed. For example, a gene isolated from Salmonella typhimurium, a bacterium, won't express in a plant unless the promoter that controls its expression is altered to function in plants. Genes which provide tolerance to herbicides through biotechnology contains two essential components: 1. A DNA sequence which encodes a specific protein which may either metabolize the herbicide or provide a target that is not supressed by the herbicide 2. A promoter sequence that will enable the expression of generated protein in specific tissues of the plant. The process of gene transfer is generally called as transformation. The cells and plants which have taken up a gene or regenerated from these cells by this method respectively are described as transformed or transgenic. To describe transgenic organisms, the term genetically modified organism (GMO) is also used, including transformed plants. Genes which are transmitted into plants are called as transgenes. The several techniques that are used to transform plants include a number of steps: Fertile plants can be regenerated through tissue culture for which a tissue is selected. The desired gene(s) are transmitted into as many cells as possible in this target tissue. To identify transformed cells selection is used. Plants are regenerated from transformed cells. Transgenic plants are analysed to confirm that the genes have been transferred, are being expressed as expected, and are having the desired effect on the phenotype of the plant. Transformation of many plants has used either Agrobacterium tumefaciens or particle bombardment in order to transfer genes into the plant. Agrobacterium tumefaciens is a soil bacterium that has the special ability to transfer DNA from the bacterium into plant cells. The bacterium uses this ability naturally to modify plant cells to secrete compounds that the bacterium can use for growth. For genetic manipulation of plants, Without the much serious efforts Agrobacterium can be used to transfer specific genes into plants normally associated with this bacterium. Until recently, Agrobacterium cannot be used for transformation of most crop plants. Hence, other methods were developed to transfer genes into these species, the most widely used being particle bombardment. Genes (pieces of DNA) are coated with metallic particles to be transferred into the plant and these DNA-coated particles are shot into the plant cells. Within the cell, the DNA will be released from the particle, reaches the nucleus, and incorporates itself into the genome. The technique of particle bombardment can transfer DNA into mostly all species, as it depends on physical methods to move DNA. Finally, the performance of transgenic is confirmed. Herbicide tolerance, involves demonstrating that transgenic plants have tolerance to field applications of the herbicide while maintaining suitable agronomic properties. The stability of the transgene to express in all plants and from generation to generation is important criteria. The time required to complete from the initial transformation of plant cells to release of a new variety will be at least 5 years. Though, once the new herbicide tolerance gene is included into elite breeding material, it can be used simply as another trait that should be followed during variety development. HRCs are classified as non-transgenic (traditional methods are used for selection of resistance character) and transgenic (genetically engineered).

Regulations of GM Crops in India (Bansal and Aurora, 2015)

In India the Environment Protection Act, 1986 implemented by Ministry of Environment, Forest and Climate Change is an umbrella legislation which provides a regulatory framework for GMO in the form of Rules, 1989.

Regulatory Framework Involved in the Development and Approval of GM Crops in India

Regulatory Approval of GM Crops

In 1989, the rules specifying the handling of genetically modified organisms (GMOs) and products were notified under Environment Protection Act, 1986 and guidelines were issued later. The Ministry of Environment and Forests (MoEF) and the Department of Biotechnology (DBT) under Ministry of Science and Technology, are responsible for implementation of the regulations. There are basically 6 authorities to handle different aspects of the regulation. These are Recombinant DNA Advisory Committee, Institutional Bio Safety Committee, Review Committee on Genetic Manipulation, Genetic Engineering Approval Committee, State Biotechnology Coordination Committee and District level Committee. Their specific responsibilities are described below:

1. Recombinant DNA Advisory Committee (RDAC): This committee is constituted by DBT to monitor the developments in biotechnology at national and international levels. RDAC submits recommendations from time to time that are acceptable for implementation for upholding the security laws in analysis and applications of GMOs and products thereof.
2. Institutional Bio Safety Committee (IBSC): This committee is constituted by organizations involved in recombinant DNA (r-DNA) research. It has

the mandate to approve low-risk (Category 1 and 2) experiments and to ensure adherence to r-DNA safety guidelines.

3. Review Committee on Genetic Manipulation (RCGM): This committee is constituted by DBT to review all ongoing projects involving high risk and controlled field experiments. RCGM regulates agricultural biotech analysis up to large scale field trials. It requests food bio safety, environmental impact and science data from candidates willing to do the analysis or conduct field trials and provides permits to import transgenic material for analysis. Its observation liquid substance analysis committee monitors field trials of transgenic crops.
4. Genetic Engineering Approval Committee (GEAC): This committee is a part of Ministry of Environment and Forests and is responsible for environmental approval of activities involving large scale use of GMOs in research, industrial production and applications. Besides food and environment safety, the GEAC also requires proof of agronomic performance and economic viability of the developed variety when cultivated by farmers.
5. State Biotechnology Coordination Committee (SBCC): This committee, constituted in each state where research and application of GMOs are contemplated, has the authority to inspect, investigate and take actions on cases of violations of the statutory provisions.
6. District level Committee (DLC): This committee is constituted at the district level to monitor the safety regulations in installations engaged in the use of GMOs in research and application.

India had approved the cultivation of *Bt* cotton with cry1Ac (Mon 531 event) by farmers in 2002 after extensive investigation for bio safety norms and agronomic performance.

Bt-cotton with is the cry1Ac first GM crop approved in India in the year 2002. Cotton production in India tripled from 13 million bales in 2003 to 37.7 million bales in 2017 (CAB, 2017). Cultivated in 11.4 million hectares (more than 93 per cent of total cotton area of the country).

BG Cotton – III (Herbicide Tolerant Cotton), contains both Round-up Ready and Round-up Flex (RRF) gene. Farmers might use the glyphosate indiscriminately, which may result in human health hazards and threat to cattle, apart from affecting cotton yield. However, herbicide-tolerant cottonseed technology has not yet received approval from the Genetic Engineering Approval Committee (GEAC) (Singh, 2019).

Brinjal is known to be infested by many insects and diseases and the most important pest is the fruit and shoot borer (FSB) and known to cause significant losses of up to 60 to 70 per cent. FSB-resistant brinjal or *Bt* brinjal was developed using similar transformation process to the one used in the development of *Bt* cotton. The cry1Ac gene expressing insecticidal protein is incorporated in *Bt* Brinjal to confer resistance against FSB. The gene source for cry1Ac gene is the soil bacterium, *Bacillus thuringiensis* (*Bt*). Maharashtra Hybrid Seeds Company (Mahyco) developed the *Bt* Brinjal. They used a DNA construct containing the cry1Ac gene, a CaMV 35S promoter and the selectable marker genes nptII and aad, to transform young cotyledons of brinjal plants. A single copy elite event, named EE-1, was selected and introduced into hybrid brinjal in Mahyco's breeding program. In the 97th meeting of GEAC held on 14th Oct 2009 has recommended the release of Bt Brinjal Event EE-1 for commercial cultivation developed indigenously by Mahyco in collaboration with the University of Agricultural Sciences (UAS), Dharwad and the Tamil Nadu Agricultural University (TNAU), Coimbatore. This is a penultimate step to commercialize *Bt* brinjal hybrids and varieties in the country (MOEF, 2009). BT brinjal was overturned in February 2010 by the then Environment Minister who assumed the role of regulator and ordered a suspension on the transgenic vegetable's cultivation.

Dhara Mustard Hybrid-11, better known as DMH - 11, is a genetically modified hybrid variety of the mustard species *Brassica juncea* resistant to glufosinate. It was developed by Professor Deepak Pental from the University of Delhi, aiming the reductio of edible oil imports by India. DMH - 11 was created through transgenic technology, primarily involving the Bar, Barnase and Barstar gene system. The Barnase gene confers male sterility, while the Barstar gene restores DMH - 11's ability to produce fertile seeds. The insertion of the third gene Bar, enables DMH - 11 to produce phosphinothricin-N- acetyl-transferase, the enzyme responsible for Glufosinate resistance. DMH - 11's Glufosinate resistance is due to an enzyme expressed by the Bar (Bialaphos resistance) gene. Derived from *Streptomyces hygroscopicus*, the cloned Bar gene in DMH-11 encodes for the synthesis of phosphinothricin-N- acetyl-transferase (PAT). This enzyme is responsible for detoxifying the active ingredient in the herbicide Glufosinate: phosphinothricin. Phosphinothricin's mechanism of action involves the inhibition of Glutamine synthetase, which prevents the detoxification of ammonia and subsequently causes toxic buildup within plant cells. Inhibition of glutamine synthetase also leads to an overall reduction in Glutamine levels. In plants, Glutamine acts as a signalling molecule, and as a major amino acid donor for nucleotide synthesis. PAT enzymes produced by the Bar gene, deactivate Bialaphos (the tripeptide precursor to phosphinothricin) through acetylation to form an inactive, non-toxic product (Peter Boger, 2002).

DMH - 11 was initially approved by GEAC for commercial cultivation, however, they retracted their approval stating that more investigations and additional data on the effect of DMH - 11 on insect pollinators, more specifically honeybees and on soil microbial diversity was needed prior to commercialisation. The Centre for Genetic Manipulation of Crop Plants at the University of Delhi applied for permission to grow DMH - 11 plots in Delhi and Punjab, to test the effects of the transgenic mustard on honeybees.

In India, Monsanto is undertaking field trials on a GM maize product which is both insect-resistant and herbicide-tolerant (MON89034 X NK603). Elsewhere, this is known by YieldGard VT/PRO brand name. The field trials on two transgenic corn hybrids with brand names 900M Gold and HiShell have been approved by the Genetic Engineering Appraisal Committee at BHU, Varanasi (U.P.), Begusarai

(Bihar); Bhagalpur (Bihar), TNAU, Coimbatore (Tamil Nadu), UAS, Dharwad (Karnataka), PJTSAU, Karimnagar (Telanagana), MPUAT, Udaipur (Rajasthan), AAU, Vadodara (Gujarat) and DWSR, Jabalpur (Madhya Pradesh). Other open-air cultivation plots have been allowed under new categories of trials called Insect Resistance Management trials in Begusarai and Bhagalpur in Bihar, in Aurangabad in Maharashtra, in Coimbatore in Tamil Nadu, Davangere in Karnataka and Warangal in Andhra Pradesh. "Seed production research trials" were also allowed in Kurnool and West Godavari districts in Andhra Pradesh apart from actual seed production allowed in Kanigel taluka in Tumkur district and Sindagi taluka of Bijapur district in Karnataka. Like all open-air trials, there is a serious threat of contamination from these trials. This GM Maize has stacked traits with cry2Ab2 and cry1A.105 Bt genes (Event MON 89034) and cp4epsps (Event NK603) gene inserted into maize, engineered for insect resistance (lepidopterans) and tolerance to the glyphosate. The toxins engineered into the plant are supposed to protect the plant from fall armyworm (*Spodoptera sp.*), black cutworm (*Agrotis ipsilon*), European corn borer (*Ostrinia nubilalis*) and the corn earworm (*Helicoverpa zea*). The genes are from soil bacteria (*Bacillus thuringiensis* and *Agrobacterium*) (Kavitha, 2011).

Conclusion

Alternative strategies need to be explored for the management of weeds. HRCs as weed management options have helped the farmers to manage weeds more effectively and economically. Increase in resistant weeds, ecological imbalance and weed species shifts were the main drawbacks observed by the continuous adoption of HRCs. The benefits are numerous like decreased cost on weed control, savings in fuel costs, positive effect on the environment. HRC's are widely adopted in developed countries like U.S.A. but in developing countries like India, they are currently under scrutiny for their effect on development resistant weeds and biodiversity.

REFERENCES

Bansal S and Arora A. (2015). Genetically Modified Crops in India: An Economic Analysis: Asia and the World Economy, pp.229-247.

Buchholtz KP. (1967). Report of the terminology committee of the Weed Science Society of America. *Weeds* 15: 388–389

CERA. 2010. GM Crop Database. Center for Environmental Risk Assessment (CERA), ILSI Research Foundation, Washington D.C. http: //cera-gmc.org/ index.php?action=gm_ crop database.

Duke SO. (2005). Taking stock of herbicide-resistant crops ten years after introduction. *Pest Management Science* 61(3): 211–18.

Green JM and Castle LA. (2010). Transitioning from single to multiple herbicide-resistant crops. In: Glyphosate Resistance in Crops and Weeds: History, Development, and Management (Nondual, V.K. Ed.), John Wiley and Sons, Inc., Hoboken, New Jersey, USA, pp. 67–92.

Green JM. (2009). Evolution of glyphosate-resistant crop technology. *Weed Science* 57: 108–17.

Green JM. (2011). Outlook on weed management in herbicide resistant crops: need for diversification. *Outlooks on Pest Management* 22: 100-4.

Green JM. (2012). The benefits of herbicide-resistant crops. *Pest Management Sci* 68: 1323–1331.

Hall JC, Donnelly Vanderloo MJ and Hume D.J. (1996). Triazine-resistant crops: The agronomic impact and physiological consequences of chloroplast mutation. In Duke, S.O. ed., Herbicide-resistant crops. Agricultural, Environmental, Economic, Regulatory and technical aspects. USA, CRC Press. pp. 107-126.

Heap I. The International Herbicide-Resistant Weed Database. Online. Monday, September 21, 2020.

Humburg NE (Ed.). (1989). Herbicide handbook, 6th ed. Weed Sci. Soc. Am., Champaign, IL. 301pp.

James C. (2001). Global GM Crop Area continues to grow and exceeds 50 million hectares for first time in 2001. Intl. Service for the Acquisition of Agri-biotech Applications. http: //www.isaaa.org/press percent20release/Global percent20Area_Jan2002.htm.

James C. (2010). Global status of commercialized biotech/GM crops: 2010. International Service for the Acquisition of AgriBiotech Applications (ISAAA) Brief 42-2010: Executive Summary. http: //www.isaaa.org/resources/ publications/ briefs/42/executivesummary/default.asp.

Kavitha Kuruganti. Briefing Paper on Herbicide Tolerant GM Crops (India), January 2011.

Peter, Böger. (2002). Herbicide Classes in Development: Mode of Action, Targets, Genetic Engineering, Chemistry. Wakabayashi, Ko., Hirai, Kenji. Berlin, Heidelberg: Springer Berlin Heidelberg. ISBN 9783642594168.

Reddy KN and Nandula VK. (2012). Herbicide resistant crops: History, development and current technologies. *Indian Journal of Agronomy* 57 (1): 1-7.

Singh A. (2019). HT Cotton in India- Current Scenario. Cotton statistics and news.

Vencill WK (Ed.). (2002). Herbicide handbook, 8th ed. Weed Sci. Soc. Am., Lawrence, KS. 493 pp.

Chapter 7

Climate Smart Agriculture Way Toward the Better Management Option for a Pest to Enhancing the Crop Production

K. Ashok, S. Pavithran and P. Dyamanagouda*

Department of Agricultural Entomology, Tamil Nadu Agricultural University, Coimbatore – 641003, Tamil Nadu

Climate change is causing global shifts in precipitation patterns, temperature as well as an increase in unpredictable, extreme weather patterns. It is already known that climate change has a significant impact on crop yields at the global level and will continue in future years. Climate shifts affect the biology, distribution, and outbreak of severe pests in a wide range of agricultural crops and across all landscapes. Climate smart way agriculture is a novel approach that aims to reduce climate induced crop losses, enhance ecosystem services, reduce the greenhouse emissions intensity per unit of food produced and increase the resilience of agricultural systems in the face of climate change. Implementation of climate smart pest management strategies involves pest forecasting models, pest diagnostics and surveillance, pest monitoring, application of chemical inputs at a low level, use of cultural, mechanical and biological control practices, biorational management practices and strengthening the farmer networking and pest management information and advice will check the invading pest populations at a low level. This review deals with the impact of climate change on insect populations and mitigating strategies to overcome pest infestation through climate smart solutions in order to enhance crop production.

Keywords: *Climate change, Crop losses, Climate smart pest management, Adaptation.*

Introduction

In recent years, overall agriculture has been influenced by the climate change *viz.,*increased temperature and carbon dioxide concentration, uncertainties in precipitation and more occurrences of drought periods which in turn have effects on insect-plant interactions in an environment (DeLucia *et al.,* 2012). Crop plants used as food by human beings are damaged by over 10,000 species of insects and cause an estimated annual loss of 13.6 per cent globally and 23.3 per cent in India (Dhaliwal *et al.,* 2010). In India, the average annual losses have been estimated to be 17.5 per cent valued at US$17.28 billion in eight major field crops (cotton, rice, maize, sugarcane, rapeseed, mustard, groundnut, pulses, coarse cereals and wheat) (Dhaliwal *et al.,* 2015).

Invertebrates are very susceptible to climate change, since they are cold blooded. Thus, many insects are affected by intense climates (droughts, heat waves or periods of cold). Tropical regions which are considered as the hotspots of biodiversity, when affected by climate change and deforestation they can dry up or fragment and many insect species are not permitted to survive (Williams *et al.,* 2003), particularly in areas with a high threat of extinction (Lewis, 2005).

Insects play a key role in ecosystem services, acting as herbivores, pollinators, predators and parasitoids, and any alterations in their abundance and diversity have the possibility to change the services they provide (Hillstrom and Lindroth, 2008).

These environmental alterations also change the life cycle of organisms, the ecological equilibrium and potential mass extinctions (Barnosky *et al.,* 2011). Plants are readily altered by a change in climatic factors, resulting in cascading effects on herbivorous insects, their predators and parasitoids (Boullis *et al.,* 2015; Rosenblatt and Schmitz, 2016).

Indirect influence includes a climate influence on atmosphere where insects emerge, plant architecture, food quality, predators, parasitoids and entomopathogens activity (Jaworski and HilszczaDœski, 2013). It is difficult to find a direct cause and effect relationship between any single factor and pest activity because the impact of meteorological factor on pests is usually compounded (Krishnaih *et al.,* 1996).

Hodkinson (1999) has reported that insect species population dynamics are largely determined by climatic factors. Climate affects the population dynamics of insects and is therefore a useful predictor of potential establishment and spread in new areas (Hance *et al.,* 2007). Numerous authors cite climatic factors to influence the changes in population size of insects over longer time periods such as several generations (Pathak, 1968). Recently frequent references have been made to rainfall, relative humidity and temperature. Pradhan (1972) suggested that cooler seasons and cooler regions had higher yields due to fewer pest problems. In temperate regions, most insects have their growth period during the warmer part of the year. Due to this, species whose niche space is defined by climatic regime will respond more predictably to climate change (Bale *et al.,* 2002). From the above, it is clear that the insect's abundance and distribution is mainly influenced by weather and climate of that particular region.

Several studies have been carried out on the population dynamics of insect pests but most considered only one species or a small region. However, at present, little is known about the temporal variation in the spatial pattern of insect pests. This chapters deals with the impact of climate change on insect populations and mitigating strategies to overcome the pest infestation through climate smart solutions in order to enhance the crop production.

Climate Smart Pest Management (CSPM)

The concept of CSPM is new and encompasses a set of interdisciplinary approaches and strategies needed for primary production to adapt, with its supporting functions (*e.g.* extension and research) and enabling environment (*e.g.* policies, infrastructure and social/human capital), to the changing climatic environment, which, together with land use characteristics, sets and influences the boundaries for geographical distributions of species (crops, pests and natural enemies). Figure 7.1 represents various approaches in Climate Smart Pest Management (CSPM) to mitigate climate change and improve food security. Implementing the CSPM concept enables more effective management of the increased threat posed by new and existing crop pests to agricultural production and ecosystem services, and therefore increases resilience of farmer livelihoods and overall local and national food security to climate change.

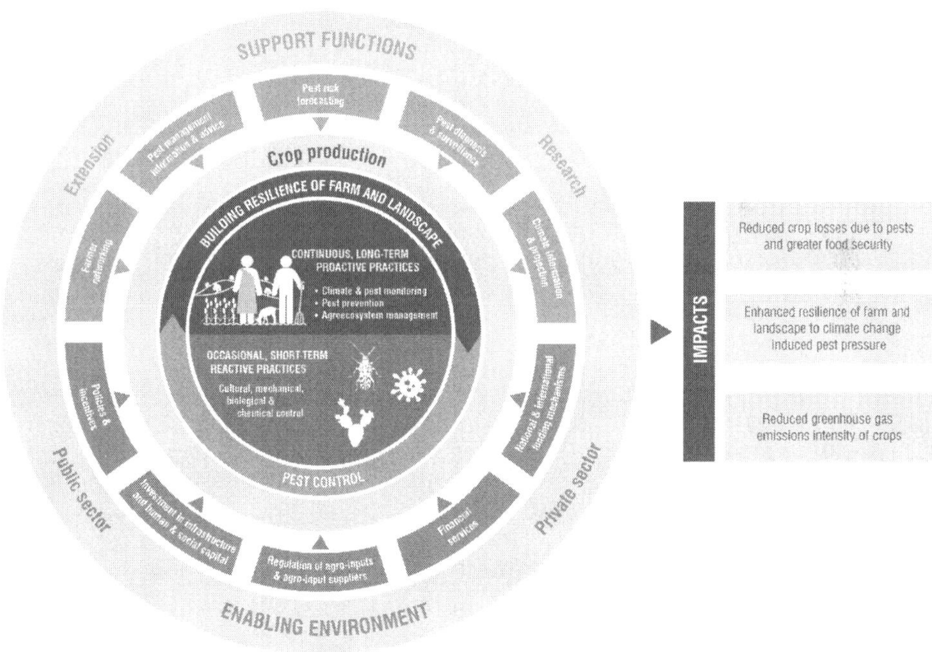

Figure 7.1: Various Approaches in Climate Smart Pest Management (CSPM) to Improve Food Security (*Source*: Heeb *et al.*, 2019).

Ecological Engineering

Odum first proposed ecological engineering in 1962, wherein his view it was "environmental manipulation by a man using small amounts of supplementary energy to control systems in which the main energy drives are still coming from natural sources"(Odum, 1962). Later, ecological engineering was synonymized with habitat manipulation, and it involves changing the agricultural field's favor for natural enemies for effective pest management. Some Ecological engineering principles involve relatively low inputs of resources such as natural processes (biocontrol agents), have developed to be consistent with ecological principles, refined by applied ecological experimentation, and contribute to theoretical and applied ecology knowledge (Gurr *et al.*, 2004). Ecological engineering mainly relies on reducing pesticide-induced pest population, cultural techniques based on the enemies hypothesis and cultural based on the resource concentration hypothesis and trap crops. Following the previous techniques will pave the way for resilience to the climate from increasing temperature, the frequency of extreme weather events, and declining biodiversity and eroding ecosystem functions. Horgan *et al.*, 2016 applied ecological engineering practices for resilient rice production systems in the Philippines. It includes floral and vegetational strips, high diversity vegetation patches and diverse vegetation patches. Following those practices increases the spider population, parasitoid and insect population, bird counts, and less pest incidence and rat damages.

Figure 7.2: Applying Ecological Engineering in the Rice Fields of Philippines (Source: Horgan *et al.*, 2016).

Planting crops such as *Sesamum indicum, Hibiscus esculentus, Hibiscus coccineus, Trida procumbens, Impatiens balsamen* will release volatiles around their surroundings, which gets attracted by egg parasitoids *Anagrus nilaparvatae* and *A.optablis*. Bund crops increase upto 70 per cent parasitism of *Anagrus nilaparvatae* on planthoppers comparing other weeds. Consequently, *Sesamum indicum, Tagetes erecta* will attract *Cyrtorhinus lividipennis* and it controls *Nilaparvatha lugens* by consuming up to 7.5 to 10.2 per adult daily. Planting of *Vetiveria zizaniodes* around the fields traps the striped stem borer *Chilo suppressalis* and it lays eggs on *V. zizaniodes*, but larvae were unable to complete its a life cycle in that plant (Lu *et al.*, 2014). Ecological engineering in rice provides two-fold way controlling pests. First, providing alternative hosts for the parasitoid during periods of rice planthopper unavailability. In eastern China, a perennial vegetable crop *Zizania caduciflora*, which is attacked by the planthopper *Saccharosydne procerus*, is often grown in rice fields. The parasitoid *A. optabilis* attacks *S. procerus* during the winter period and then moves to rice crops in spring to parasitize rice planthoppers (Lu 2003; Zheng *et al.*, 2003). A second and complementary approach is the use of flowering plants to provide nectar to parasitoids. Other than rice production systems, ecological engineering has been successfully applied to pest management practices. For example, the planting of buckweed, *Fagopyrum esculentum*, as a cover crop in vineyards and Alyssum, *Lobularia maritima* between rows of vegetables provide resources for predators and parasitoids, resulting in reduced herbivore damage (Berndt *et al.*, 2006).

Figure 7.3: Planting Trap Crop along the Rice Bunds for Attracting Stem Borer in China (Lu *et al.*, 2014).

Push-Pull Strategy

The term push-pull was first introduced as a strategy for management of insect pest by Pyke in Australia in 1987 and later Miller and Cowles, termed this strategy to stimulo-deterrent diversion (Pyke *et al.*, 1987; Miller and Cowles, 1990). Push-pull strategy involves use of repellent and attractant stimuli to manipulate the distribution and abundance of pest and/or beneficial insects for pest management. The pests are repelled or deterred away from this resource (push) by using stimuli that mask host apparency or are repellent or deterrent. The pests are simultaneously attracted (pull), using highly distinct and attractive stimuli, to other areas such as trap crops where they are concentrated, facilitating their elimination (Cook *et al.*, 2007). The push pull strategy is an alternative for chemical pesticides and other modern technologies that are economical and practical for poor and small scale farmers.

Table 7.1: Components of Push-pull Strategy

Push Strategy	*Pull Strategy*
☆ Visual clue	☆ Visual stimulant
☆ Synthetic repellants	☆ Host volatiles
☆ Non host volatiles	☆ Sex and aggregation pheromones
☆ Host derived semiochemicals	☆ Gustatory stimulants
☆ Antiaggregation pheromones	☆ Ovipositional stimulants
☆ Alarm pheromones	
☆ Antiffedants	
☆ Oviposition deterrants	

Figure 7.3: Climate Smart Push Pull Strategy: Green leaf *Desmodium* is planted between the rows of sorghum, while *Brachiaria cv. mulato* is planted around this intercrop (Murage *et al.*, 2015).

Table 7.2: Case Studies of Push-pull Strategy that are Under Development or Used in Practice in the Significant Areas of Insect Pest Control

Crop	*Pest*	*Push (Repellent)*	*Pull (Attractant)*	*References*
Maize, Sorghum	*Chilo partellus, Eldana saccharina, Busseola fusca, and Sesamia calamistis*	Molasses grass (*M. minutiflora*), silverleaf desmodium (*D. uncinatum*), or greenleaf desmodium (*D. intortum*)	Napier grass (*Pennisetum purpureum*) or Sudan grass (*Sorghum vulgare sudanense*)	Khan *et al.*, 1997
Sorghum	Stem borers and Striga	*Desmodium uncinatum*	*Pennisetum purpureum*	Murage *et al.*, 2015
Cotton	*Helicoverpa armigera and H. punctigera*	Main crop with neem extracts	Pigeon pea (*Cajanus cajan*) or maize (*Z. mays*)	Pyke *et al.*, 1987
Beans	*Sitona lineatus*	Neem extract	4-methyl-3,5-heptanedione	Smart *et al.*, 1994
Potato	*Leptinotarsa decemlineata*	Antifeedants	Aggregation pheromone ((S)-3,7-dimethyl-2-oxo-6-octene-1,3-diol)	Martel *et al.*, 2005
Oilseed rape	*Meligethes aeneus*	Lavender, (*Lavandula angustifolia*)	Turnip rape (*Brassica rapa*)	Cook *et al.*, 2006
Onion	*Delia antiqua*	Cinnamaldehyde	Onion culls	Cowles *et al.*, 1992
Chrysan-themum	*Frankliniella occidentalis*	Rosemary (*Rosmarinus officinalis*)	Aggregation pheromone ((R)-lavandulyl acetate and neryl (S)-2-methylbutanoate)	Bennison *et al.*, 2001

Protected Cultivation

Protected cultivation involves modified form of agroecosystem where, unlike other agroecosystems. "Protected cultivation practices can be defined as a cropping technique wherein the microenvironment surrounding the plant body is controlled partially/fully as per the plants' need during their period of growth to maximize the yield and resource saving" (Reddy, 2016). Protected cultivation is becoming increasingly important to modern climate smart way agriculture because of its capacity to produce all year round irrespective of cropping seasons. The greenhouse system is the main factor in protected cultivation, which gives a maximum advantage of the natural resources offered by the area in certain latitudes while minimizing the disadvantages. Protected cultivation involves three levels of sustainability *viz.* (Pascale and Maggio, 2004)

1. Use of natural resources,
2. Use of non-renewable resources (CO_2 enrichment, Pesticides),
3. Pollution levels

Crop Protection in Protected Cultivation

A controlled environment, high humidity and a warm temperature make the greenhouses suitable for insects and pathogens to attack. The primary purpose of greenhouse cultivation is to extend their cropping season and survive them from adverse environmental factors and maintain the quality of output resources. So crop protection in greenhouses involves specific management practices without altering the quality of produces.

i) Insect Screening

Screening involves the prevention of entry of arthropods into the greenhouses. It also provides light, temperature and humidity to crops. *e.g.*, UV absorbing plastic sheets, whitefly exclusion and thrips exclusion sheets.

ii) Biorational Pest Management

It integrates two words, biological and rational, which means the use of naturally derived pesticides which has no adverse effects on the environment. *e.g.*, Botanicals, Insect growth regulators for insects. Acibenzolar-S-methyl (Downy mildew), Copper sulfate (Bacterial blight), Cottonseed oil (Powdery mildew), Neem oil (Anthracnose) and Saponins (Bacterial diseases) (Reddy, 2016).

iii) Chemigation

Application of insecticides and fungicides along with drip irrigation or sprinkler irrigation for the management of soil invading arthropods (Weevils, root grubs, Cucumber fly) and certain plant diseases such as *Phytophthora capsici* (Ristaino and Johnston, 1999; Kuhar *et al.*, 2002).

iv) Grafting

Cultivation of grafted vegetables and fruits were common in the greenhouse system. The purpose of grafting in protected cultivation is to reduce soil borne pathogens like *Fusarium, Verticillium, Phytophthora, Ralstonia* and viral diseases such as tomato spotted wilt virus, Pepino mosaic virus and tomato leaf curl virus (King *et al.*, 2008; Louws *et al.*, 2010).

Conclusion

Food and Agricultural Organization (FAO) estimates that global food production must increase by 60 per cent by mid-century in order to satisfy the needs of the growing world population and their changing food habits. However, climate change is already having impacts on agriculture, including through its effects on the biology, distribution and outbreak potential of pests across all land uses and landscapes. At the same time, agriculture is also a major climate change driver, since it remains the world's second largest emitter of GHG. Reducing the impact of pests is more important than ever to ensure global food security, reduced application of pesticides and decreased GHG emissions intensity per unit of food produced. Climate smart agriculture through climate smart pest management is the only way toward the better management option for a pest to enhance the crop production and to the improve the food security.

REFERENCES

Bale, J., G. Masters, I. Hodkinson, C. Awmack, T.M. Jnbezemer, V.K. Brown, J. Butterfield, A. Buse, J.C. Coulson, J. Farrar, J.G. Good, R. Harrington, S. Hartley, T.H. Jones, L. Lindroth, M. Press, I. Mrnioudis, A. Watt and A. Whittaker, (2002). Herbivory in global climate change research: direct effects of rising temperature on insect herbivores. Global Change Biology, 8: 1-16.

Barnosky, A. D., Matzke, N., Tomiya, S., Wogan, G. O., Swartz, B., Quental, T. B. and Ferrer, E. A. (2011). Has the Earth's sixth mass extinction already arrived? *Nature*, *471*(7336): 51-57.

Bennison, J., Maulden, K., Dewhirst, S., Pow, E., Slatter, P., and Wadhams, L. (2002). Towards the development of a push-pull strategy for improving biological control of western flower thrips on chrysanthemum. In Proceedings of the 7th international symposium on Thysanoptera: Thrips and Tospoviruses. Reggio Calabria, Italy: Australian National Insect Collection, Canberra, pp. 199-206.

Berndt, L.A, Wratten, S.D, and Scarratt, S. L. (2006). The influence of floral resource subsidies on parasitism rates of leafrollers (Lepidoptera: Tortricidae) in New Zealand vineyards. Biological Control, 37(1): 50-55.

Boullis, A., Francis, F., and Verheggen, F. J. (2015). Climate change and tritrophic interactions: will modifications to greenhouse gas emissions increase the vulnerability of herbivorous insects to natural enemies? Environmental entomology, 44(2): 277-286

Cook, S. M., Khan, Z. R., and Pickett, J. A. (2007). The use of push-pull strategies in integrated pest management. Annual review of entomology, 52.

Cook, S. M., Smart, L. E., Martin, J. L., Murray, D. A., Watts, N. P., and Williams, I. H. (2006). Exploitation of host plant preferences in pest management strategies for oilseed rape (Brassica napus). Entomologia experimentalis et applicata, 119(3): 221-229.

Cowles, R. S., and Miller, J. R. (1992). Diverting Delia antiqua (Diptera: Anthomyiidae) oviposition with cull onions: field studies on planting depth and a greenhouse test of the stimulo-deterrent concept. Environmental entomology, 21(3): 453-460.

De Pascale, S., and Maggio, A. (2004). Sustainable protected cultivation at a Mediterranean climate. Perspectives and challenges. In International Conference on Sustainable Greenhouse Systems-Greensys, pp. 29-42.

DeLucia, E. H., Nabity, P. D., Zavala, J. A. and Berenbaum, M. R. (2012). Climate change: resetting plant-insect interactions. *Plant physiology*, *160*(4): 1677-1685.

Dhaliwal, G. S., Jindal, V. and Dhawan, A. K. (2010). Insect pest problems and crop losses: changing trends. Indian Journal of Ecology, 37(1): 1-7.

Dhaliwal, G. S., Jindal, V. and Mohindru, B. (2015). Crop losses due to insect pests: global and Indian scenario. Indian Journal of Entomology, 77(2): 165-168.

Gurr, G.M, Wratten, S.D. and Altieri M.A. (2004). Ecological engineering for pest management: advances in habitat manipulation for arthropods. Australia: CSIRO Publishing 238pp.

Hance, T., van Baaren, J., Vernon, P., and Boivin, G. (2007). Impact of extreme temperatures on parasitoids in a climate change perspective. Annual review of entomology, 52.

Hillstrom, M. L. and Lindroth, R. L. (2008). Elevated atmospheric carbon dioxide and ozone alter forest insect abundance and community composition. *Insect Conservation and Diversity, 1*(4): 233-241.

Hodkinson, I. D. (1999). Species response to global environmental change or why ecophysiological models are important: a reply to Davis *et al.*, Journal of Animal Ecology, 68(6): 1259-1262.

Horgan, F.G, Ramal, A.F, Bernal, C.C, Villegas, J.M., Stuart, A.M., and Almazan, M.L. (2016). Applying ecological engineering for sustainable and resilient rice production systems. Procedia food science, 6: 7-15.

Jaworski, T., and Hilszczañski, J. (2013). The effect of temperature and humidity changes on insects evelopment their impact on forest ecosystems in the expected climate change. Forest Research Papers, *74*(4): 345-355.

Khan, Z. R., Ampong-Nyarko, K., Chiliswa, P., Hassanali, A., Kimani, S., Lwande, W. and Woodcock, C. M. (1997). Intercropping increases parasitism of pests. Nature, 388(6643): 631-632.

King, S. R., Davis, A. R., Liu, W., and Levi, A. (2008). Grafting for disease resistance. HortScience, 43(6), 1673-1676.

Krishnaih, K., Pasalu, I.C., Krishnaiah, N.V. and Bentur, I.C. (1996). Integrated pest management in rice, In: Plant Protection and environment,Reddy D V S (eds) Plant Association of Indian, Hyderabad, India, pp. 94-104

Kuhar, T. P., Speese, J., and Hohlt, H. E. (2002). Evaluation of Drip Line-Injected And Foliar Insecticides For Controlling Cucumber Beetle In Melons. Arthropod Management Tests, 27(1).

Lewis, O. T. (2006). Climate change, species–area curves and the extinction crisis. Philosophical Transactions of the Royal Society B: Biological Sciences, 361(1465): 163-171.

Louws, F. J., Rivard, C. L., and Kubota, C. (2010). Grafting fruiting vegetables to manage soilborne pathogens, foliar pathogens, arthropods and weeds. Scientia Horticulturae, 127(2): 127-146.

Lu, B.R. (2003). Exploring sustainable production model of Jiaobai (*Zizania caduciflora* L.) through strategic biodiversity deployments. Acta Agriculturae Zhejiangensis, 15(3): 118-123.

Lu, Z. Zhu, P. Gurr, G.M, Zheng, X, Chen, G, and Heong, K. L. (2015). Rice pest management by ecological engineering: a pioneering attempt in China. In Rice Planthoppers. Springer, Dordrecht. 161-178pp.

Martel, J. W., Alfored, A. R., and Dickens, J. C. (2005). Synthetic host volatiles increase efficacy of trap cropping for management of Colorado potato beetle, Leptinotarsa decemlineata (Say). Agric. For. Entomol.7: 79–86.

Miller, J. R., and Cowles, R. S. (1990). Stimulo-deterrent diversion: a concept and its possible application to onion maggot control. Journal of Chemical Ecology, 16(11): 3197-3212.

Murage, A. W., Midega, C. A. O., Pittchar, J. O., Pickett, J. A., and Khan, Z. R. (2015). Determinants of adoption of climate smart push-pull technology for enhanced food security through integrated pest management in eastern Africa. Food Security, 7(3): 709-724.

Odum, H.T. (1962). Man in the ecosystem. In: Proceedings Lockwood Conference on the Suburban Forest and Ecology. Bulletin of the Connecticut Agricultural Station 652, pp. 57–75.

Pathak, M. D. (1968). Ecology of common insect pests of rice. Annual Review of Entomology, 13(1): 257-294.

Pradhan, S. (1972). Paddy yield higher in cooler regions. Entomology Newsletter, 2(10): 63-64.

Pyke, B., Rice, M., Sabine, B., and Zalucki, M. P. (1987). The push-pull strategy-behavioural control of Heliothis. Australian Cotton Grower, 9(1): 7-9.

Reddy, P. P. (2016). Sustainable crop protection under protected cultivation (No. BOOK). Singapore: Springer.

Ristaino, J. B., and Johnston, S. A. (1999). Ecologically based approaches to management of Phytophthora blight on bell pepper. Plant Disease, 83(12): 1080-1089.

Rosenblatt, A. E., and Schmitz, O. J. (2016). Climate change, nutrition, and bottom-up and top-down food web processes. Trends in Ecology and Evolution, 31(12): 965-975.

Smart, L. E., Blight, M. M., Pickett, J. A., and Pye, B. J. (1994). Development of field strategies incorporating semiochemicals for the control of the pea and bean weevil, Sitona lineatus L. Crop Protection, 13(2): 127-135.

Williams, S. E., Bolitho, E. E. and Fox, S. (2003). Climate change in Australian tropical rainforests: an impending environmental catastrophe. Proceedings of the Royal Society of London. Series B: Biological Sciences, 270(1527): 1887-1892.

Zheng, X., Yu, X. and Lu, Z. (2003). Parasitization adaptability of *Anagrus optabilis* on *Nilaparvata lugens*. Chinese Journal of Biological Control, 19(3): 136-138.

Chapter 8

Climate Change Impacts on Vegetable Crop Production and Adaptation Strategies

Khyati Singh[1]* *and Mukesh Kumar Mehla*[2]

[1]*Department of Vegetable Science, CCS Haryana Agricultural University, Hisar – 125004, Haryana*
[2]*Department of Soil and Water Engineering, Maharana Pratap University of Agriculture and Technology, Udaipur – 313001, Rajasthan*

Vegetable production is highly sensitive to environmental changes. Adverse climatic conditions are the primary cause of low production, along with crop failure, loss of yield quality, increased incidence of insect pests and diseases that make vegetable production a risky and unprofitable job. Moreover, enzymatic activities and physiological processes of plants are largely affected because of their temperature-dependent nature. These changes influence the occurrence of diseases and pest, host-pathogen interactions, ecology, distribution, migration and overwintering capacity of insects. Adopting climate smart agriculture will improve crop productivity and mitigate the direct and indirect impacts of adverse climate on vegetables. This approach includes activities to promote sustainable crop production technologies such as adjustment based on hot and dry weather condition to increase water use efficiency, application of mulches to conserve soil moisture level. Heavy rainfall causes flood-like conditions in the field, so to avoid excessive soil moisture, crops should be grown on raised beds. The development of tolerant genotypes against several biotic and abiotic stress like salinity, moisture stress and high temperature using conventional methods, breeding techniques or genomics and biotechnology is a primary focus of crop improvement programme to overcome these challenges. The objective of this chapter is to review the impact of changing climate conditions and adoption of climate smart agriculture strategies in vegetable crop production.

Keywords: *Climate smart agriculture, Climate change, Abiotic factors, Host-pathogen interaction.*

Introduction

With rise in the global population at consistent rate *i.e.*, estimated to be 1.1 per cent per year, the consumption of resources like food, feed, fiber, energy, land and freshwater have also increased significantly leading to decease in per capita availability. As per an estimate given by the UNESCO world's population will increase to 9.7 billion by 2050 (UNESCO, 2011), to feed this large population, food production must also increase by leaps and bounds. Considering population growth and food consumption trends, agricultural production should increase by 60 per cent by 2050, most notably in developing countries (FAO, 2012). Therefore, it is necessary that agriculture in developing countries undergo sufficient changes to overcome the challenges of food insecurity and climate change. The changes in climatic parameters such as temperature, relative humidity, rainfall, the composition of atmospheric gases, *etc.* over the large geographical area for a long period due to natural variability or human intervention is considered as 'climate change'. The ability of any cropping system to withstand these changing conditions depends on its level of sensitivity to avoid the adverse effects of climate change. Agriculture is a climate dependent sector and is sensitive to any changes in its climate or growing environment, showing both direct and indirect negative effects on production systems. A study by Rosenzweig *et al.* (2001) predicts that global temperatures will rise by 1.5-5.8°C at the end of the 21st century. Agriculture sector accounts for nearly 70 per cent of global freshwater use (Adamides, 2020). Himalayan glaciers that provide large amount of water required for agriculture in Asia, which is expected to decline by 20 per cent by the year 2030. Increased climate variability in the coming decades will increase the frequency and severity of floods and droughts which will pose increased production risks for agriculture especially crop production and reduce their coping ability (Thornton and Gerber, 2010).

The term 'vegetable' is defined as any plant product like leaves, petiole, flower, fruit, seeds and roots which is edible as raw or can be used for culinary purpose. Based on plant parts used as vegetable, these crops can be classified majorly into fruit vegetables such as tomato, brinjal, bottle gourd, cucumber, pea; root vegetables such as potato, carrot, radish, sweet potato, elephant foot yam; green leafy vegetables such as cabbage, celery, amaranths, curry leaf and bulb vegetables such as garlic and onion. As we know, vegetables play important role in our balanced diet and are known for their richness of vitamins, mineral, carbohydrate and protein, also help in overcoming micronutrient deficiency and when grown along with field crops in the intercropping system provide additional income to farmers. Therefore, it is required to simultaneously increase the vegetable crops productivity and minimize yield variability over time under unfavorable climatic conditions. Tropical zones are highly influenced by variation in climatic parameters, including a rise in atmospheric CO_2 level, temperature, high UV radiation, changing rainfall patterns and frequent occurrence of extreme weather events such as droughts and floods. In East-Asian countries, yields are highest due to temperate and sub-temperate climate which is most suitable for most vegetable production. However, being sensitive to the extreme environmental conditions, the physiological and biochemical processes of vegetables like photosynthetic enzymatic and other metabolic activity, pollination, fruit set, *etc.*

are greatly affected by the changing climate. Shifting in climatic patterns as a result of changing weather conditions has affected the crop production due to extreme weather events, variability in temperature and rainfall regimes and shockwaves during important phenological stages (IPCC, 2012).

The factors like short growing period and insufficient rainfall affect the growth and development due to terminal heat stress and less availability of water which leads to a reduction in the production. The variation in rainfall intensity and days of precipitation has a major impact on rainfed agriculture. Moreover, climate change influences the time of occurrence of diseases and pest, host-pathogen interactions, ecology, distribution, migration and overwintering capacity of insects and fostering the evolution of new strains of pathogen (insect-pest, bacterial, fungal and viral diseases). To maintain sustainability and production potential of horticultural crops under degradation of natural resources like land, water and changing climate is the severe challenge which needs to be overcome by agricultural institutes, breeders and farmers to meet the growing demands.

To sustain under the climatic vagaries, we require a climate smart approach in production system for vegetable, which will be location-specific, knowledge and technology-intensive for improving productivity in the challenging environmental situation (Malhotra and Srivastava, 2015). The objective of this chapter is to review the impact of changing climate conditions and its management practice in vegetable crops for sustainable intensification of production.

Climate Smart Agriculture (CSA)

Climate Smart Agriculture is an integrated approach for developing a sustainable agricultural system that can address the challenges posed by climate change by developing better policies, technical and investment conditions. CSA aims to transform, reorient and develop better crop production systems by application of latest technologies for increasing global food security by following climate change adaptation and mitigation efforts (FAO, 2010; Hellin and Fisher, 2019).

CSA should not be considered a set of practices that can be universally adopted, but rather an approach of different practices that can be adopted in local context. It deals with all on-farm and off-farm activities which involve appropriate policies, technologies, roles of various institutions and investments. The concept of CSA is based on three pillars of sustainable development as follows:

I. **Economic**: It focuses on sustainably improving agricultural productivity and development, provisions for doubling farmers' incomes and providing food security.

II. **Social**: The concept helps in enhancing adaptive capabilities and resilience of agricultural and food security systems to climate change at multiple levels.

III. **Environmental**: It provides a system to diminish the emissions of greenhouse gas (like CO_2, CFC's, CH_4, N_2O, *etc.*) from agriculture and allied activities including crops, livestock and fisheries.

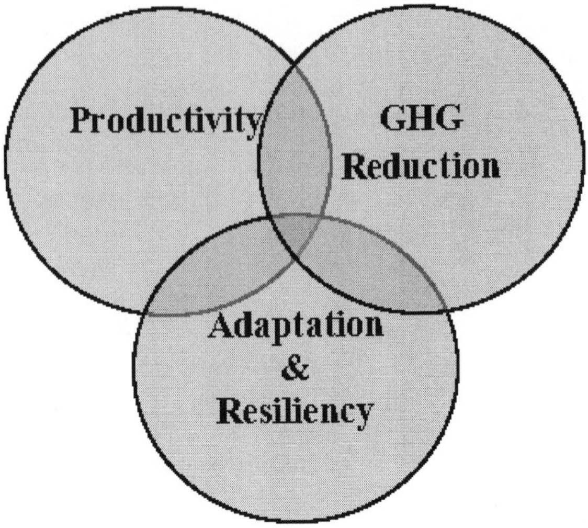

Figure 8.1: Three Main Pillars of Climate Smart Agriculture.

The approach includes various practices so as to better utilize and manage natural resources (land, freshwater, sunlight) which maintain a balance between ecosystem and landscape. It also includes the application of various digital services to ensure execution of significant changes (Zecca, 2019) for example adoption of smart farming technologies like precision farming (Gebbers and Adamchuk, 2010), Internet of Things (IoT) in agriculture (Brewster *et al.*, 2017) or robotics (Edan, 1999).

The use of climatic, crop and soil data for agricultural planning can help us decrease the uncertainties involved with crop production and increase farmers ability to plan and take early action, allocate resources effectively and reduce associated risks. For example, developing early warning systems for predicting extreme weather (*e.g.*, drought, flood, haél), pest and disease monitoring and occurrence. It can be a solution to climate change impacts and food insecurity (Tantely *et al.*, 2015).

Major approach for climate change adaptation includes:

1. Integrated planning of various resources at local, watershed and regional scales.
2. Promoting sustainable agriculture practices that can increase carbon storage, reduce water use and improve soil fertility.
3. Reducing emissions from agriculture caused from fertilizer application, livestock and methane emissions.
4. Promote diversified agriculture, improved cultural practices and improve genetic traits of crops for sustainable agricultural practices.
5. Developing risk insurance and risk management strategies that can provide safety nets to even the poorest farmers.

6. Disseminating climatic and other important information to farmers and monitors outcomes of different actions. To build important technological intervention for sustaining crop production systems under chaning climatic conditions without adverse impact on ecosystems.

Impact of Climatic Changes on Vegetable Production

Vegetable crops, like other field crops, are sensitive to environmental extremes. Climate changes affect the cultivation and management practices of vegetable crops and are the primary reason for low production in most of the horticultural crops. Moreover, the variations in climatic conditions may lead to crop failure or reduction in the quality of produce, which makes the vegetable production unprofitable. The enzymatic activities and physiological processes of plants are largely affected because of their temperature-dependent nature. The rise in temperature leads to two major constraints, drought and salinity, which limits the possibility of sustainable vegetable cultivation. Studies have revealed that continuous variations in weather and climatic conditions due to the rising temperature, erratic rainfall pattern, increase the demand of water and rise in disease incidence which will affect the production potential of various vegetable crops.

The vegetable crop productivity may get reduced because of climatic changes at the global level, such as unpredictable rainfall pattern and high-temperature spells. Yusuf (2012) reported negative effects on the production of tomato due to environmental factors. It was reported that water availability greatly affects yield and quality of vegetables; especially tomato where the productivity is reduced under excess water as well as drought conditions. The important environmental factors that affect vegetable crop production are studied below.

Rise in Atmospheric Concentration of Carbon Dioxide

Plant growth is highly affected by the elevated CO_2 levels and plants adapt to mechanisms against water stress conditions. Plants adopt mechanisms such as photosynthesis, accumulation of biomass and economic yield increases. Under such conditions, there is partial closure of stomata which reduces the loss of water by transpiration and the water use efficiency improves. The study of effects of elevated CO_2 in *Amaranthus bicolor* reported an increase in total dry matter production (0.99 g), phenol content (25.46 mg/g), SOD activity (1.65 $g^{-1}minute^{-1}$) and decrease starch (3.22 mg/g) and protein (13.90 mg/g) content (Chatti and Manju, 2017). Increasing atmospheric CO_2 levels has less effect on C_4 plants because they have a mechanism to increase CO_2 levels within the leaves before the onset of photosynthesis under normal conditions. But it has been reported that the yield of these plants is relatively less affected in water stress conditions (Simpson, 2017). In tuber crops, like potatoes, there is an increase in tuber yield of about 10 per cent for every 100 ppm increase in CO_2 concentration. This happens due to advances in tuber initiation and flowering, rise in mean tuber weight and number of tubers produced but the disadvantage is that there occurs reduction of chlorophyll in later stages. The photosynthetic activity in C_3 plants depends on the carbon dioxide concentration naturally present in the atmosphere. Although in vegetables like cucumber, yield may increase by 34 per cent with enhanced carbon dioxide level but in most of the C3 plants, the nutritional value of plant (protein, minerals and trace elements, like iron and zinc) declines at elevated carbon dioxide levels (Taub *et al.*, 2008; Loladze, 2014).

Temperature

The majority of plants' metabolic, physical and biochemical activities are temperature dependent due to which the fluctuations in daily average temperature are the primary factor which influences vegetable production, especially in tropical and arid areas where situations of high temperature prevail. The morphological, physiological, biochemical and molecular response of plant is significantly altered under high temperature which results in poor plant growth, development and yield. Rise in temperature also increases the rate of growing cycle of plants (Hatfield *et al.*, 2011) and this may give less time to reproduce causing reproductive failures and lower yields (Craufurd and Wheeler, 2009).

Hazra *et al.* (2007) summarized the symptoms causing major effects of high temperatures are reported in tomato where temperature above 25°C reduces pollination and the occurrence of fruit set failure is commonly found due to reproductive abnormalities like poor pollen production, dehiscence and viability, abnormal flower development, ovule abortion, poor viability, bud drop and reduced carbohydrate availability. Moreover, high temperature can reduce tomato fruit size, lower quality, fruit loss owing to sunscald and ultimately lower productivity. Lycopene which is responsible for red color in tomato fruit starts degrading at a temperature higher than 27°C and its production is completely inhibited at 40°C. Even temperatures below the optimum level can cause problems such as poor germination, developed growth, chlorosis, wilting, acute dehydration by freezing that can lead to plant death. In pepper, exposure of high temperature at post-pollination stage inhibits fruit set but the pistil and stamen viability remain unaffected at the pre-anthesis stage, signifying the sensitivity of plants against fertilization under high temperature conditions. High temperature also affects red colour development in chilli at the ripened stage and is also responsible for poor fruit set, ovule abortion, flower and fruit drop.

Temperature equal or higher than 42°C, hampers the germination of watermelon, cucumber, winter squash, summer squash and pumpkin seeds and temperature higher than 45°C suppresses the germination of melon seeds, may also delay fruits ripening and lowers the sweetness in fruits. The female flower production in cucurbits like, bottle gourd, ash gourd and pumpkin are affected by warm humid climate resulting in low yield. In cucumber, maleness increases due to rise in temperature. Cole crops are also affected by high temperature that causes early bolting in plants which are cultivated for vegetable purpose. In leguminous vegetable like pea temperature above 25.5°C during blooming reduces pod sets, number of flowers, pods and yield. In bulb crops, 1 per cent rise in temperature will decrease bulb yield by 3.5-15 per cent.

Water Stress

It is a major problem in arid and semi-arid regions, which causes more than

50 per cent crop loss worldwide (Sivakumar *et al.*, 2016). Drought stress conditions may occur due to scarce rainfall or deficient soil moisture which stimulate many physiological, biochemicals and genetic responses in plants and may restrict crop growth. The incidence of drought conditions has adverse effects on different vegetable crops, for examples, it causes a reduction in the germination of onion and okra seeds, sprouting of potato tubers and induces flower abscission, nearly 50 per cent reduction in fruit yield of tomato. In legumes, water stress condition increases the ratio of pod drop to number of pods and seed abortion in pods by reducing water content and cell turgor.

The physiology of plants is greatly affected by water stress condition as described below:

- The rate of photosynthesis and photosynthate assimilation due to reducing stomatal conductance is greatly affected when water stress befalls at flowering stage and there may be an increase in the rate of abscission.
- It leads to an increase in solute concentration in the soil, which causes loss of water from plant cells due to osmotic flow. This loss of water in plant cells triggers the inhibition of physiological and biochemical processes like respiration, photosynthesis, *etc.*
- The sucrose phosphate synthase (SPS) is important for the resynthesis of sucrose and assist the assimilatory carbon flux from source to sink. But drought stress impairs the availability and utilization of sucrose by decreasing SPS and invertase activities which may affect the utilization of sucrose and as consequence reduces ovary growth and concentration of hexoses.

Salinity

The productivity of many horticultural crops gets reduced in the salt affected area, especially vegetables are more sensitive to excessive salt concentration. The high salt concentration in soil and irrigation water may affect the physiological and metabolic processes of plants, leading to cell growth reduction. Physiologically, high solute concentrations in soil impose an initial water deficit and alter K^+/Na^+ ratios which leads to high Na^+ and Cl^- concentrations causing an injurious effect to plants. Salt stress, on the other hand, causes problems like reduction in growth, leaf abscission, loss of cell turgor, decreased respiration and photosynthesis, tissue necrosis, wilting and finally plant death. The susceptibility of vegetable crops varies with the range of salt concentration, for example, onions are highly susceptible to salinity, while cucumber, pepper, eggplant and tomato are moderately sensitive to salinity. In cabbage, salinity may result in a significant loss in rate and percentage of germination, length and fresh weight of roots and shoots.

Salinity under high temperature conditions results in the accumulation of salt in young expanding potato leaves causing a reduction in vegetative growth recovery and leaf area index. In chilli, reduction in dry matter production, leaf area, relative growth rate and net assimilation rate is reported when cultivated in saline soils (Lopez *et al.*, 2011). In all cucurbits, the number of seeds per plant and fruit weight is affected by salinity due to decrease in total chlorophyll and relative water content. Fruit yield in muskmelon decreases by 14.3 per cent with per unit increase in salinity. Salt stress result into suppression of growth, transpiration, cell water potential, changes in stomata conductivity and photosynthesis, number and size in bean plants, while, higher salt levels reduce flowering in peas, tomato and brinjal.

Flooding

Almost all vegetable crops are flood sensitive which creates a disadvantage for the growth and yield of crops. The excess of water in the soil normally causes oxygen (O_2) depletion in the rhizosphere (root zone) caused by a slow gaseous exchange in water, thus low O_2 consumption by plant roots and microorganisms or inhibition of aerobic activities. The severity of the effect of flooding on different vegetables depends on the developmental stage of plant and the period of stress. Tomato is highly sensitive to flooding with a low genetic variation for this character. Flooding in tomato crops causes damage to the plants by an accumulation of endogenous ethylene. The characteristic response of tomato plants under waterlogged conditions is the epinastic growth of leaves. While even a short period of flooding under high temperature conditions may lead to more severe damage like rapid wilting and finally death of tomato plants. The sensitivity of onion to waterlogging at bulb development stage causes a huge loss of yield up to 30-40 per cent. The physiology of vegetable plants is highly influenced by excess water conditions. Some of these significant changes in plants are described as under:

- ☆ The reduction in stomatal conduction is one of major plant physiological responses of plants to flooding which is responsible for an increase in leaf water potential, and subsequent reduction in the carbon exchange rate and increase in internal CO_2 (C_i) concentration.
- ☆ Flooding also harms vegetative growth and reproduction stages of plants due to damaging influences on physiological functioning.
- ☆ Plant damaging symptoms such as the reduction in shoot and root growth, accumulation of dry matter content, wilting, leaf chlorosis and decline in total plant yield are also caused under excessive water conditions.
- ☆ It can also cause spread of various water-borne pathogens which can infect the plants under droughts and heatwaves, followed by wind-borne dispersal of spores.

Pests and Diseases Responses to Climatic Change

Changes in temperature and precipitation regimes may alter the ecology and biology of insect pests. The rise in temperature may cause an increase in fecundity in some group of insects such as aphids and diamond back moth that means the rate of reproduction increases and the insect pest would be capable of producing more generations in a year than usual. Being stenotherms (cold-blooded), insects are sensitivity to temperature change. The insect outbreaks, developmental and oviposition rates, and invasive species introductions increase under high temperature condition, which becomes the reason for the decrease in reliability of economic threshold levels and lessens the efficiency of bio-control measures by fungi and parasitism (Das *et al.*, 2011). Temperature rises in higher latitude influence the migration of insect species.

Generally, the development rate of insects between the two generation increases with the rise in temperature, for example, european corn borer (*Ostrinia nubilalis*), *onion* maggot (*Delia antiqua*), cabbage maggot (*Delia radicum*) and colorado potato beetle (*Leptinotarsa decemlineata*), mealy bugs in crops like onion, potato, brinjal, tomato and okra. There is incidence of hadda beetle (*Epilachna vigintioctopunctata*) in bitter gourd which was previously found in brinjal. New race of cabbage butterfly (*Pieris brassicae*) has been found to infect radish, knol-khol in addition to cabbage, cauliflower, and mustard. The other influences of high temperature on insect pests includes increase in dispersal capabilities as in aphids and moths; accelerated metabolic rates and depletion of stored nutrient resources leading to sooner breakage of insect diapauses. Rise in temperature in winter up to 1-5°C and early summers may upsurge survival rate of insect pests with reduced winter mortality, increased population multiplication, early infestations and crop damage under global warming scenario.

Climate variation not only influences the pathogenicity of disease-causing agents, their growth and development rate but also affects the resistance of host plant. Under low temperature conditions, the activity, survival and reproduction rate of most agro-pathogens is reduced during the growing season of crops. Hence, the attack of pathogens is relatively higher in the warm season of vegetable crop production. The air borne pathogens are more likely to spread, complete their life cycle faster and increase their survival due to higher temperatures and absence of frost. With a rise in temperature during winters and reduction of frost, there is observed an early appearance of insect vectors of viral diseases in crops like sugarbeet and potato but under such environment, there is less incidence of pathogens such as *Fusarium*.

Climate Smart Vegetable Production and Management Practices

Cultural Management Practices

The vegetable growers should follow recommended production systems for stress management to adapt to hot and dry conditions and to improve water-use efficiency of crop. The escaping defence mechanism like early sowing or planting can be adopted to avoid the forecasted water stress or rise in temperature during the crop-growing season. The nutrient availability in the soil can be increased by modified application of fertilizer and different soil amendments can also be used to improve soil fertility and nutrient uptake by plants (Srivastava *et al.*, 2014). Irrigation should essentially be provided at critical stages of crop growth like initial vegetative growth, flowering and fruiting. Conservation of soil moisture reserves is an important intervention which involves practices like mulching either with crop residues or plastic mulches. Under heavy rainfall situation, raised beds or clear plastic rain shelters should be used to grow vegetable crops to avoid excess soil moisture or waterlogging which may lead to increase in yield by improving drainage and prevent the development of anoxic stress to root system.

Grafting

The soil-borne diseases like fusarium wilt was major issue in vegetable crop production in the 20th century. To mitigate the problem, the concept of grafting vegetables was introduced in East Asia, which can help in increasing production of vegetables such as tomato, pepper, eggplant, and cucurbits (watermelon, pumpkin and cucumber). Grafting is an efficient method of crop improvement and can substitute the relatively slow breeding methods designed for development of stress tolerance in different cultivars of vegetable crops, hence these days it has become quite popular in Asia and Europe such as Japan and Korea. Grafting makes it possible to improve the abiotic stress tolerance caused due to various environmental factors such as low and high temperatures, salinity, drought and flooding by modifying the root system of the otherwise good cultivar.

The introduction of grafted vegetable plants was done in Japan and Korea in the late 1920s where watermelon was grafted onto gourd rootstock to improve yield and quality of watermelon. Some reported more salt tolerance in melons grafted onto hybrid squash than the non-grafted melons. Salt tolerance depends on the species to be used as rootstocks, for example, *Cucurbita* spp. rootstock provides more tolerance against salt than *Lagenaria siceraria* rootstock. Drought tolerant genotypes of eggplants are used as rootstock to improve plant adaptation to drought condition. About 10 per cent increase in fruit yield can be obtained by using a heat sensitive eggplants cultivar (*S. melongena cv. Yuanqie*) grafted onto a heat-tolerant rootstock (*cv. Nianmaoquie*). Under high temperature conditions, temperature-sensitive tomato grafted onto resistant rootstock cultivars has been useful to protect plants against heat-stress.

Developing Climate Resilient Vegetables

Many improved vegetable germplasms have been effective to control the crop losses due to changing climate. The development of improved cultivar is intensive process, requires high input and can be expensive initially but this method is most cost-effective for cultivation of different vegetable cultivars for subsequent years. However, most of the vegetables have limited genetic variability, especially for abiotic stress tolerance. Superior varieties which are more adaptable to challenging environmental conditions can be developed from novel genetic variation responsible for tolerance/resistance to biotic and abiotic stresses. Tolerant cultivars, wild relatives and other related species can be used as a source of tolerance and improved selection techniques can be adopted to identify these superior genotypes and associated traits. Plants originating from certain climate zones easily acclimatize to varying atmospheric conditions and can be a source of genes or gene combinations that confer such resilience.

Biotechnology

The traditional methods can be substituted by advanced technology like biotechnology which can aid to increase crop productivity under unfavourable environmental conditions. Genes responsible for particular traits have been discovered and their functions understood. This has made possible the genetic manipulation of genes linked with tolerance to stresses. These methods are more rapid, efficient, and provide higher returns but requires a large initial investment. The identification of QTLs underlying stress tolerance is processed *via* analysis of molecular marker but its application is still limited. In tomato, QTLs for drought tolerance have been identified. Three QTLs associated with water use efficiency were identified in *S. pennellii* based on 13C composition, while four QTLs associated with seed germination and drought tolerance were identified from salt tolerant species, *S. pimpinellifolium* (Foolad *et al.*, 2010). QTL mapping reveals the quantitative inheritance of salt tolerance. Several studies have described the quantitative inheritance of stress tolerance and often tolerance is affected by the developmental stage of plants. Being a complex trait, several genes may be involved in environmental stress tolerance (Wang *et al.*, 2003).

Prospects of Work

With changing climatic conditions in recent times, the sustainability of vegetable crop production is the need of the hour. It can be done based on the agro-ecological zones and susceptibility of the individual crops, the integrated system with all available resources should be developed which supports the crop adaptability under extreme environmental conditions. Due to poor understanding of strategies to analyze the impact of climate change, it is difficult to develop advanced adaptation measures in vegetable crops. To prepare for many climate changes and create a well-managed plan, we must identify research issues based on the perspectives of farmers, scientists, industries, business and policy planners. It is essential to consider the weather changes which are forecasted for next seasons and analyze their effect on growth and development of vegetable crops, based on which the research or innovative technologies can be adopted to mitigate the challenges of climate change.

- Emphasis should be given to climate change research, education and development
- More stress should be on education, research and development should be given for increasing acclimatization of vegetable crops to climate variability
- Short and long-term policies should be formed to mitigate the effects of changing climate on vegetables
- Improved techniques for identification and selection for the source of resistance and development of resistant vegetable crop varieties should be adopted
- An appropriate method of water harvesting and its judicious utilization in the form of drip, sprinkler or mist irrigation techniques should be implemented to mitigate drought stress
- Adopting soil moisture conservation practices like mulching with crop residues or plastic mulch, which can effectively improve the soil moisture level and availability of water to plants
- Development of parthenocarpic cultivars by utilizing plant growth hormones like auxin in tomato, eggplant and cucumber

- ✰ Grafting of a plant with otherwise good performance (scion) on to plant with high heat, drought, salt and flood stress tolerance (rootstock) can improve the yield of crops
- ✰ Extension activities can be useful in generating awareness among growers to modify the present vegetable cultivation practices and higher acceptability of greenhouse technology

Summary and Conclusions

With a growing population and urbanization, agriculture especially vegetable production is facing challenges of food/nutritional security. Sufficient production to meet growing food demands has become a very difficult task due to the reduction of agricultural land. The problem is intensified due to global warming which causes a decline in quality of the environment and aggravates different biotic and abiotic stresses. Vegetable crops being highly sensitive to extreme temperature, drought and flooding. This calls for urgent need to study the effect of climate change on growth, development, yield and quality of crops and implement climate smart agriculture approach.

The mitigation strategies should involve the development of production systems to increase the adaptability of crops to adverse environmental conditions so that the productivity and quality of vegetable crops can be improved. Under heavy rainfall situations, crops should be cultivated on raised beds. Mulching can be done using crop residues or plastic mulch which help in conserving soil moisture and also maintains soil temperature. The best and cheapest method to avoid abiotic and biotic stresses is growing of varieties or hybrids tolerant/resistant against heat, flooding, drought, diseases and insect pest. The genetic tools for gene isolation, characterization, and genetic engineering have proved to be effective for generating stress resistant and high-quality vegetable cultivars. Furthermore, several agronomic practices are useful in water conservation and to protect crops from unfavorable environmental conditions. Education and capacity building considering the socio-economic and political systems through effective extension strategies can play a key role for sustainable adaptation approach to climate change.

REFERENCES

Adamides, G. (2020). A Review of Climate Smart Agriculture Applications in Cyprus. Atmosphere, 11: 898.

Brewster, C., Roussaki, I., Kalatzis, N., Doolin, K., Ellis, K. (2017). IoT in agriculture: Designing a Europe-wide large-scale pilot. IEEE Communications Magazine, 55: 26-33.

Chatti, D., Manju, R.V. (2017). Carbon dioxide enrichment induced drought tolerance responses in amaranthus (*Amaranthus tricolor* L.). Chemical Science Review and Letters, 6(23): 2019-2024.

Craufurd, P.Q. and Wheeler, T.R. (2009). Climate change and the flowering time of annual crops. Journal of Experimental Botany, 60: 2529-2539.

Das, D.K., Singh, J., Vennila, S. (2011). Emerging crop pest scenario under the impact of climate change: a brief review. Journal of Agricultural Physics, 11: 13-20.

Edan, Y. (1999). Food and Agriculture Robotics. In: Handbook of Industrial Robotics, 2nd ed.; Nof, S.Y., Ed.; John Wiley and Sons, Inc.: Hoboken, NJ, USA. Thornton and Gerber, 2010.

FAO. (2010). Climate Smart Agriculture: Policies, Practices and Financing for Food Security, Adaptation and Mitigation. Food and Agriculture Organization of the United Nations, Roma, Italy.

FAO. (2012). Using marginal abatement cost curves to realize the economic appraisal of climate smart agriculture: policy options.

Foolad, M.R., Zhang, L.P., Subbiah, P. (2010). Genetics of drought tolerance during seed germination in tomato: inheritance and QTL mapping. Genome, 46: 536-545.

Gebbers, R., Adamchuk, V.I. (2010). Precision agriculture and food security. Science, 327, 828-83.

Hatfield, J.L., Boote, K.J., Kimball, B.A., Ziska, L.H., Izaurralde, R.C., Ort, D., Thomson, A.M. and Wolfe, D.W. (2011). Climate impacts on agriculture: Implications for crop production. Agronomy Journal, 103: 351-370.

Hazra, P., Samsul, H.A., Sikder, D., Peter, K.V. (2007). Breeding tomato (*Lycopersicon esculentum* Mill) resistant to high temperature stress. International Journal of Plant Breeding, 1: 31-40.

Hellin, J. and Fisher, E. (2019). Climate smart agriculture and non-agricultural livelihood transformation. Climate, 7: 48.

IPCC. (2012). Managing the risks of extreme events and disasters to advance climate change adaptation.

Loladze, I. (2014). Hidden shift of the ionome of plants exposed to elevated CO_2 depletes minerals at the base of human nutrition. Life, 3: e02245, 2014.

Lopez, M.A.H., Ulery, A.L., Samani, Z., Picchioni, G., Flynn, R.P. (2011). Response of chile pepper (*Capsicum annuum* L.) to salt stress and organic and inorganic nitrogen sources: growth and yield. Tropical and Subtropical Agroecosystems, 14: 137-147.

Malhotra, S.K., Srivastava, A.K. (2015). Fertilizer requirement of Indian horticulture. Indian Journal of Fertilizers, 11: 16-25.

Rosenzweig, C., Iglesias, A., Yang, X., Epstein, P.R. and Chivian, E. (2001). Climate change and extreme weather events; implications for food production, plant diseases, and pests. Global Change and Human Health, 2(2): 90-104.

Simpson, B.M. (2017). Preparing Smallholder Farm Families to Adapt to Climate Change. Pocket Guide 2: Managing crop resources. Catholic Relief Services: Baltimore, MD, USA.

Sivakumar, R., Nandhitha, G.K., Boominathan, P. (2016). Impact of drought on growth characters and yield of contrasting tomato genotypes. Madras Agricultural Journal, 103: 78-82.

Srivastava, A.K., Das, S.N., Malhotra, S.K., Majumdar, K. (2014). SSNM-based rationale of fertilizer uses in perennial crops: A review. Indian Journal of Agricultural Sciences, 84: 3-17.

Tantely, R., Herintsitohaina, R., Tovonarivo, R., Narindra, R., Tiana, S., Andry, A., Marie Lucia, F., Stéphan, R., Sylvain, D., Falinirina, V., Bernard, L., Masse, D., Alain, A. (2015). Climate smart agriculture-Climate smart practices impact soil organic carbon storage in Madagascar. In Proceeding: Third global science conference on climate smart agriculture, 16-18 March 2015, Montpellier France.

Taub, D., Miller, B., Allen, H. (2008). Effects of elevated CO_2 on the protein concentration of food crops: a meta-analysis. Global Change Biology, 14: 565-575.

UNESCO (United Nations Educational, Scientific and Cultural Organization). (2011). Climate change starter's guidebook places de Fontenoy, 75352 Paris 07 SP, France and the United Nations environment programme 15 rue de Milan.

Wang, W., Vinocur, B., Altman, A. (2003). Plant responses to drought, salinity and extreme temperatures: towards genetic engineering for stress tolerance. Planta, 218: 1-14.

Yusuf, R.O. (2012). Coping with environmentally induced change in tomato production in rural settlement of Zuru local government area of Kebbi state. Environmental Issues, 5: 47-54.

Zecca, F. (2019). The use of internet of things for the sustainability of the agricultural sector: The case of climate smart agriculture. International Journal of Civil Engineering and Technology, 10: 494-50.

Chapter 9

Prospects of Agroforestry as Climate Smart Agricultural Strategy in India

M.R. Nayak[1], P.J. Mishra[2] and L.M. Garnayak[2]*

[1]*Krishi Vigyan Kendra, Koraput (OUAT), Odisha*
[2]*Directorate of Extension Education, OUAT, , Bhubaneswar, Odisha*

Indian agriculture is very much weather dependent. Sudden changes in Climate severely reduce the agricultural production. Modern cultivation systems and climate change are unified with each other. Agriculture is the cause as well as sufferer of climate change. The adverse effect of climate change leads to food insecurity in rural population. Climate Smart Agriculture is increasingly being promoted by the internationally to help farmers get used to climate change and lift them out of poverty. An essential technique that is promoted under the climate mitigating smart agriculture policy package to reduce forest loss is agroforestry- the planting of woody plants or trees into farming systems. Achieving food security and responding to the challenges of climate change are two goals that must be achieved collectively. That's why agriculture and forestry in developing countries must endure a significant transformation for the future of world. This transformation must be accomplished without depletion of the natural resource base. Therefore, tree farming, conservation agriculture, minimum tillage, natural resource management and silvipastoral are gathered under single umbrella of Climate smart agriculture, where agroforestry stands as a solution for all the problems. Hence the aim of this chapter is to have glimpse of relationships between climate Smart Agriculture and Agroforestry, different aspects of agroforestry technology and area covered by Agroforestry. Climate Smart agriculture with agroforestry will act as an effective instrument to increase the resilience of farmers in the era of climate change.

Keywords: *Agroforestry, Carbon sequestration, Climate smart agriculture and ecosystem.*

Introduction

For low and medium income countries agriculture is the prime source of employment, approximately 45 per cent of total labor force (FAO 2015). Climate is the key factor governing crop and other farming performances. Change is predestined; a low of nature, which is acceptable if ascribed to nature. Likewise climate change is also to be expected and refers to the variation in climate over time which leads to environmental contrariety. There are some interrelationships between agriculture and climate change, *i.e.* in one hand agriculture is a supplier to climate change and in the other hand agriculture is a sufferer too. Climate change already became a dare to zero hunger and food security. In this menace, the system of food production from small farm to global level must be very smart, efficient and resilient to maintain sustainability.

Nearly 50 per cent population of developing countries (Asian and African) depends on agriculture for their primary income generation. Food security is bound to fulfill four dimensions simultaneously *viz.*, food supply, food access or marketing, food utilization and food system stability (FAO 2012). Uncertainty of any one of four dimensions marks food vulnerability. Climate change transforms agricultural production through species distribution, reduced availability of good quality water in time, changing length and time of crop growing season, nutrient cycling and weather unpredictability. It also has indirect impact on storage, marketing, transport and human health which governs food accessibility (household income, price of food and market availability) and digestibility (food utilization). Adverse effect of climate change on agriculture leads to food insecurity especially among rural and marginal population as their livelihood depends largely on highly vulnerable agriculture, forestry and fishery. We can depict an idea about uncertainty of agriculture in relation to climate change from Figure 9.1.

Climate smart agriculture is being widely debated at present and global recognition is needed for the transformation of agriculture. However, the total need to feed a population of over 9 billion by 2050 remains, and it is important that the aspirations of climate smart agriculture to increase yields and food availability are not conquered to climate priorities (Newaj *et al.*, 2015). An immediate suggestion is that there may well be a tension between the growth of cropland to meet growing food demands which will increase greenhouse gas emission and the maintenance of other land uses such as forests that maintain ecosystem services including sequestration and emissions evasion. Another and equally important move towards should be through the introduction and expansion of tree based agricultural systems. Agroforestry has the capacity to meet many of the aspirations of climate smart agriculture. Agroforestry is a dynamic, ecologically based natural resource management system which the integration of trees on farm and in agricultural landscapes, or through the production of agricultural products in forests, agroforestry diversifies and sustains production for increased economic, social and environmental benefits for land users at both the farm and landscape levels (Alao and Shuaibu, 2013). This makes an important and cost-effective climate smart production system. Trees on farms and in landscapes help to improve food security. Fruits and leaves contribute to human nutrition and fodder from trees

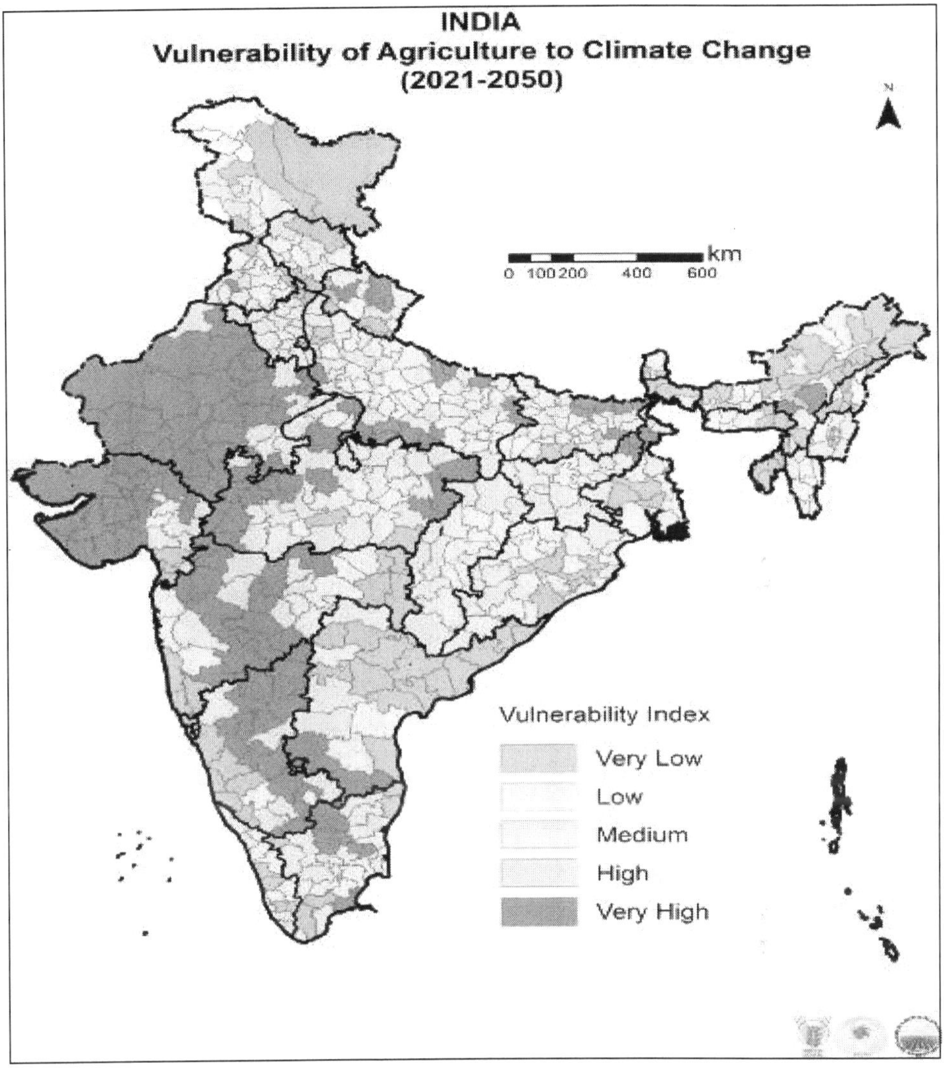

Figure 9.1: Vulnerability of Agriculture to Climate Change (2021-2050) (*Source*: https://www.google.com).

feeds livestock. Appropriate trees, if well managed in crop fields can also increase the fertility and organic matter content of the soils and also increase crop yields. Timber and fuel wood are harvests that can be used or sold, fuel wood still being the most significant source of cooking fuel in many rural communities. In rural India, 70 per cent of people are dependent on forest and common trees for fuel wood (Chauhan *et al.*, 2015). Trees on farms and in extensive grazing lands also contribute to a broad range of ecosystem services that landscapes provide, such as improving biodiversity, soil health and water conservation. Trees also contribute to mitigation, as they sequester carbon in above and below ground biomass (Chavan *et al.*, 2015).

In summary, trees are part of sustainable increase approaches as well as within extensive systems produce abundant products including food, medicines, fencing, timber and fuel. Agroforestry can rapidly help to make agriculture climate smart.

With recent world food price spikes, anthropogenic climate change concerns, and the challenge of a rising global human population, the roles of farms and forests in following food availability and nutritional security have returned centre stage in politics and development. In order to afford context for the on-going discussions concerning the importance of different possible interventions to support food and nutritional security, in this working Paper we consider the direct and indirect roles of agroforestry. In the following section we discuss agroforestry importance in providing food directly, in providing incomes to support access to food, in providing fuel for cooking, and through ecosystem service provision. Despite the increasing emphasis and promotion of agroforestry as an essential component of the climate smart agriculture approach of reducing forest loss, our understanding of its potentials and barriers from the perspective of farmers is still limited research on this is critical because whether agroforestry will work or not depends on the participation of farmers in the various climate smart agricultural programmes that are being promoted by various institutions in various communities. The aim of this chapter is, therefore, to gather some perspectives about the extent to which agroforestry can be successful as a climate smart agricultural strategy.

Principles of Climate Smart Agriculture

As earlier mentioned climate smart agriculture (CSA) has three focal areas *i.e.* sustainably increasing agricultural productivity and incomes, adapting and building resilience to climate change, reducing greenhouse gases emissions, where possible. Adapting and building resilience to climate change is especially important since it ensures food satisfactoriness despite unsuitable conditions. Resilience is called as the capability of a social or ecological system to absorb disturbances while retaining the same basic structure and ways of functioning, the capacity for self organization, and the capacity to adapt to stress and change (IPCC, 2007). The Rockefeller Foundation (2009) went on to specifically define climate resilience as the capacity of an individual, community, or institution to dynamically and effectively respond to shifting climate impact conditions while continuing to role at an acceptable level. This is achieved through several soil management practices that sequester carbon in the soil, reduce greenhouse gas emissions and aid intensifies production (FAO, 2013). Above all the practices have to improve the natural resource base. Therefore, the most important premise of CSA is the building of healthy soils (Stabinshy and Ching, 2012) through increasing the soil organic matter status of the soil (Blanco-Canqui and Lal, 2004). Soil management practices for CSA include; direct seeding under no/reduced tillage (Zheng *et al.*, 2014); improved protective soil cover through cover crops, crop residues or mulch and crop diversification (Muzangwa *et al.*, 2013). In addition, integrated soil fertility management, which includes both inorganic and organic sources and considers combining inputs of organic matter *i.e.* mulch, compost, crop residues and green manure with fertilizers to address or prevent macro and micro nutrient deficiencies should be carefully considered (FAO, 2013).

Farmer's Perception about Agroforestry

All most farmers think about tangible benefit and expect for early return like agriculture crop. They never worried on intangible benefit long turn advantage. The farmers have common perception that tree component reduce the yield of agriculture crop or main crop in following way:

- ☆ Tree component interfere with agriculture operation.
- ☆ Tree exhausts water and nutrient from field which is essential for main crop.
- ☆ Tree is home of certain pest and insect which is harmful for concerned crop.
- ☆ Tree provide shed which interfere the growth of agriculture crop.

They also have thought that planting of tree incurred additional cost. Majority of farmer's view that the Agroforestry systems would have no effect on farm income and boost their extra work and care for land (Matthews *et al.*, 1993). These are common observation prevalent in farming community about Agroforestry.

Lack of Research and Extension Activities in Field of Agroforestry

Already recognized field of agriculture and there are many technology for getting better the production. In Agroforestry, however, there are only a few research validated technologies. Many promising Agroforestry technologies, whether found in existing farmers' practice or newly conceived by researchers, require considerably more research attention before they can confidently be recommended for wider adoption (ICRAF 1986). In Indian context, in this field the research began 1983 by the started as ICAR (Indian Council of Agriculture Research) with the help of All India Co-ordinated Research Project on Agroforestry (AICRAF). The National Research Centre for Agroforestry is establishment in 1988, which has now been upgraded as the Central Agroforestry Research Institute (CAFRI, 2014), Jhansi (U.P., India). However, in India less concerned was given in the field of Agroforestry with respect to research and extension activities compare to agriculture. The major constraint on Agroforestry is that most of the system is on low level of management, primarily resulting from shortage of technical support (Kumar *et al.*, 2017). There is many Krishi Vigyan Kendras running in country but these all are mainly concerned with agriculture activities. Unlike agriculture, in forestry either none field specific hybrid or variety invent or if invented it not reach up to farmer field. So the program of **LAB TO LAND** had revealed failure in case of forestry sector.

There is no doubt that some renowned scientist of forestry developed hybrid and variety in forestry sector but due to lack of extension activities this research or technology bound up to lab only. So there is an urgent need to open **VAN VIGHYAN KENDRA** as early as possible. There are some policies and mission is implemented by government of India and some institute also actively involved in research and extension activities associated to Agroforestry.

Status of Agroforestry in India

India had started organized research in agroforestry world-wide about 25 years ago. The agroforestry research was initiated in early 1970s through industry participation in plantation of commercial tree species through Indian Council of Agricultural Research (ICAR) institutes. As action, the All India Co-ordinated Research Project (AICRP) on Agroforestry was established in 1983 by the ICAR for conducting research work in collaboration with State Agricultural Universities situated in different agro-climatic zones. Since then huge data has been generated about both traditional as well as improved agroforestry systems and practices. Now, about 25 million ha area in the country (8.2 per cent of the total reported geographical area), is reported under agroforestry in both irrigated and rainfed agriculture which also includes trees outside forests and scattered trees on and off the agricultural fields (Dhyani *et al.*, 2013). Considering the country's unique land-use, demographic, political, and socio-cultural characteristics as well as its strong record in agricultural and forestry research, agroforestry has received much attention in India from researchers, policymakers and others for its supposed ability to contribute appreciably to economic growth, poverty alleviation and environmental quality, so that today agroforestry is an imperative part of the 'evergreen revolution' movement in the country. Twenty-five years of investments in research have clearly demonstrated the potential of agroforestry in many parts of the country, and some practices have been widely adopted. But the vast potential remains principally underexploited, and many technologies have not been widely adopted. This situation is a result of the interaction of several complex factors. The understanding of the biophysical issues related to productivity, water resource partaking, soil fertility, and plant interactions in mixed communities is imperfect and insufficient, mainly because research has mostly been observational in nature rather than process oriented. Methods to value and measure the social, cultural and economic benefits of a variety of tangible and non-tangible benefits of agroforestry are not available, and the socio-economic processes concerned in the success and failure of agroforestry have not been investigated. On the other hand, the success stories of wasteland reclamation, and poplar-based agroforestry show that the technologies are widely adopted when their scientific principles are understood and socio-economic benefits are convincing. An assessment of the impact of agroforestry technology generation and acceptance in different parts of the country highlights the major role of smallholders as agroforestry producers of the future. It is vital that progressive lawful and institutional policies are created to avoid the historical dichotomy between agriculture and forestry and encourage integrated land use systems. Government policies hold the key to agroforestry adoption.

India has announced the National Agroforestry Policy, 2014 that will mainstream the growing of trees on farms to gather a wide range of developmental and environmental goals. For planning and functioning, the Policy has suggested developing a sound database and information system on agroforestry. However, at present unlike forest cover and area under crops, exact estimates for area under agroforestry are not available. The accurate estimation of area under agroforestry can be done with the help of geospatial technologies as manual methods of

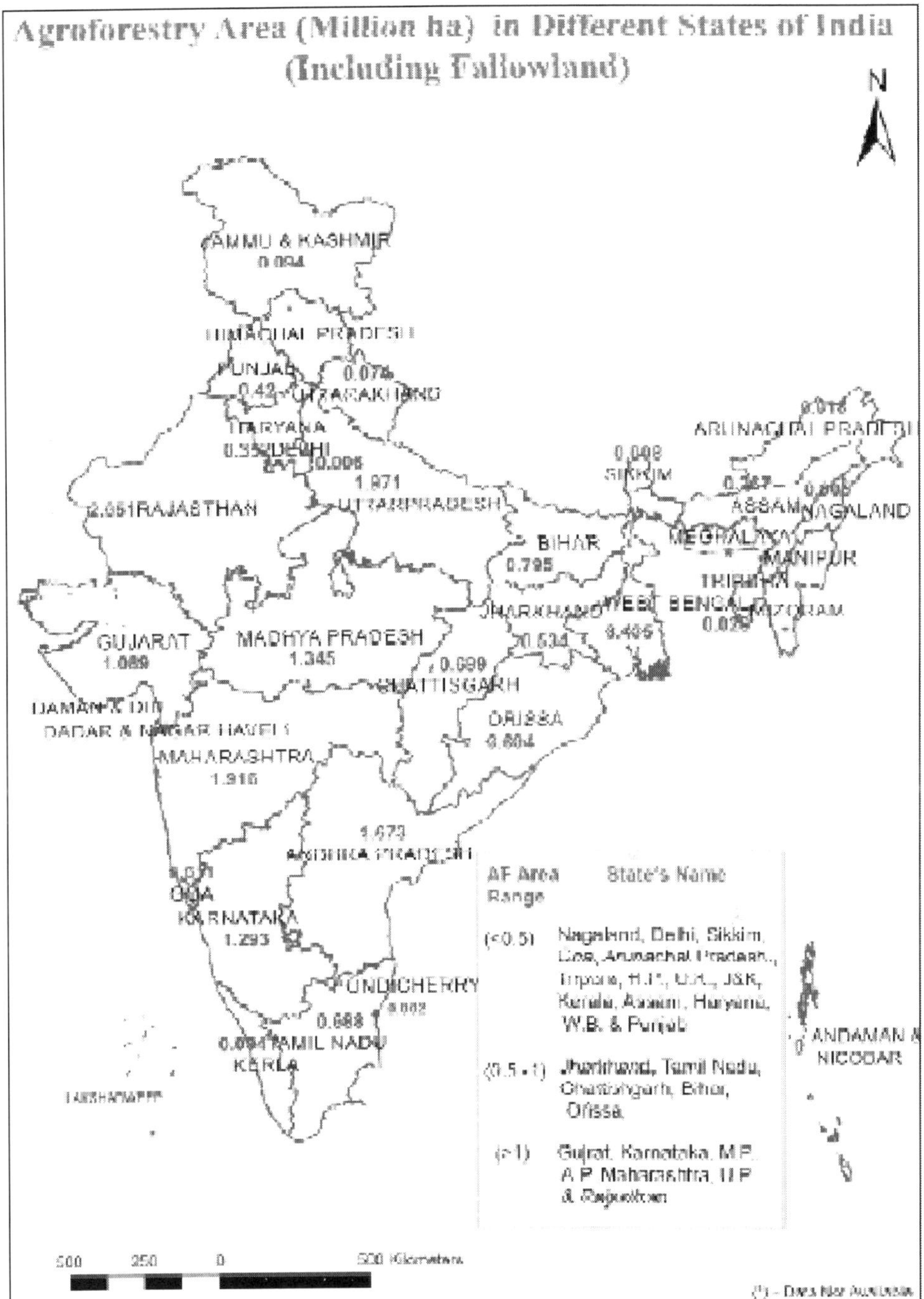

Figure 9.2: Agroforestry Area in different State of India (*Source*: Rizvi *et al.*, 2014).

mapping take a relatively long time and high cost (Chavan *et al.*, 2015). The geo spatial technologies are the integrated use of Geographical Information System (GIS), Remote Sensing (RS) and Geographical Positioning System (GPS), and all three are used for different principle during the area evaluation. GIS is used for geo referencing, masking of RS data and estimation of area; GPS is used for collecting reference data on agroforestry which is later on used for signature formation and accuracy assessment; and RS data is used for land use and land cover study and delineation of desired features (Dhyani, 2014). Use of geospatial technologies to approximation agroforestry area has been initiated by NRC for Agroforestry, Jhansi using medium resolution data. The agroforestry area in the country is approximate 17.45 mha through remote sensing (Rizvi *et al.*, 2014). Figure 9.2 states the area of agroforestry in India. The methodology will be further advanced for accurate assessment of area under agroforestry in the country.

Agroforestry for Higher Biomass Production and Improving Rural Livelihood

Agroforestry has the high biomass production potential with diverse products resulting financial benefits over the monoculture of crops. Product diversification reduces the risk associated with farming of single crop and gives the year round income. There are different uses of tree products (timber, fuel, fodder and bioenergy) which can reduce the inputs and enhance the eco-efficiency of agroforestry systems (AFS). At present agroforestry meets almost half of the command of fuel wood, two thirds of the small timber, 70-80 per cent wood for plywood, 60 per cent raw material for paper pulp and 9-11 per cent of the green fodder requirement of livestock, besides meeting the subsistence needs of households for food, fruit, fibre, medicine *etc.* However, at present the biomass efficiency per unit area is less than 2 t $ha^{-1}yr^{-1}$. Agroforestry practices have demonstrated that this could be safely increased to 10 t $ha^{-1}yr^{-1}$ by carefully selecting tree-crop combinations (Dev *et al.*, 2017)

Agroforestry: Smart Practice for Climate Change Mitigation and Adaptation

Developing countries like India are going to more sufferers due to negative impacts of climate change especially of agriculture. Mitigation efforts will therefore only provide a partial softening of the effects of climate change (Verchot *et al.*, 2007). Agroforestry provides unique opportunity to unite the twin objectives of climate change adaptation and mitigation. Agroforestry systems store carbon in the soils and woody biomass, and these also reduce greenhouse gas emissions from soils (Dhyani *et al.*, 2016). Recent studies under a variety of agroforestry systems in varied ecological conditions emphasized that agroforestry systems increase and store carbon stocks in above ground biomass and in soil and also has an vital role in increasing livelihood security and reducing vulnerability to climate change.

Out of all landuse analyzed, agroforestry has a great potential for sequestration of carbon from the atmosphere (Figure 9.3). India recently submitted voluntary pledges for action on mitigating climate change in the form of the Intended Nationally Determined Contributions (INDCs) to the United Nations Framework

Convention on Climate Change (UNFCCC). India targets a 33-35 per cent reduction of emission intensity of its GDP by 2030 from 2005 levels. The target will be achieved through increasing area under trees through afforestation, agroforestry and increase in share of renewable source of energy like biofuel, bioenergy, solar energy *etc.*

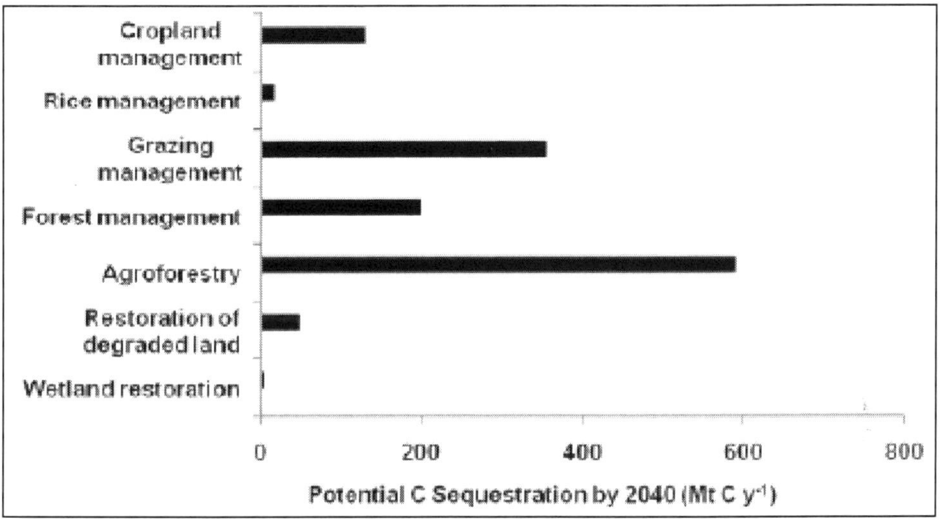

Figure 9.3: C-sequestration Potential of different Landuse and Management Options (adapted from IPCC 2000).

Based upon numerous studies it can be revealed that the overall biomass productivity, soil fertility improvement, soil conservation, nutrient cycling, microclimate improvement, and carbon sequestration potential of an agroforestry system is generally greater than that of an annual cropping systems (Dhyani *et al.*, 2009). Agroforestry has the potential to enhance above- and below-ground carbon stocks to mitigate climate change.

Agroforestry for Ecosystem Services

Different studies have shown the importance of agroforestry in conserving the biodiversity, sustainable use of natural resources and providing various ecosystem services (Alam *et al.*, 2014). The Millennium Ecosystem Assessment (2005) has characterized the ecosystem services into provisioning service, regulating service, supporting service, and cultural service (Table 9.1). Ecosystem services and benefits provided by the agroforestry systems are not confined to in-situ but occur over a wide range of spatial and temporal scale.

Climate Smart Agriculture and Agroforestry

In relating tensions between maximizing global agricultural productivity, increasing resilience of agricultural systems in the face of climate change and mitigating greenhouse gas (GHG) emissions from agriculture, the term climate smart agricultural development was first used in 2009 (FAO, 2009). FAO coined the

Table 9.1: Categories of Ecosystem Services Provided by Agroforestry

Ecosystem	*Services Definition*
Provisioning services	Goods, such as food or freshwater, that ecosystems provide andhumans consume or use, fuel-wood, fodder, timber, poles *etc.*
Regulatory	Services such as flood reduction and water purification, otherhydrological benefits, micro-climatic modifications provided byagroforestry systems.
Cultural	Intangible benefits, such as aesthetic enjoyment or religiousinspiration and recreation.
Supporting	Basic processes and functions, such as soil formation and nutrientcycling and biodiversity conservation that are critical to theprovision of the first three types of ecosystem services

term CSA in the background document prepared for the Hague Conference on Food Security, Agriculture and Climate Change in 2010. The CSA concept was developed with a strong focus on food security, for now and the future, counting adaptation to climate change. The CSA concept now has wide possession among governments, regional and international agencies, civil society and private sector. CSA is not a set of practices that can be universally functional, but rather an approach that involves different elements entrenched in local contexts. CSA relates to performance both on-farm and beyond the farm, and incorporates technologies, policies, institutions and investment (CCAFS and UNFAO, 2014). Linkages between CSA and agroforestry systems (AFS) are an important and closely connected aspect particularly on soil health, carbon sequestration, resources distribution, diversity, water management and food security *etc.* Agroforestry is a judicious integration of tree species with agricultural crops and/or animals which has been practised since ancient times across the world in both the tropics and temperate regions (Chavan *et al.*, 2015).

Agroforestry systems may also enhance ecosystems by storing carbon, preventing deforestation, increasing biodiversity, protecting water resources and reducing erosion. In addition, when applied deliberately on a large scale, agroforestry enables agricultural lands to withstand weather events, such as floods and drought, and climate change. Agroforestry practices were evolved long ago to tackle the different issues, which resemble as a natural forest on a farm. CSA is moreover a broader term of agroforestry and it has close relationship among them.

The Indian agricultural production system faces the daunting task of having to feed 17.5 per cent of the global population with only 2.4 per cent of land and 4 per cent of the water resources at its disposal. With the continuously degrading natural resource base compounded further by global warming and associated climate changes resulting in increased frequency and intensity of extreme weather events, "business as usual" approach will not be able to guarantee food and nutrition security to the vast population as well as environmental security (the need of the hour). The challenge is alarming because more has to be produced with reduced carbon and water footprints. To achieve this task of paving the way for climate smart agriculture, we need to take several measures that will have enabling policies, institutions and infrastructure in place and the farming community be better informed and empowered with necessary resources. The major benefits CSA and AFS are as follows:

1. CSA and AFS helps to increase soil health and improve water conservation and availability on farms, with practices such as composting and other soil management techniques, harvesting and retention systems.
2. CSA and AFS promotes ways to reduce greenhouse gas emissions, such as: reducing encroachment release of farms or ranches into forested areas, encouraging tree-based systems with added planted trees and promoting more efficient use of fertilizer to help avoid methane emissions
3. Diversifying crops through CSA and AFS can help address food security and create alternative income opportunities and sustainable economic livelihoods for farmers.
4. CSA and AFS can also help reduce their vulnerability to crop losses in the face of climate change weather impacts.
5. Integrated planning of land, agriculture, forests, fisheries and water at local, watershed and regional scales, to ensure synergies are properly captured.
6. Promoting activities that increase carbon storage, combine animal husbandry and trees with food production, and are geared towards improving soil fertility.
7. Reducing a variety of emissions from agriculture such as nitrous oxygen from fertilizer application, livestock emissions and methane from rice cultivation.
8. Exploring carbon finance as a "lever" to promote sustainable agricultural practices that have many other direct benefits for smallholder farmers and the environment.

Conclusion

There is both direct and indirect influence of climate change on agriculture, direct influence by regulating temperature, rainfall, humidity and indirect by changing soil physical, biological and chemical properties. Climate smart pathway combines both mitigation and adoption which increases damage resistance and ensures speedy recovery to a variety of risks. Evidences are clear to suggest that agroforesry is desirable, both for its beneficial effects on climate change adaptation and mitigation, and for sustaining farm income. The combination of trees with crop gives not only timber, fuel, fodder and food but also reduces CO_2 from the atmosphere at acceptable level. Agroforestry can improve the resilience of agricultural production to current climate variability as long-term climate through the use of trees for intensification and diversification and buffering of faming systems. It helps to increase yields (property reduction and food security), make yield more resilient in the face of weather extremes and slow onset climate changes, and make a farm as solution to climate change rather than being part of the problem (mitigation), which is basic requirement of CSA. Tree based production systems often produce crops of higher value than row crops. Thus, diversifying the production system to include as significant tree component may buffer against income risk associated with climate variability. Besides the potential of agroforestry systems to providing various ecosystem services, these systems provide the unique opportunity to increase the tree cover to a level of 33 per cent in India.

REFERENCES

Alam, M., Olivier, A., Paquette, A., Dupras, J., Reveret, J.P., Messier, C. (2014). A general framework for the quantification and valuation of ecosystem services of tree-based intercropping systems. *Agroforest Syst*, 88: 679–691.

Alao, J. S. and Shuaibu, R. B. (2013). Agroforestry practices and concepts in sustainable land use systems in Nigeria. *Journal of horticulture and forestry*, 5(10): 156-159.

Blanco-Canqui, H. And Lal, R. (2004). Mechanisms of Carbon Sequestration in Soil Aggregates. *Critical Reviews in Plant Science*, 23: 481-504.

CCAFS and UNFAO (2014). Questions and Answers: Knowledge on Climate Smart Agric ulture. United Nations Food and Agriculture Organisation (UNFAO), Rome.

Chavan, S. B., Keerthika, A, Dhyani, S. K. Handa, A. K., Ram Newaj and Rajarajan K. (2015). National Agroforestry Policy in India: A low hanging fruit. *Current Science*, 108(10): 1826-1834.

Chouhan, D. S., Chavan, S. B. Keerthika, A., Jha, Ankur, Dwivedi, R. P. and Kumar, A. (2015). Socio-economic fuelwood survey in Bundelkhand region of central India. In: Energy Research and Environmental management: An innovative approach (Eds) G.C. Mishra and S.K. Yadav. Krishi Sanskriti Publications, New Delhi.

Dhyani S K, Newaj R and Sharma A R (2009). Agroforestry: its relation with agronomy, challenges and opportunities. *Indian Journal of Agroforestry*, 54(3): 249-266.

Dhyani, S. K., Karemulla, K. Ajit and Handa, A. K. (2009). Agroforestry potential and scope for development across agroclimatic zones of India. *Indian J Forestry*, 32(2): 181-190.

Dhyani, S. K. (2014). Agroforestry Research in India: experience of hundred years with focus on organized national programme. *In:* World Agroforestry Congress on Agroforestry held at New Delhi during 9th-14th February.

Dhyani, S. K., Asha Ram and Dev I. (2016). Potential of agroforestry systems in carbon sequestration in India. *Indian Journal of Agricultural Sciences*, 86 (9): 1103–12.

FAO (2012). Impacts of climate change on food security, climate change and food security.

FAO (2015) The state of Food Insecurity in the World. Meeting the 2015 International hunger targets: Taking stock of uneven progress (http: //www.fao.org/3/a-i4646e.pdf).

FAO (2013). Climate Smart Agriculture Source book. Food and Agriculture Organization of the United Nations, Rome.

Food and Agriculture Organization of the United Nations (2009). Food Security and Agricultural Mitigation in Developing Countries: Options for Capturing Synergies. Rome.

https: //www.google.com

Dev, I., Asha Ram, Srivastava, M., Chaturvedi, O.P., Kumar N. and Kumar, D. (2017). Agroforestry: A Climate Smart Land Use System. PROCEEDINGS of International Symposium on Eco – Efficiency in Agriculture and Allied Research held at BCKV, Kalyani, Nadia, West Bengal, India, Pp, 95-97.

IPCC (2007). "Contribution of Working Group II to the Fourth Assessment Report of the Intergovernmental Panel on Climate Change", M.L. Parry, O.F. Canziani, J.P. Palutikof, P.J. van der Linden and C. E. Hanson (eds), in Climate Change 2007, Fourth Assessment Report, Intergovernmental panel on Climate Change, Cambridge University Press, Cambridge, United Kingdom.

Kumar, Y., Thakur, T. K. and Thakur, A. (2017). Socio-Cultural Paradigm of Agroforestry in India. *Int.J.Curr.Microbiol.App.Sci.* 6(6): 1371-1377.

Matthews, S., Pease, S.M., Gordon, A.M. and Williams. P.A. (1993). Landowner perceptions and the adoption of Agroforestry practices in southern Ontario, Canada. *Agrofor. System*, 21(2): 159–168.

Millennium Ecosystem Assessment (2005). Ecosystems and human wellbeing-volume 4-multiscale assessments: findings of the sub-global assessments working group. Island Press, Washington DC, 388p.

Muzangwa, L., Chiduza, C. and Muchaonyerwa, P. (2013). Feasibility of Winter Cover Crop Production under Rainfed Conditions in the Eastern Cape Province of South Africa. *African Crop Science Journal*, 21: 73-184.

Newaj, R., Chavan, S.B. and Prasad, R. (2015) Climate smart agriculture with special reference to agroforestry. *Indian Journal of Agroforestry. 17 (1): 96-108.*

Rizvi, R. H., Dhyani, S. K., Newaj, R., Karmakar, P. S. and Saxena, A. (2014). Mapping agroforestry area in India through remote sensing and preliminary estimates. *Indian Farming* 63(11): 62-64.

Rockefeller Foundation (2009). Building Climate Change Resilience. Rockfeller Foundation White Paper.

Stabinsky, D. and Ching, L. (2012). Ecological Agriculture, Climate Resilience and a Roadmap to Get There. TWN Environment and Development Series No. 14, Third World Network, Penang.

Verchot, L. V., Noordwijk, M. V., Kandji, S., Tomich, T., Ong, C. Albrecht, A., Mackensen, J. Bantilan, C., Anupama, K.V. and Palm, C. (2007). Climate change: linking adaptation and mitigation through agroforestry. *Mitig Adapt Strat Glob Change* DOI 10.1007/s11027-007-9105-6.

Zheng, C., Jiang, Y., Chen, C., Sun, Y., Feng, J., Deng, A. and Song, Z. (2014). The impacts of Conservation Agriculture on Crop Yield in China Depends on Specific Practices, Crops and Cropping Regions. *The Crop Journal*, 2: 289-296.

Chapter 10

Agriculture Extension for Climate Smart Agriculture

K. Manobharathi[1], Kethavath Naresh[2] and Shijaatt Hussain Bhat[2]*

[1]Department of Agricultural Extension, Bidhan Chandra Krishi Viswavidyalaya, Mohanpur, Nadia (Dist) – 741252, West Bengal
[2]Department of Agriculture Extension and Communication, Sher-e-Kashmir University of Agricultural Sciences and Technology of Kashmir, Jammu and Kashmir – 193201

Agriculture is the foundation of most countries' economic structures. Agricultural production has been closely related with variations in weather. It is expected that climate change will have a major effect on conditions affecting agriculture including temperature, carbon dioxide etc. Agriculture extension is a out of school education involves out stretching of skills, knowledge to farm people. It also includes transfer of information regarding Climate smart agriculture. In general aspect, CSA is one that increases the socio-economic background of farmers economically by adopting farming with suitable climatic conditions. By utilizing extension approaches such as FFS, ICT networks, F2F strategy etc., CSA is getting adopted by most of the people. Gender-specific consequences in the sense of climate smart farming vary by the degree to which women can equally access resources such as land or livestock, business opportunities etc. Institutions like NGOs, Research stations, other civil societies etc. Reforming Governance in supply, demand, administration, structure and legally are necessary for effective Extension delivery. Some of the climate related financing agents worldwide and in national side were involved in sponsering climate related services and also various schemes for sustainable livelihood for climate smart agriculture. Thus, Agriculture Extension is playing a crucial part in supporting and improving new innovative technologies introduced in agriculture. Future role of Agriculture Extension is vitally important for developing every farmers livelihood and their socio-economic conditions in the society. Farmers also need to initiate using new innovative ideas like CSA in order to lead a sustainable life.

Keywords: *Agricultural extension, Climate smart agriculture, Communication technology and information.*

Introduction

Agriculture is the foundation of most countries' economic structures. In addition to food resources, agriculture also provides vast communities with better job opportunities. As this sector is inherently susceptible to climate conditions and considered as a vulnerable sector at risk and effects of global climate change, climate change directly affects agricultural production (Parry *et al.*, 2005). Agricultural production has always been closely linked with variations in weather. It is expected that climate change will have a substantial effect on agricultural conditions including temperature, precipitation *etc.* and association of these factors. In agriculture, knowledge and decision-making capacity determine how production factors – soil, water and capital – are utilized. In the formulation and distribution of information, and in teaching farm people to be professional decision makers, agricultural extension is essential. Therefore, agriculture extension plays the essential role in most of agricultural development projects.

Maunder (1973) described extension as a service that helps farmers in improving farming metods and techniques through educational procedures, increasing production efficiency and income, raising their living standards and raising their social and educational standards through educational procedures and policies. It is called agricultural extension when it is extended to agriculture. The key objective of agriculture extension is to help farming families to shift their development and marketing strategies to rapidly evolving social, political and economic circumstances in order to change their lives based on their personal and community preferences in the long term. Agriculture Extension is generally described as a process and a system in which knowledge, skills, attitudes relating to farming practices are transmitted through various channels and methods to its clients.

Agricultural Extension is mainly consists of following main aims:

1. Transfer of useful agricultural knowledge that includes improved seeds, fertilisers, tools and implements, pesticides, information related to dairy, poultry, climate change *etc.*
2. The practical application of acquired knowledge in farm and home.
3. Ultimately developing all ways of life of rural people in the particular framework of national, economic and social policies involving the population as a whole.

To meet the aim of Agriculture Extension, Agricultural Extension Functionaries are needed. Agriculture Extension Functionaries are responsible for the technology transfer and information dissemination. Today meaning and use of extension is beyond transfer of technology to facilitation and to help farmers in grouped way in order to deal with marketing issues and maintainin partenering with service dealers and various agencies. More recently, extension began to be seen as a bridge between research and farmers -a bridge between newly developed agricultural technologies and policy directives which are going to be utilized by farmers. The role of Agricultural extension functionaries is to translate these technologies and advisories from research system to farmers into an understandable form so that farmers could understand easily and extension personnel also provides a technological feedback to research system to improve a particular technology.

Climate Smart Agriculture (CSA) and its Objectives

CSA is characterized as an approach to transform and reorient agricultural production (Lipper *et al.*, 2014). It is defined as "farming that sustainably increasing productivity, improving resilience, reducing effect of Greenhouse Gases and improving the achievement of food security of nation and developmental goals" (FAO, 2013). Productivity, adaptation and mitigation are three interrelated pillars in the development of CSAs food security and growth.

Objectives

1. Sustainable improvement in agricultural productivity to encourage equitable rises in farm income, food security and better personal growth;
2. Adapting and constructing multi level resilience of agricultural and food security systems to climate change;
3. Reducing agricultural (including crops, poultry, livestock, fisheries *etc.*) greenhouse gas emissions.

Extension Approaches for CSA

There are several ways in which CSA can be contributed by extension systems. The philosophy used, however, (*e.g.* demand vs supply led, individual vs mass contact) and specialized methods that fits different types of farmers' messages and offer various possibilities for collecting information from the fields of farmers. Mass media also suits simplistic messages, while intense interactions can be more successful for complex information through farmers' field schools. The suitability and choice of approach combinations can influence the ability of extension services to contribute to food security and income, adaptation and resilience, and climate change mitigation.

As shown in Figure 10.1, Sala *et al.* (2016) concluded that two different indicators (reach and impact potential) were negatively correlated and they differ for various Extension approaches. If reach potential is higher, impact potential is lower and vice-versa.

Various Extension approaces used for CSA were:

a. Campaigns, Exhibitions and Awareness Programmes for CSA

The campaign is an intensive educational activity to inspire and organize a society to take action, to solve an issue or to fulfill a need that it desperately feels.

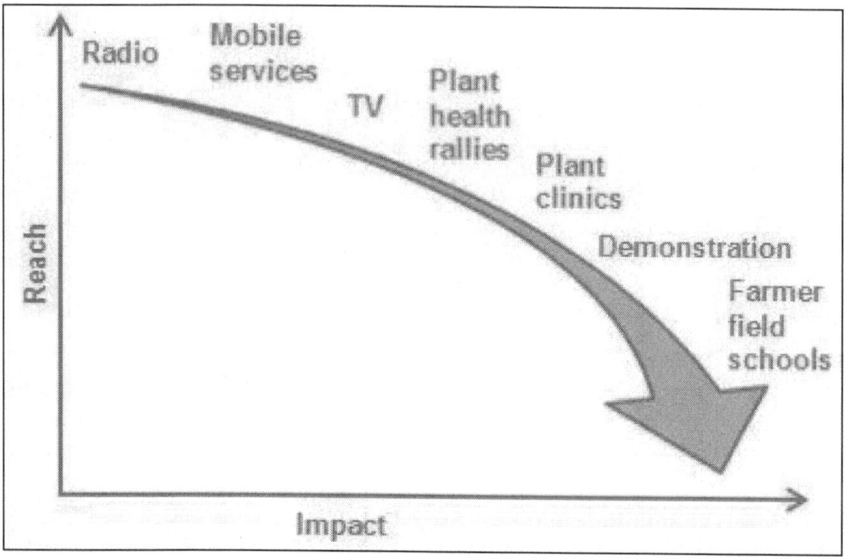

Figure 10.1: Impact of Extension Approaches (Sala *et al.*, 2016).

It is used to create mass awareness about an important problem or felt need of the community and encourage them to solve it. Exhibition is a systematic show of models, specimens, posters *etc.* in a series around a theme to build awareness and interest in the group (Ray G.L. 1991). As we know, the major constraints of national extension systems are availability of less number of extension functionaries and limited resources to reach wide range of farm people spread across geographical areas. To tackle these constraints, extension can be more efficiently performed using mass media. A participatory mass extension campaign in Malawi, for example, has shown positive impact of radio messages on farmers' understanding of climate change along with its effects, about producing better quality compost or essentiality of compost in constructing resilience and adaptation (Mloza Banda, 2014). The above model clearly shows that extension approaches which are suited for awareness raising have the potential to contribute to climate mitigation, adaptation and increase food security.

b. Trainings, Clinics and Rallies for Promoting CSA

Training of extension functionaries is very essential for updating their awareness related to climate change, impacts and implications on various adaptation and mitiation strategies. As we know, climate smart agriculture is a new and emerging issue so extension functionaries should be trained on this. The major component in which training can be given is agro-meteorology, and extension officials should be well versed with local climatic conditions and local terms and well versed in reading the scientific data or interpret it. Extension intermediaries should be closest to farm people and operate exclusively at field level. Farm clinic is a facility developed and extended to the farmers for diagnosis and treatment of farm problems and to provide some specialist advice to individual farmers (Ray G.L. 1991). Plant clinics are the medium to facilitate face-to-face exchange and two-way flow of information between farmers and extension functionaries and linking of all other components of plant system (Boa *et al.*, 2015). Different type of plant issues that brought to clinics can be either abiotic or biotic factors. Crop doctors should be well versed and knowledgable about specific farm people and farm conditions and local dialects should be well known and understandable. Plant doctors are usually trained extension workers and remain employed by their current organisations.

c. Farmer Field Schools

Farmer Field School (FFS) is non-formal educational activity. It is a group activity and field based. Empowers farmers to solve their field problems by themselves. It includes fosters participation, interaction and joint decision making. In FFS, farmers learn by carrying out activities through constant observation. FFS is a type of adult education that developed from concept in which farm people learn directly from the field observation and through field experience. FFS was initially developed in order to help farmers to learn about Integrated Pest Management (IPM) practices for varied ecological and climatic conditions. Initially it was started in 1989 in Indonesia to promote IPM in rice. (Anonymous 2015). Climate Field Schools in Indonesia raised climatic change awareness and promoted solutions to cope with changing rainfall patterns, such as recording and interpretation of on-farm rainfall measurements and in-field water harvesting (Winarto *et al.*, 2008).

d. Contingency Crop Planning

It is a document that includes the recommendations across the key aspects of crop management and cultural practices. This form of calendar is very useful in terms of crop planting, irrigation scheduling and plant protection measures for farmers. It is done on the bases of local weather conditions for the local crops grown in that particular area. It is participatory in nature because the local knowledge of farmers about the crops or various agricultural practices and scientific knowledge of scientist both plays important role in crop planning. It is developed along with the support of agro-meteorological department.

e. Information and Communication Technology (ICT) Networks

ICT stands for "Information and communication technology". ICT is defined as technologies which provide access to information connected through telecommunication. ICT includes internet, mobile phones, wireless networks and other mediums of communication. ICT can be used in teacher training programmes effectively to improve educational quality. The term ICT played an vital role as a medium of information as well as communication in climate change awareness, adaptation and mitigation strategies. Usage of mobiles, radios *etc.* addressed the problem of change in climate by creating awareness among the farmers about the availability of different adaptation and mitigation strategies. e-Arik was an ICT related project initiated in 2007 in Arunachal Pradesh of India, disseminating 'Climate smart agriculture practices' and at achieving food security.

The e-Arik project used variety of different ICTs in different ways. Thus, mobile technologies were used to record from the field. Radio and TV were used as a channel for raising general awareness about climate and agricultural issues but not for specific guidance. Videos were sometimes shown via laptop actually in the field in order to communicate specific details of adopting new agricultural technologies. Physical publications were kept in a village library for use when power outages prevented ICTs from being used. Finally, physical display of organic farm inputs at the village knowledge centre was used to stimulate interest and awareness with 90 per cent of visitors recorded as having enquired about availability of the input. The project was successful in demonstrating application of ICTs in promoting climate smart agriculture practices, new approaches to farming that require few external inputs and which are organic. Such projects must necessarily be driven by the needs and interests of the farmer beneficiaries. For them, climate is an important issue and they identify the keys of change in climate. However, their overriding priority and the main aspect that will contribute to the resilience in aspect of change in climate is increased incomes. (Source- Saravanan, 2014).

f. Farmer-to-Farmer Extension Approach (F2FE)

F2FF is a provision of farmers' training to farmers, mostly through the establishment of a farmer-promoter and farmer-trainer structure (Scarborough *et al.*, 1997). F2FF also provides a great promise to extend climate smart agriculture effectively. The approach empowers farmers as change agents and helps to increase adoption because farmers are highly willing to learn from their colleagues than from extension staff (Franzel *et al.*, 2015).

g. Climate Smart Villages

Climate smart villages are established villages or local action models which uarantee food security, foster adaptation and create resilience to climatic stress. The four essential components of CSV are village development plans, climate smart technology, climate information services and local knowledge and institutions. Based on its risk profile of climate and the willingness of farmers and local governments to participate in project, location of CSV is chosen. No standardized package of treatments or a one-size-fits-all approach is available. The emphasis is on customizing the portfolio of strategies that complement each other and fit local circumstances.

Gender Issues in Climate Smart Agriculture

Understanding gender issues is crucial to effectively deal with climate change and achieve climate smart agriculture. Due to different gender roles and the related constraints, both women and men may have various perspectives and awareness about the climate risk, how it may affect their livelihoods and how to respond to this challenge. They have different access to the resources and services needed to adopt climate smart practices. Gender productivity gaps exist in agriculture. Due to traditional gender-based discrimination, women have fewer privileges, entitlements and endowments. Generally women face more challenges than men nowadays in accessing, using and controlling productive resources and services, such as land, water, credit, inputs, technologies, information, knowledge, education, extension and other rural advisory services, markets and weather and climate information. This affects their adaptive capacity to climate threats. Gender-specific consequences in the sense of climate smart farming vary by the degree to which women can equally access resources such as land or livestock, services, employment and business opportunities (World Bank, FAO and IFAD, 2015). It is reported that closeness of gender gap in farming is expected to reduce hungry people by 100-150 million (FAO, 2011). However, climate change exacerbates the existing barriers that women face.

Institutional Support to Promote Climate Smart Agriculture

Strong institutional support is required to improve the dissemination of information, provide financial support and access to markets, and coordinate the work of a large number of farmers over a wide area. In order to effectively develop the requisite CSA partnerships, needs, goals and drivers of individual organizations must be identified and handled appropriately. Institutional synergies and knowledge sharing networks are essential to CSA programmes being developed, sustained and scaled up. The innovations needed to enforce the CSA can seem daunting to smallholder farmers without adequate institutional frameworks in place. Some of the improvements expected by CSA from institutional support include:

Adapting Management Practices

Competition for natural resources will grow as populations increase. Moreover, environmental and climatic variables are starting to degrade these services. For this purpose, CSA is based on seeking new approaches for soil, water and land management. In several areas, water tables drops drastically (*e.g.* in parts of India). The techniques of water harvesting and water conservation need to be able to cope with the possibility of rising fluctuations in rainfall. Land loss and soil fertility decreases are widespeard concerns. To improve soil quality and restore deraded land property, adoption of new strategies and approaches is important.

Climate Smart Crop Varieties

One major CSA approach to adjusting to new circumstances is to change crops as climatic change alters rainfall patterns. The switch to new crops may entail planting varieties with a shorter growing period that can withstand saline soils and severe weather events better (droughts and floods) are more resistant to pests and diseases.

Changing Farming Schedules

Climate change is pushing farmers to change the pace of their normal agricultural operations. For example, in East Africa, many farmers have already started planting sooner than they have istorically done because of changing rainfall patterns (Kristjanson *et al.*, 2012).

Mitigating while Adapting to Climate Change

Planting trees on farms and enhancing livestock and rangeland management will also be essential CSA practices as the need to reduce cabon emissions increases

(FOA, 2010). These activities are typically followed by farmers mainly because they help to increase and diversify incomes, not because they reduce emissions. An additional advantage is combating climatic change. When establishing CSA, local organizations and institutions can play positive roles in three areas:

a. Production and Transfer of Technical Knowledge

- ☆ Identify the key problems and most vulnerable houseolds in the local agriculture and food systems.
- ☆ Consider and pick from collection of possible climatic smart practices the most locally relevant technologies (both institutional and technical).
- ☆ New and recent information on weather, environment and agricultural options is rapidly relayed throughout nation.
- ☆ Improve the consistency and relevance of CSA research by local involvement.

b. Providing Credit Services and Access to Markets

- ☆ To kick-start new practices, innovations and behaviours among farmers, channel micro-finance effectively.
- ☆ Provide credit, insurance, social protection nets, and environmental service payments or incentives.
- ☆ Stimulating local markets, building ties with national and international markets, and improving smallholder farmers' market literacy.

c. Supporting Coordination of Collaborative Action

- ☆ Foster new cultural standards for agriculture, food production and household food management activities.
- ☆ Shift the emphasis of agricultural extension from providing technology to working to establish solutions in collaboration with local farmers.
- ☆ Include especially disadvantaged social groups in benefit of climate smart agriculture, such as women, youth or immigrants.

i) Public Sector Institutions

The institutional structures through which such assistance is delivered stretch from national ministries (primarily Ministries of Agriculture) and Department of Agriculture in national to regional, subregional and local offices. Public sector institutions are generally top-down in structure whereas in formulating needs of specific reigonal farmers, Public sector institutions make use of bottom-up idea. National Adaptation Prorammes of Actions (NAPAs), National Adaptation Plans (NAPs) and National Appropriate Mitigation Actions (NAMAs), are intended to provide national frameworks to guide government actions on mitigation and adaptation under a future international climate agreement. These voluntary proposals for the United Nations Framework Convention on Climate Change (UNFCCC) are still being created, with the result that their particular functions are still developing.

ii) Civil Societies

Civil society organizations' include non-governmental organizations (NGOs). They also include trade unions, professional associations, research and educational faculties, private foundations, religious organizations, issue groups, the media and similar organizations reflecting civic interests, values and concerns. Such institutions, acting independently of the state.

iii) Non-Governmental Organizations

NGOs are typically strong in providing social services, especially when supporting poor communities (Swanson and Rajalahti, 2010). For example, building social capital by organizing farmer groups to articulate their own needs and formulate solutions collaboratively, brings immediate positive effects at the local level. NGOs can also be valuable partners for implementing projects, conducting research and carrying out outreach activities.

iv) Universities and Research Centers

Universities and research institutes are good providers of pre-training and in-service training for extension workers. As faculty members often conduct agricultural research of their own, they are obvious focal points for field staff in need of technical and management backstopping (Swanson and Rajalahti, 2010).

v) Private Foundations

Private foundations rarely aim to organize popular movements. However the funding they offer and the expertise they provide in specific activities are usually not influenced by the public sector and for this reason they often have scope for greater innovation (Uphoff and Krishna, 2004).

vi) The Media

Mass media can reach large audiences with messages about CSA. Broadcast and electronic media include radio, television, film and the internet; print media includes newspapers, journals, magazines; outdoor media includes posters placed in public view. An important channel for disseminating informative and educational contents, media can both reflect public opinion and shape it. In recent years, more and more people, even in rural areas have gained access to media hardware such as radios, televisions, computers and smart phones.

vii) Religious Communities and Institutions

An awareness of public opinions and values is therefore important when formulating a CSA strategy. While religious institutions or issue groups are unlikely to provide CSA-related advisory services, they can offer indispensable insights into prevailing beliefs and values. In addition to these institutions, some other institutions include private sector institutions, individual farmer support, producer cooperatives, commercial consultancies, national and international aribusinesses, banks, credit and saving institutions and other local stakeholders are also responsible and playing a key role for promoting Climate Smart Agriculture.

Reforming Governance to Improve Extension Service Delivery

One of the major problems worldwide is enhancing public service delivery. It is considered to be a core feature of government to coordinate public service provision. World Bank (2002) defined governance as "the way in which power is exercised in management of the country's social and economic resources for the development". It is the brain that creates various aspects of society and identifies functions, whereas administration is the act of implementing the result of governance. Extension is a public service in most developing countries and its failure to effectively support rural producers has led to its reform. When we refer to reforms, it is evident that there are both supply side reforms as well as demand side reforms.

1. **Supply side reforms** related in provision and funding of public, private and third sector services, administrative and fiscal decentralization, capacity development and budgeting.
2. **Demand side reforms** related in participatory planning and implementation principles. Proper reforms relating to institutional reforms, structural reforms and legal reforms need to be implemented to ensure good governance.
3. **Administrative reforms** seek to focus on issues related to people and systems/processes. Administrative reforms are defined as comprising reforms in three important elements, *viz.* Human Resources Development and Personnel Management; Internal Systems and Processes; and Citizen Interface Systems and Processes.
4. **Structural reforms:** As a consequence of economic or technological changes, structural reforms refer to changes in the structure and operation of organizations.
5. **Legal reforms** refer to changes in role of institutions as a consequence of changes in policy or law or legislation.

Climate Financing Options for CSA

Financing options specifically targeting CSA are still limited, necessitating a strategic use and combination of existing funding sources. The basis for any CSA activity should be the identification of a country's opportunities and vulnerabilities, corresponding needs and preferred options for CSA activities. After national priorities have been defined, a strategic approach to sources of international finance based on an understanding of available channels will not only increase the chances for approval, but also enhance the fit between the finance option and the country's overall approach to climate change in agriculture.

List of categories of important climate finance options at international level provided by:

1. Financing mechanisms directly under United Nations Framework Convention on Climate Change (UNFCCC);
2. United Nations (UN) organizations or programmes;
3. Multilateral Development Banks (MDBs);

4. Bilateral public financing channels;
5. Compliance and voluntary carbon markets; and
6. Private sector actors and philanthropy.

On 30 June 2008, the Government of India unveiled the National Action Plan on Climate Change (NAPCC), outlining eight national climate change missions. These include:

I. National Water Mission
II. National Mission for Sustaining the Himalayan Eco-system
III. National Mission for Sustainable Agriculture
IV. National Solar Mission
V. National Mission for Enhanced Energy Efficiency
VI. National Mission on Strategic Knowledge for Climatic Change
VII. National Mission on Sustainable Habitat
VIII. National Mission for a Green India

It was assigned to the Department of Science and Technology, Ministry of Science and Technology, to manage two of these eight national climate change missions. These are:

I. National Mission on Strategic Knowledge for Climate Change (NMSKCC) and
II. National Mission for Sustaining Himalayan Ecosystem (NMSHE)

Programmes Related to Climate Smart Agriculture Promotion in India

1. For Sustainable Livelihoods

i) National Innovations in Climate Resilient Agriculture (NICRA)

NICRA was launched by Indian Council of Agricultural Research (ICAR) in February 2011, with funding from the Ministry of Agriculture and Farmers' Welfare, Government of India.

Objectives

- ☆ To improve Indian agricultures' resilience to change in climate by designing and implementing enhanced development and risk management technologies covering crops, livestock and fisheries.
- ☆ To demonstrate site-specific application packages in the fields of farmers to resolve existing climatic risks.
- ☆ To enhance the capacity building and deployment of climatic resilient agricultural research by scientists and other stakeholders.

ii) National Mission on Sustainable Agriculture (NMSA)

NMSA has been formulated to enhance agricultural productivity, with a focus on integrated agriculture, water quality, soil health management and resource conservation synergies, especially in rainfed areas. NMSA addresses key aspects of water use efficiency, nutrient management and diversification of livelihood by following path of sustainble development by gradually moving to environmentally friendly integrated farming *etc.* NMSA aims to promote specific enhanced agronomic practices.

Objectives

i. By supporting place specific integrated/composite farming systems, to make agriculture more efficient, sustainable and climate resilient.

ii. To preserve natural resources by suitable steps to conserve soil and moisture.

iii. To follow systematic practices of soil health management based on soil fertility maps, application of macro and micro nutrients based on soil testing, judicious application of fertilizers *etc.*

iv. Optimizing the use of water supplies through successful water management to increase coverage to achieve 'more crop per drop'.

v. To promote capacity building of stakeholders and farmers with on-going missions in order to follow mitigation measures and climatic change adaptation.

vi. To create effective and efficient inter and intra departmental coordination under NAPCC for achievement of key deliverables of NMSA.

NMSA has three essential components such as:

- ☆ Rainfed Area Development (RAD).
- ☆ Soil Health Management (SHM).
- ☆ Paramparagat Krishi Vikas Yojana (PKVY): The state is bestowed with the potential of organic farming due to its huge natural resource, biodiversity, varied and rain fed agro climatic conditions. To promote organic farming, Indian Government has recently launched a new scheme, PKVY under the NMSA for adoption of organic villages.

2. For Watershed Development

Pradhan Mantri Krishi Sinchai Yojana (PMKSY)

Government of India has started a new scheme, PMKSY. The main motto of this scheme will be micro irrigation projects ("Har Khet Ko Pani") and end to end irrigation solutions. The big concept is "More crop per drop".

Objectives

i. Achieving convergance of irrigation investments at field level.

ii. Expansion of cultivable region under guaranteed irrigation.

iii. Improving water use efficiency to reduce water wastage.
iv. Improving the adoption of precision agriculture and other innovations to conserve water.

3. For Crop Insurance

National Crop Insurance Scheme/Pradhan Mantri Fasal Bima Yojana (PMFBY)

The state governament. has introduced this scheme from Rabi, 1999-2000 season. Crops that covered under this scheme are wheat, barley, maize, paddy, potato and ginger. 50 per cent subsidy is being provided to the small holding and marginal farmers. For loaned farmers, Scheme is obligatory and for non-loaned farmers, it is optional. The framework offers comprehensive risk insurance against yield losses, in other words, drought, hailstorm, floods, pests and diseases. The scheme is being introduced by Agriculture Insurance Corporation of India (AICI). The farmers of the state will benefit from this scheme. The government of state and India share the losses equally. The premium subsidy has been increased from 10 to 50 per cent for small as well as marginal farmers since Rabi 2007-2008.

Conclusion

This chapter clearly explained about Climate smart agriculture and role of Agricultural Extension in Climate smart agriculture. As per the reports, most of the positions of agricultural officers are vacant in many states. At natinal level, the ratio of extension staff to farmers is 1:1162, as compared to the recommended ratio of 1:750 (Anonymous, 2018). Eventhough Government implemented schemes in national as well as international level is giving satisfactory results in implementation level, due to low ratio of extension workers, sometimes schemes are not reaching farmers effectively and not utilised efficiently. Extension approaches shows us regarding the methods, methodologies and technologies used in CSA. This chapter also tells us about institutional support for promoting Climate smart farming. Thus, Agriculture Extension plays the vital role in supporting and improving new innovative technologies introduced in agriculture. Future role of Agriculture Extension is vitally important for developing every farmers livelihood and their socio-economic conditions in the society. Farmers also need to initiate using new innovative ideas like CSA in order to lead a sustainable life.

REFERENCES

Anonymous, (2015). Agritech.tnau.ac.in.

Anonymous,(2018).HansIndia,https: //www.thehansindia.com/posts/index/NewsAnalysis/2018-07-24/Reform-agri-extension-to-boost-ryots-income/400625.

Anonymous,(2020). dst.gov.in/climate change programme. Ministry of Science and Technology.

Boa, E. Danielsen, S. and Haesen, S. (2015). Better Together: Identifying the Benefit of a Closer Integration Between Plant Health, Agriculture and One Health. In:

One Health: The Theory and Practice of Integrated Health Approaches, 22: pp. 258-259.

FAO. (2010). Climate smart agriculture: policies, practices and financing for food security, adaptation, and mitigation. Rome. 49 pp.

FAO. (2011). The State of Food and Agriculture. Women in Agriculture. Closing the gender gap for development. Rome, FAO. 160 pp.

FAO. (2013). Climate Smart Agriculture: Sourcebook. Food and Agriculture Organisation of The United Nations, Rome, Italy. 570 pp.

FAO. (2018). Climate smart agriculture training manual – A reference manual for agricultural extension agents. Rome. 106 pp.

Franzel, S. Degrande, A. and Kiptot, E. (2015). Farmer-to-farmer extension. Global forum for rural advisory services good practice notes for extension and advisory services, Note. 7. Global Forum for Rural Advisory Services, Lindau, Switzerland. 4 pp.

Kristjanson, P. Neufeldt, H. Gassner, A. Mango, J. Kyazze, F. Desta, S. Sayula, G. Thiede, B. Förch, W. Thornton, P.K. and Coe, R. (2012). Are food insecure smallholder households making changes in their farming practices? Evidence from East Africa. Food Security, 4: 381-397.

Lipper, L. Thornton, P. Campbell, B.M. and Torquebiau, E.F. (2014). Climate –smart agriculture for food security. Nature Climate Change, 4: 1068-1072.

Manobharathi, K. (2020). Job Satisfaction of Agricultural Extension Functionaries for Dissemination of Information in District Baramulla of Jammu and Kashmir, M.Sc. (Agri) thesis. Sher-e-Kashmir University of Agricultural Sciences and Technology, Srinagar, Kashmir: 145 pp.

Maunder, A.H. (1973). Agricultural Extension: A Reference Manual, Abridged Edition, Rome, 374 pp.

Mloza Banda, B.C. (2014). E-adaptation to Climate Change in Malawi, ICT Update, 78 pp.

Neeta Sabar. (2015). A study on the level of Job satisfaction of agriculture extension functionaries of Nuapada district in Odisha, M. Sc. (Agri.) Thesis, Department of Extension Education, Orissa University of Agriculture and Technology Bhubaneswar, Odisha. 122 pp.

Parry, M. C. Rosenzweig, and M. Livermore. (2005). Climate change, global food supply and risk of hunger. Phil. Trans. Roy. Soc. B, 360: 2125-2138.

Rasheed Sulaiman, V. Onima, V T. Nimisha Mittal. Athira, E. (2019). Taking Stock and Shaping the Future: Conversations on Extension. Agricultural Extension in South Asia (AESA), Centre for Research on Innovation and Science Policy (CRISP), Hyderabad, India, 572 pp.

Ratheeswari, K. (2018). Information Communication Technology in Education. Journal of Applied and Advanced Research. In Proceedings of the Conference on "Recent Trend of Teaching Methods in Education" Organised by Sri Sai Bharath College of Education Dindigul, Tamil Nadu, India.

Ray, G.L. (1991). Extension communication and management. Naya Prakash, Calcutta, India. ISBN: 8185421072. 358 pp.

Rupan, R. Saravanan, R. and Suchiradipta, B. (2018). Climate Smart Agriculture and Advisory Services: Approaches and Implication for Future. National Institute of Agricultural Extension Management, Hyderabad, India, 54 pp.

Sala, S. Rossi, F. David, S. (2016). Supporting agricultural extension towards Climate Smart Agriculture: An overview of existing tools. Global Alliance for Climate Smart Agriculture FAO, Italy, 90 pp.

Scarbourough, V. S. Killough, D.A. Johnson and J. Farrington (eds.). (1997). Farmer-led Extension: Concepts and Practices. Intermediate Technology Publications, London, 214 pp.

Swanson, B.E. and Rajalahti, R. (2010). Strengthening agricultural extension and advisory systems: procedures for assessing, transforming, and evaluating extension systems. World Bank Agriculture and Rural Development Discussion Paper No. 45, 206 pp.

Uphoff, N. and Krishna, A. (2004). Civil society and public sector institutions: more than a zero-sum relationship. Public Administration and Development, 24: 357-372.

Winarto, Y.T. Stitger, K. Anantasari, E. And Hidayah, S.N. (2008). Climate Field Schools in Indonesia: Improving "response farming" to climate change. Leisa Magazine, Indonesia, 24 pp.

World Bank. (2002). Reforming Public Institutions and Strengthening Governance: A World Bank Strategy- Implementation Update, USA, 239 pp.

World Bank, FAO and International Fund for Agricultural Development (IFAD). (2015). Gender in Climate Smart Agriculture: Module 18 for the Gender in Agriculture Sourcebook. Agriculture global practice. World Bank Group, Washington, DC, 96 pp.

Chapter 11

Protected Cultivation for Climate Smart Agriculture

*Mukesh Kumar Mehla[1] *, Khyati Singh[2], Charan Singh[3] and Ravi Kumar[4]*

[1]Department of Soil and Water Engineering, College of Technology and Engineering, MPUAT, Udaipur – 313001, Rajasthan
[2]Department of Vegetable Science, COA, CCSHAU, Hisar – 125004, Haryana
[3]Department of Processing and Food Engineering, College of Agricultural Engineering and Technology, JAU, Junagadh – 362001, Gujarat
[4]Department of Processing and Food Engineering, College of Agricultural Engineering and Technology, CCSHAU, Hisar – 125004, Haryana

Agriculture plays very important role in agricultural growth and economic prosperity of our country. Protected Cultivation has emerged as a good alternative to traditional crop cultivation practices specially under climate change, it can help overcome biotic and abiotic stresses. Climate change has impacted agriculture adversely with occurrence of erratic seasonal behaviours of rainfall and extreme events like drought and flood threatening sustainability of crop production ad food security. Protected cultivation can help in breaking the seasonal barrier and open doors to cultivation of offseason crops which fetch much higher prices. Round the year cultivation of high-value vegetables crop, flowers etc. especially, during the off-season is quite profitable. Higher yield under protected cultivation can be achieved by maintaining proper microclimate, better irrigation and fertigation management practices. They also give greater net return and have high B: C in comparison with open field condition. Poly-sheet cladding prevents entry of harmful ultraviolet light, conserves CO_2 and enhances plant growth and development. Temperature and moisture can be properly maintained inside the structure, which *improves the rate of photosynthesis and supports uniform plant growth. The construction of protected structures is quite expensive as it requires high initial cost but government is encouraging farmers to adopt them by providing subsidies and technical support.*

***Keywords**: Protected cultivation, Polyhouse, Climate smart agriculture and Climate change.*

Introduction

Injudicious use and climate change have significantly worsened the availability of precious land and water natural resources. There has been consistent adverse effect of climate change on agriculture especially in recent times with happenings of extreme events like drought and flood which has added to the misery of farmers and put a big question marks on sustainability of crop production ad food security under climate change which is worsening day by day (Srinivasarao *et al.*, 2020). Indian agriculture is consistently struggling with problems due to population explosion, low productivity, depleting water tables, loss of fertile lands, pollution of land and water resources due to indiscriminate use of both resources and decreasing of these due to expansion of industrial and urban area (Kumar *et al.*, 2020). Use of protected cultivation systems and following recommend adaptations in terms of sufficient cooling, improved natural and additional light for winter production, innovations like the semi-/closed greenhouse with mechanical cooling and dehumidification, structural solutions like passively ventilated greenhouses and screenhouses can help cultivation systems to cope with climate change (Gruda *et al.*, 2019).

Protected cultivation is a concept that involves modifying and controlling the natural environment for optimum plant growth. This involves modifying and controlling the temperature, RH, air velocity, concentration of carbon dioxide in atmosphere and root zone oxygen, nutrients and moisture supply to enhance crop growth and yield. This allows crop cultivation to be independent of external climate and weather conditions. Polyhouse has become a staple structure that provides affordable means of environmental control for providing opportunity to cultivate high value and offseason crops with full or partial control on the growing environment. Protected cultivation consists of all sorts of systems and practices of cultivation ranging from hot beds, plastic low tunnels, hydroponics, aquaculture, soilless cultivation, polyhouse and growth rooms to completely enclosed factory type setting with complete control over the environment along with artificially lighting. Mostly vegetable crops are environment sensitive, and thus very high or very low temperatures and deficient soil moisture constitute the major causes of low yields. India is second largest producer of vegetables in the world contributing 14 per cent in the world vegetable market. India's vegetable production is 184.39 million tonnes with 10.25 million hectares area under production and an average productivity of 17.97 MT/ha. Several methods can be adopted to increase the production of quality produce and consumption of vegetables such as increasing area under vegetable hybrids, using improved agro-techniques and cultivation of vegetable crops under protected conditions. The off-season cultivation of vegetable crops and improved storage infrastructures should be promoted to avoid the flooding of vegetables in the market in natural season and generating higher profits.

India is a diverse country both culturally as well as in terms of agro climatic variations, it has a wide range of diverse agro-climatic conditions. Most cultivation practices are generally restricted to regional and seasonal needs. Even though the production of vegetables has increased upto 184.39 million tonnes from an area of about 10.25 million hectares, but still the technology used and practices followed by the farmers are pre-dominantly traditional which consequently results in low productivity, poor quality and high losses in the produce.

Most parts of India have non-constraints of soil fertility, especially the northern plains where the soils are highly fertile but extremes variability of temperature ranging between 0 to 48°C around the year does not allow year-round cultivation vegetables outdoor due such biotic (mainly during rainy and post rainy season) and abiotic stresses.

When climate prevents or reduces the choices for year-round out door crop production, crops are produced in protected environments. Polyhouse has become widely accepted and most efficient means to overcome climatic diversity. Polyhouse vegetable production systems makes the use of most recent and advance technology to control the environment for maximizing crop productivity and increasing the quality of the vegetable crops. Although, contribution of polyhouse technology in terms of area under cultivation is very small part of worldwide agricultural operations but the income generated from polyhouse cultivation by growing off season crop, enhanced quality and prospects of foreign exchange earnings from exports. India is still new in the area of protected vegetable cultivation and the total area under protected vegetable production is still very less. India has a vast landscape with diverse and extreme agro-climatic conditions, the protected vegetable cultivation technology has great potential. It can be utilized for year-round, or production of virus free high-quality seedlings, production of off-season vegetables, hybrid seed production of high value vegetables and as a tool for diseases resistance breeding programs (Murthy *et al.*, 2009).

India moving with a steadily pace to expand the area under protected cultivation and improve develop new technologies. This pace has increased many folds by keen interest of the government which continues to support protected cultivation efforts strongly in the country. Polyhouse cultivation of horticultural crops is considered to be the next logical step to the open field agriculture as small area under polyhouse cultivation gives higher production and productivity of vegetables. Also, has potential for generating additional jobs in the rural sector where unemployment problem is more acute. This has a tremendous scope in peri-urban areas by providing local production requirements of urban consumer vegetable demands for off-season and high-quality produce of different horticultural crops. Polyhouse cultivators in area nearing to metro cities and good transport facilities have tremendous potential to go ahead with vegetable production under polyhouse for high productivity and high-quality vegetables to take the advantage of the available upmarket in those cities and get good price of their produce. Next consideration should be which crops should be grown under polyhouse the choice of the vegetable crop or variety to be grown in a polyhouse is made on the basis of the physical size of the structure and market available for that produce. Vegetables crops like tomato, cherry tomato, coloured sweet peppers, slicing cucumbers, off season muskmelon, summer squash including some rare vegetables, medicinal and ornamental plants especially offseason and under adverse agro-climatic conditions can be very profitable. Nursery raising of vegetables seedling and providing disease free quality planting material has become good business, it is very specialized and technical job and its importance has increased with increasing susceptibility of vegetables to various biotic and abiotic stresses and very high cost of hybrid seeds. Keeping in view these critical issues, this chapter presents important aspects related to protected cultivation for climate smart agriculture.

Protected Structure for Climate Smart Agriculture

Protected cultivation systems are used for farming under adverse agro-climatic conditions, especially offseason. Plants grown in a controlled environment, have good potential for producing more and better-quality produce with higher productivity. Crops require suitable climate conditions for successful cultivation crop but with constantly changing and abnormal weather patterns the climate has changed. Protected cultivation can be used grow crops which are remarkably difficult to grow in open field conditions, which are otherwise very susceptible to climatic conditions even in harsh winters by controlling the micro-climate in the protected structure. Most modern polyhouse are either Normally Ventilated Polyhouses or the Fan-and-Pad Polyhouses. All kind of protected technologies may not be economical and useful to the farmers in India, because of their very high initial, running and maintenance cost, but some protected technologies are simple, suitable and highly profitable under Indian conditions and more specifically under peri-urban areas, which can be adopted by Indian farmers for production of different horticultural crops.

Table 11.1: Major Climatic Factors for Consideration in Protected Structures

Sl.No.	Climate Factors	Desirable Ranges and Values	Remarks
1.	Air/wind movement	Inflow: outflow air circulation ration (1:19	Affects temperature, Relative Humidity and CO_2 concentration
2.	Carbon dioxide CO_2	350 – 1000 ppm	Required for good photosynthesis
3.	Radiation/light	50000 lux	Photosynthesis, photo morphogenesis, photoperiodism
4.	Temperature	18 – 25°C	Cell division, elongation, respiration, photosynthesis, water uptake, transpiration
5.	Relative humidity (RH, per cent)	60 – 80 per cent	Plant stability

Protected structures have cladding which protects crops from UV and harmful infrared radiation. Also, it protects the plants from extreme temperatures and maintains air and soil moisture (Maikhuri *et al.*, 2007). Even simple structures like shade nets were utilized for the protection of important crops against abundance daylight, cold, ice, wind, and insect/birds (Takte *et al.*, 2003). Walk-in tunnel with lower initial cost covered with UV plastic film can be used to grow reasonably wide range of crops, flowers and vegetables resulting increase yield and farmers income. Early crops in protected structures using low-cost and simple greenhouses can enhance income of vegetable growers in even cold areas (Sanwal *et al.*, 2004).

Components of a Protected Structure

Major components of a protected structure are as follows:

Roof: It is the transparent cover of a Polyhouse over the protected structure.

Gable: It is the transparent wall of a protected structure.

Cladding material: It is the transparent material mounted on the walls and roof.

- ☆ **Rigid cladding material:** It is the cladding material with high degree of rigidity, where any deformation may result in permanent damage to it. *e.g.* Glass.
- ☆ **Flexible cladding material:** It is the cladding material with a degree of flexibility such that there is no permanent structural damage due to deformation. *e.g.* Plastic film.

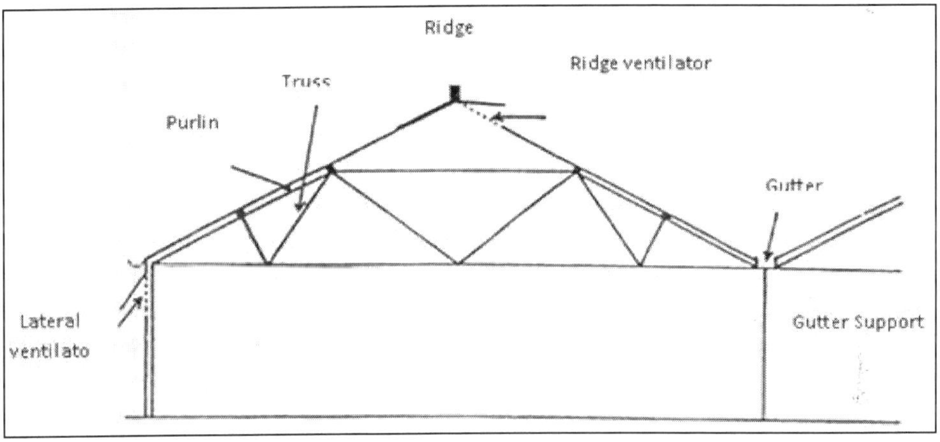

Figure 111.: Major Components of a Protected Structure.

Girder: It is the horizontal structure member, connecting columns on gutter height.

Bracings: These are sued to support the structure against wind.

Gutter: It collects and drains rain water and snow which is placed at an elevated level between two spans.

Column: It is the vertical structure member carrying the protected structure.

Ridge: It is the highest horizontal section on the top of the roof.

Arches: These are the members supporting covering materials.

Purlin: It is a member which connects cladding supporting bars to the columns.

Foundation pipe: It is the connection between the structure and ground.

Span width: It is the center to center distance of the gutters in multispan houses.

Polyhouse length: It is the dimension of the protected structure in the direction of gable.

Polyhouse width: It is the dimension of the protected structure in the direction of the gutter.

Designs and Classification of Polyhouse

Polyhouse structure consists of base frame of wooden, plastic, galvanized iron (G.I), steel or other metal material covered with a transparent cladding material in which crops are grown under controlled environment conditions. Polyhouse cultivations basic advantage revolves around creating favourable micro-climates, which favours the crop production and crops can be possible throughout the year. It is generally associated with the off-season production of ornamentals plants and high value crops in cold climate areas where outdoor production is not possible. The temperature and moisture are primary environmental parameter that are traditionally controlled in a protected structure, usually providing heat to overcome extreme cold conditions and increasing moisture and RH by misting. Greenhouse vegetable is particularly useful for early crop production keeping in mind the high energy efficiency in cold areas. The development of a temperature integration approach along with low pre-night temperature (PNT) increases earliness, improves energy use efficiency and reduces energy consumption (Zhang *et al.*, 2010). Environmental control includes cooling to mitigate excessive temperatures, light control either shading or adding supplemental light, carbon dioxide levels, relative humidity, water, plant nutrients and pest control.

Table 11.2: Cost (per sq. meter) for different Types Polyhouse (Singh *et al.*, 2015)

Sl.No.	*Type of structure*	*Cost (¹) per sq. meter*
1.	Less expensive polyhouse without fan and pad	300 to 500/m^2
2.	Medium cost polyhouse with pad and fan system without automation	800 to 1100/m^2
3.	Expensive polyhouse with fully automatic control system	2000 to 3500/m^2

Classification of Polyhouse Based on Suitability and Cost

a) Low Cost or Low Tech Polyhouse

Low cost polyhouse is a simple structure constructed with material like bamboo, timber *etc.* which are locally available and cladding of the ultra violet (UV) film is used. There is no specific control device that can be used for regulating environmental unlike conventional or hi-tech where control over different environmental parameters inside the polyhouse are provided. But there are simple techniques that can be used somewhat increasing or decreasing the temperature and humidity. Light intensity can be reduced by using net type or reflecting type shading materials. The temperature can be maintained during summer by opening the side walls. Such structures are suitable for cold climatic conditions.

b) Medium-tech Polyhouse

These type of polyhouses are suitable for dry climatic regions. A polyhouse users prefers to have some degree of control over the environment by having manually or semiautomatic control arrangement with minimum investment. This type of polyhouse is constructed using galvanized iron (G.I) pipes. Exhaust fans with thermostat are available for temperature control. Evaporative cooling pads and misting arrangements are also made to maintain a favourable humidity inside the Polyhouse. But as it is still only semi-automatic a lot of attention is required to operate and maintain the system successfully. It is very difficult to maintain uniform environment throughout the cropping period.

c) Hi-tech Polyhouse

Hi-tech Polyhouse overcome the difficulties faced in medium-tech polyhouse. In a hi-tech polyhouse there is network of sensors and controllers that automate the environmental control process. These type of polyhouses are suitable under various climatic regions.

Other Classifications

The polyhouse are classified based on type of structures, type of glazing, number of spans, environmental control *etc.* The various types are as follows.

Classification Based on the Type of Structure

a. Quonset type
b. Curved roof type
c. Gable roof type

Classification Based on the Type of Glazing

a. Glass glazing
b. Fiberglass reinforced plastic glazing
    i. Plain sheet
    ii. Corrugated sheet
c. Plastic film
    i. Ultra violet stabilized low density poly ethylene
    ii. Silpaulin

Classification Based on Number of Spans

a. Free standing or single span
b. Multispan or ridge and furrow or gutter connected

Classification Based on Environmental Control

a. Naturally ventilated
b. Passive ventilation

Nursery Raising in Protected Structures

It is easier to get higher production and disease-free seedlings under protected structures even in the offseason of a year to get the early crop under protected conditions or/and open field conditions. Raising nursery of the vegetable and flower crops under protected structures has many advantages like easy handling, early germination, and security environment free from different biotic and abiotic stresses (Singh and Sirohi, 2002).

Table 11.3: Different Protected Structures and Vegetable Crops Suitable for Seed Production

Sl.No.	*Protected Structure*	*Suitable Vegetable Crops for Seed Production*
1.	Climate controlled polyhouse	Tomato, cherry tomato, sweet pepper (one crop)
		Cucumber (three crops)
2.	Semi-climate controlled Polyhouse	Tomato, cherry tomato, sweet pepper (one crop)
		Cucumber (three crops)
3.	Naturally ventilated polyhouse	Tomato (one crop)
		Sweet pepper (one crop)
		Cucumber (three crops)
		Muskmelon (two crops)
		Summer squash (two crops)
4.	Insect proof net-houses	Sweet pepper (one crop)
		Tomato (one crop)
		Cucumber (two crops)
5.	Walk-in-tunnels	Muskmelon (one crop), Watermelon (one crop) and other cucurbits
6.	Plastic low tunnels and Rain shelters	All cucurbits

To ensure high productivity and high quality of the produce, raising of high-quality seedlings through use of good quality seeds at right time and at an appropriate place is one of the most important opportunity that can be captured to reap multiple benefits. Most of the Indian farmers are raising their vegetable seedlings under open field conditions, which are easily infected by diseases and always inferior in quality especially during rainy and post rainy season. On one side soil borne fungus and nematodes create severe problem for raising the seedlings in soil media in open fields during hot summers and rainy season but on the other hand the very high cost of hybrid seeds in vegetables has also warranted the farmers to change their traditional nursery raising method to increase the productivity and quality of vegetables. Under this system seedling are raised in multi-celled plastic pro-trays in artificial soil-less media in especially designed polyhouse or other protected structures. Large number of virus free healthy seedlings of different vegetables can be raised in a small area of polyhouse in multicelled plastic pro-trays by using soil-less media for growing vegetables either in main season or during their off season. With the use of this technology it is now almost possible to even raise healthy, vigorous seedlings of different cucurbits, otherwise it was not possible in the traditional system of nursery raising. This provides great opportunity for farmers or unemployed youth of our country to successfully start nursery raising as a small-scale industry and with proper marketing it can turn into good business in major vegetable growing pockets of the country.

Flower Cultivation in Protected Structures

Growing ornamental and flower in protected structure have significant economic advantages. Cost for protected cultivation structures can be easily recovered in 3-5 years, depending upon the crops grown. Carnation development has been the most beneficial activity as compared to other type of crop production (Punera *et al.*, 2017). Profitability in cut-flowers under poly house conditions was found to be maximum in orchid and minimum in rose (Wadkar *et al.*, 2006). Major flower crops like rose, carnation, Lilium, gerbera, and chrysanthemum have found to be having great commercial feasibility in the low-cost, naturally ventilated, galvanized iron-framed poly houses.

Off-season Cultivation of Crops under Polyhouse and Plastic Low Tunnels

In most parts of our country the farmers are growing various vegetables during their main season of cultivation but in these seasons due to high supply during the harvesting period the prices of vegetables go down and sometimes farmers are even not able to breakeven cost of cultivation. But the process of same vegetables goes up during the off season owing to low supplies and poor cold storage infrastructure, they are sold at very high price during their off-season. The demand of off-season vegetables is increasing day by day and this provides good scope for protected vegetable cultivation through off-season cultivation of vegetables in polyhouses and low tunnels (Sirohi *et al.*, 2002). Plastic low tunnel technology is a simple, suitable and profitable technology for off-season cultivation during the winter season. Cucurbits are warm season crops, generally grown in last week of March to April under open field conditions when night temperature is around 18-20°C. However, their seedlings can be cultivated in polythene bags during December-January in polyhouse and can be transplanted during February-March in the field. This allows yield to be obtained at least one month before direct sowing and a bonus profit can be obtained in off-season (Sanwal *et al.*, 2004). Crops like summer squash can be grown as a complete off-season crop, whereas other cucurbits like muskmelon, bottle gourd, cucumber, bitter gourd, watermelon can be advanced by 30-40 days over their normal growing season (5, 9, and 10). Plastic low tunnels are flexible transparent coverings that are installed over single or multiple rows of vegetables to enhance the plant growth by warming the air around the plants in the open field during winter season normally when the temperature is below 10°C. Plastic low tunnels are often used to promote the growth of plants during the period of winter season. Plastic low tunnels are supported above the plants by using hoops of GI wire and a clear or transparent plastic of 30-50 micron is covered or stretched over the hoops and the sides are secured by placing in soil. Farmers fetches very high price during off-season and several crops can be grown successfully by using the plastic low tunnel technology. This technology is suitable and profitable for the farmers of northern plains of India.

Use of Insect Proof Net Houses for Production of Vegetable Crops

Usually the farmers are growing their vegetable crops like tomato, chilli, sweet pepper, okra, *etc.* under open fields. But during rainy and post rainy season it is very difficult to grow these crops successfully due to leaf curl and yellow vein mosaic and other viruses, respectively. These viruses are mainly spread by insect vectors like whitefly or aphids. The population of white fly after on start of monsoons is very high. The farmers are spraying several harmful insecticides to control these vectors, but still they could not control these vectors and their tomatoes, brinjal, sweet pepper, chilli or okra crops are highly infected with viruses. The second most common and most severe problem in tomato, brinjal and okra is the fruit borer against which also the growers are using huge amount of insecticide even then they are unable to control this insect perfectly. The only way to control the virus and fruit borer is to put a mechanical barrier between the crops and open environment and this is possible with the use of insect proof nylon nets of 40 or 50 mesh density in form of net houses or insect proof net covered walk in tunnels. This reduce the dependence on insecticides, the farmers can erect these insect proof net houses by using half inch size GI pipes after bending them in half circle shape. Other insect proof net houses can also be made by covering all sides and top with insect proof net of 40 or 50 mesh, but the nylon net should be UV stabilized. Under these net house crops like sweet pepper, tomato chilli, brinjal and okra can be grown successfully without infestation of viruses or other insects like fruit borer *etc.*

Cultivation of Vegetable Crops under Naturally Ventilated Polyhouse

Naturally ventilated polyhouse are the protected structures where no heating or cooling devices are provided for climate control. These are simple and medium cost polyhouse which can be erected with a cost of 500-600 [1] per m^2 and these polyhouses can be used successfully and efficiently for growing year-round parthenocarpic slicing cucumber, off season muskmelon, tomato and sweet pepper crops. These structures are having a manually operated cross ventilation system as and when required. This is one of the most suitable and profitable way for crop cultivation under naturally ventilated polyhouse for higher yield and good economics returns.

Vegetable Production under Climate and Semi-climate Controlled Polyhouse

High value vegetables like slicing tomatoes, cherry tomatoes and coloured peppers are three crops which can be grown for long duration (10-12 months' period) under climate control polyhouse conditions. Near metro cities there is year round demand of these high quality vegetables in the up markets. Farmers can get good prices for the high-quality produce. Although the initial and running cost of the climate controlled Polyhouse is very high, which restrict the adoption of this technology. But one can use the greenhouse technology for cultivation of high value vegetables for high profits and produce can also be exported to other countries.

Table 11.4: Comparison of Cost of Production of Vegetables under different Protected Structures (Singh *et al.*, 2015)

Sl. No.	*Type of Structure*	*Cost (Rs/m²)*	*Suitable crops*	*Duration of Crops (Days)*	*Yield (t/1000m²)*	*Break-even Cost (Rs/kg)*
1.	Climate controlled polyhouse	3200-3500	Tomato	320	25.0	22-22
			Sweet pepper	300	6.0	45-50
			Cucumber	360 (4 crops)	18.0	20-25
2.	Semi-climate controlled polyhouse	1400-1500	Tomato	300	20.0	15-18
			Sweet pepper	270	5.0	35-40
			Cucumber	280 (3 crops)	15.0	15-18
3.	Naturally ventilated Polyhouse	500-600	Tomato	270	16.0	8-9
			Sweet pepper	240	4.0	18-20
			Cucumber	280	15:0	7-8
4.	Insect proof net house	80-100	Sweet pepper	220-240	4.0	15-16
5.	Plastic low tunnels	5-6	Summer squash	90-95	5.0-6.0	1.5-2.0
			Muskmelon	110-120	2.0-3.0	3.0-4.0
			Bitter gourd	110-120	1.0-1.5	4.0-5.0
			Bottle gourd	80-90	2.0-3.0	3.0-4.0

Use of Plant Growing Structures/Containers in Polyhouse Production

The duration of crop in polyhouse is the key factor to make this technology profitable and it should be short. For this use of containers in polyhouse production have a greater significance. There are some innate advantages of using containers in polyhouse production. They can increase production capacity by reducing crop time. We get high quality of the polyhouse product. There uniformity in plant growth with good vigour. Provide quick take off with little or no transplanting shock. Easy maintenance, easy to handle, grade and shift or for transportation. Better water drainage and aeration in pot media. Easy to monitor chemical characteristics and plant nutrition with advanced irrigation systems like drips. Different types of the containers are used for various activities in polyhouse production like:

- Raising of seedlings in the nursery
- Hybrid seed production of flowers
- Growing plants for cut flower production
- Growing potted ornamental plants

Table 11.5: Advantages and Disadvantages of different Types of Plant Growing Containers

Type of Containers	Advantages	Disadvantages
Clay pot	Low costEasy water management	Slow to work with pots and dry out fastThey are heavy to handle
Fiber block	Easy to handle	Slow root penetrationShort life
Fiber tray	Minimum use of space	Hard to handle when wet
Single peat pallet	No media preparationLow storage requirement	Requires individual handling Limited sizes can be handled
Prespaced peat pallet	No media preparationLimited to small sizes	
Single peat	Good root penetration	Difficult to separate
Pot	Easy to handle in field available (square/round) in large sizes	
Strip peat pot	Good root penetration	Slow to separate
Protrays	Easy to handle Reusable	May be limited in sizes
Plastic pack	Easy to handle	Roots may grow out of container
Plastic pot	ReusableGood root penetration	Requires handling as single plant
Polyurethane foam	Easy to handle Requires less medium Reusable	Requires regular fertilization
Soil band	Good root penetration	Requires extensive labour
Soil block	Excellent root penetration	Expensive machinery
Perforated	Easy to handle	Requires regular fertigation
Plastic tray	Requires less medium Available in many sizes reusable	Roots may grow out of the container
Perforated	Less expensive	Less durable
Polyethylene	Reusable bags	Requires less storage space

Media Preparation and Fumigation

Soil mixes used for Polyhouse production of potted plants and cut flowers are highly modified mixtures of soil, organic and inorganic materials. When top soil is included as a portion of the mixture, it is generally combined with other materials to improve the water holding capacity and aeration of the potting soil. Many polyhouse do not use topsoil as an additive to the soil mixes, but rather use a combination of these organic and inorganic components as an artificial soil mix. When managed properly as to watering and fertilization practices, these artificial mixes grow crops that are equal to those grown in top soil.

Conclusion

The protected cultivation structure offers a good solution for sustainable crop production under varying climates to ensure food security and water savings. Protected cultivation can be used to grow crops which are remarkably difficult to grow in open field conditions with ease and with higher production by controlling environment. Growing vegetables, flowers and seedlings under protected structures has great potential for improving farmers income. High initial cost is one of the major constraints in adoptions of this technology by the poor and less informed cultivators who lack technical understanding and thinks of this as a risky business. The other constraints prolonged subsidies and loan approval system, high expense of production, lack of cold storage and transportation infrastructures, non-availability of proper seedlings and marketing constraints are some major issues with protected cultivation.

REFERENCES

Gruda, N., Bisbis, M. and Tanny, J. (2019). Influence of climate change on protected cultivation: Impacts and sustainable adaptation strategies - A review. Journal of Cleaner Production. 225. 481-495.

Kumar, K., Bhattacharjee, S., Vaikuntapu, PR., Sharma, CL., Jayaswal, D., Sharma, R. and Sundaram, R.M. (2020). Climate Change Mitigation and Adaptation Through Biotechnological Interventions. In: Ch. Srinivasarao *et al.* (Eds). Climate Change and Indian Agriculture: Challenges and Adaptation Strategies, ICAR-National Academy of Agricultural Research Management, Hyderabad, Telangana, India. pp-1-22

Maikhuri, R. K., Rawat, L. S., Negi, V. and Purohit, V. K. (2007). Eco-friendly appropriate technologies for sustainable development of rural ecosystems in Central Himalaya. GB Pant Institute of Himalayan Environment and Development, 45.

Murthy, D. S., Prabhakar, B. S., Hebbar, S. S., Srinivas, V. and Prabhakar, M. (2009). Economic feasibility of vegetable production under polyhouse: A case study of capsicum and tomato. Journal of horticultural sciences, 4(2), 148-152.

Punera, B., Pal, S., Jha, G. K. and Kumar, P. (2017). Economics and institutional aspects of protected cultivation of carnation in Himachal Pradesh. Agricultural Economics Research Review, 30(347-2017-2038), 73-80.

Sanwal, S. K., Patel, K. K., and Yadav, D. S. (2004). Vegetable production under protected conditions in NEH region: Problems and prospects. ENVIS Bulletin: Himalayan Ecology, 12(2), 7.

Singh, A., Dhankhar, S.S. and Dahiya, K.K. (2015). Protected Cultivation of Horticultural Crops. CCS HAU, Hisar, 2015, pp 37-42.

Singh, B. and Sirohi, N.P.S. (2002) High tech nursery raising technology for vegetables. Indian Hort. 47 (3): 22-24.

Sirohi, N.P.S, Neubauer, E. and Singh, B. (2002). Growing vegetables under protected conditions. In Proceedings: International conference on vegetables, November 11-14, 2002. Bangalore pp. 207-212.

Srinivasarao, C., Srinivas, T., Rao, R. V. S., Rao, N. S., Vinayagam, S. S., and Krishnan, P. (2020). Climate Change and Indian Agriculture: Challenges and Adaptation Strategies, ICAR-National Academy of Agricultural Research Management, Hyderabad, Telangana, India.

Takte, R. L., Ambad, S. N., Kadam, U. S., and Dhawale, B. C. (2003). Greenhouse clodding material shade nets ventilation *etc.* Proc. of All India on Sem. Potential and Prospects for Protective Cultivation, 12-13.

Wadkar, S.S., Bagade, S.R. and Veerkar, P.D. (2006). A study on the economics of cut flower production under polyhouse. Agricultural Economics Research Review 19: 214

Zhang, L., Hao, X., Li, Y. and Jiang, G. (2010). Response of Greenhouse Tomato to Varied Low Pre-night Temperatures at the Same Daily Integrated Temperature, Hortscience, 45(11): 1654–1661.

Chapter 12

Climate Smart Agriculture Way Toward the Better Management Option for a Pest and Diseases to Enhancing the Crop Production

N. Ponnusamy[1], V. Prithiv Raj[2], V. Prabakaran[3] and P. Ramesh[4]*

[1]Department of Agricultural Entomology, [4]Department of Plant Pathology, Dr. Rajendra Prasad Central Agricultural University, Pusa – 848125, Bihar
[2]Department of Agricultural Entomology, Sardarkrushinagar Dantiwada Agricultural University, Sardarkrushinagar – 385506, Gujarat
[3]Department of Agricultural Entomology, Anand Agricultural University, Anand – 388110, Gujarat

Climate is emerging as a prominent issue in the world nowadays. Changing climatic conditions most severely affect agriculture because it depends on local weather factors viz., rainfall, temperature and relative humidity etc. Modifications in climate has major impact on productivity and production; also changes in agricultural inputs viz., herbicides, insecticides, fertilizers and environmental effects. Modifications in climate are having impacts on outbreak, biology and distribution of significant pests on wide crop range. Early diagnosis of outbreak also management are essential for successful mitigation of pests, permitting mitigation plans are introduced prior to establishment of pests and economic losses accrue. Modern visualizing methods could diagnose and henceforth are used as non-invasive crop monitoring methods. The combined actuation and sensing and methods are placed in the instrument passing into fields, on smaller drones and unmanned driving vehicles. Changing disease scenarios could need for good agronomical methods also utilization of eco friendly methods in management for *sustainable crop production. The choice of crop management practices like weather based monitoring, inoculum monitoring, especially for soil borne diseases and rapid diagnostics would play an important role. Integrated disease management strategies and multiple sustainable methods including seeds with healthy and instinctive forms of durable and broad resistance to disease and inter-crop methods in which refuge fostering for naturally occurring biological mitigation system. Climate smart agriculture is a cross-sectorial method which focuses in reducing disease and pest stimulated losses, improved services in ecosystem, decreasing CO_2 level per food production unit and developed the agriculture flexibility systems related to weather modifications.*

Keywords: *Climate smart agriculture, Eco-friendly method, Management and monitoring.*

Introduction

Climate is the entire portion of human civilization, agriculture being the most important outcome of civilization that has undergone transformation with the changes in climate and human food habits. Climate is bound to change as the earth is ageing naturally, however, the most unexpected and unprecedented changes in pace are due to anthropogenic activities like invasion of alien pests, pollution in environment and urbanization, *etc.* In upcoming modern era, weather modifications are having chances of further exacerbates the risk of disasters. More regular and rapid floods, sustainable droughts and also the storms could cause erosion in prevailing ecosystem managing abilities for recovering and responding to such hazardous function. The hazardous function even has the effect to production in agriculture directly (livestock maintenance and farming of crops) and security of livelihoods.

Modifications in climate are the major concern for agricultural communities worldwide. The cultivation methods contains host, pathogen and environmental situations, in which the connection in those are the major factor for initiation of infection on its absence, while weather modification had severe impact in this whole conditions. Modifications in weather conditions have severe impact on pest longevity also susceptibility and crop production. Weather modification effect the susceptibility of disease and pests where in affecting the health of crop also those modifications could lead to alternation in agronomic practices as to cope with the effects of these changes and to prevent a decline in productivity. Farming is the activity in where climate was greatly dependent also for the fiber and food production essential for living organism survival and sustainability. Farming is anyhow the practice in where highly affecting due to weather modification. The impacts on agriculture due to change in climate were categorized into several uncertainty forms. Primarily, the uncertainty in looking about the magnitude and rate of change in weather happen it. Secondarily, the uncertainty cases in responding on outputs regarding farming-based like CO_2 fertilizer application. Tertiary, this uncertainty happens in which community how they are responding or else even though the competency of responding with those expecting constraints. Few research perspectives on weather modifications are restricted with those following those uncertainty things. Many of this uncertainty couldn't are measurable, creating ignorance of our particular understanding range on weather modifications in future. This was mostly probable in which modification in climate has affecting the security of food at regional, local and global range. Weather modifications could affect and decreasing the availability of food also in lessening the quality of food. For instance, reductions in water availability, changes in precipitation patterns, modifications on severe weather events and peak in temperature range can all resulting with reduction in agricultural productivity. Occurrence of severe climatic regulations could interfere in delivery of food also resulted raise in cost of food because of less distribution severe condition afterwards, these were probably highly more predominant in the upcoming years. Also, the temperature raise could partake to contamination and spoilage.

The farming system is on-going the changing through modern methods, where this looks like more trusting as this enables the first system for moving into further range of profitability and productivity. The agriculture subtlety, containing the inputs application (needed one) and needed on where and when, had becoming the tertiary thing for revolution of new agriculture farming (mechanization was the primarily one and green revolution as secondary along with modification genetically); presently this was highly stimulated along with developed knowledgeable systems in farm because of greater data range availability. The USDA (United States Department of Agriculture) in 2016 has revealed already in which the agriculture subtlety methods has raised the operating benefits and net returns. Moreover, in environment consideration, modern methods improved were implemented into crop fields for maintaining the production stability of farm. Anyhow, this technology adoption involved the trade-offs and v uncertainty. According with marketing sector, the conditions in which carry out the implementation of stable cultivation technology like good training and giving knowledge to crop growers, information sharing, ready financial resource availability, also developing the organic food buyer demand. While implementing these modern practices, the crop data retrieving challenges are came along with few thing valuable and coherent, as this data itself were not in need, as even images or numbers. Crop fields which decided as modern driven method in few ways are showing worthy benefits, like work and money saving; in cost reduction along with minimized effort, or raised production also the food with quality production in most of eco-friendly methods.

Pest defines that any organisms microbes, insects, animals, plants and other factors that are abiotic which caused crop damage. They are almost a diseases, nuisance and destructive to crops, damaging property or make lives difficult to humans. It depends with the condition, considering as pests particularly the insects are destructive on particular time period also at other side as beneficial. Therefore an insect is usually considered a pest when it is in competition with humans for some resources and when a significant number is presents. Farming is worker profound also even covered along with difficult hard work manually. Farming which in case as one such modern industrialization in which the utilization of automated machines and robotic technology. Dissimilar mining and manufacturing industrial sectors were embracing modern technological tools more quickly because of continuous dependence on the machinery throughout the year, the use of equipment in agriculture is astringent or else involved on particular agricultural methods like planting, ploughing, harvesting, pesticides application and so on.

In present process in agriculture, real-time machines or robotics were used commonly in the plant factory concepts, but these require large investment and modern facilities. Crop growers are deeply involved the path which makes those toil easily. Most of the less worker consuming instruments also normally operating devices on farming are created themselves by crop growers. Crop cultivation with quality produce and economical yield are greatly normal approach as this could be more amend through assisting of support technologically. Modern techniques played the constantly developing range in world action particularly this are related with farming instrument, collection of survey data also the assessment also getting prominent tidings relating to those cropping area.

Direct and Indirect Effects of Climate Change on Pest Scenario

Weather modification initiated environmental changes are having major impact on demography, distribution also many species life history, the insects in particular. This was moreover inducing the pest phenology as emergence and arrival time of certain level of pests. Those deviations are and also would result in livelihoods constraints of human, like more dispersal of disease and pest of major significant crop plants. The following were the noticeable impacts on population dynamics of pests and pest scenario due to changes in weather.

a. Increases in Temperature

It increases the higher injury level of damage due to pests in such crops like cereals, oilseed rape and potato. Those developing pressures and populations of pests could result into mostly higher pesticide applications, in which temperature raised level resulting into greater populace of pests has leads on higher pesticide utilization in comparison with, higher and cooler altitude areas (Barzman *et al.*, 2015). The particular instance for those in which the spray application of pesticides for mitigation of maize (sweet corn) pests (Lepidoptera) were highly great (15-32 applications/year) on Florida as compared with Delaware (4-8 applications/year) and New York (0–5 applications/year) (Hatfield *et al.*, 2011). This high utilization in application of pesticides could results negatively significant impacts on human health and environment (Larsen *et al.*, 2017). The temperature increases also leads to such reduction in pesticide effectiveness, for instance, the toxicity to *Ostrinia nubilalis* with spinosad (spinosyn) and lambda-cyhalothrin and bifenthrin (synthetic pyrethroids) was revealed as reduction with increased after exposing temperature (Musser and Shelton, 2005). Even little modifications on temperature regulation has reported also in influencing the parasitoid potential on mitigating the species of pests also revealing the traits which are defensive against their larval parasitoids utilized by the pests (Iltis *et al.*, 2018). The increased temperatures could affect positively the production of biomass, the herbivores relative and absolute biomass were disproportionately increased also the parasitoids didn't showed characteristic responding with temperature change (De Sassi and Tylianakis, 2012).

b. Changing Precipitation (Excessive or Insufficient)

It could have real impacts in interactions with pest-crop, due to several pests were influenced with humid and warm regulations, also includes the pathogens, as greatly responding to rainfall and humidity (Hatfield *et al.*, 2011). Along with the plants which are suffered due to stressed with water were highly dangerous for injury caused by insect pests, which reported during 1993 on Iowa floods (Rosenzweig *et al.*, 2001). The desert locust occurrence were increased with events of precipitation changes, which are proved during 2004 with nearly 10 countries invasion on northern and western parts of Africa followed by high rains as compared with normal; these revealed that characteristic shortage of food and losses to crop. Outbreak of locusts was majorly general with becoming raise in events of precipitation severity and frequency in upcoming predictions (Masters and Norgrove, 2010).

c. Increasing CO_2 Levels

It could lead mainly with increased yield of crop, only which the yield profit might be getting down entirely or partially due to plant feeding pest damage (Coakley *et al.*, 1999). For instance in North America, *Trichoplusia ni* (cabbage loopers) was reported in consuming the leaves with higher quantity in raised CO_2 range, where found as conclusion in cabbage leaves with decreased levels of nitrogen under cultivating those levels (Trumble and Butler, 2009).

d. Extreme Weather Events

This are influenced with pests and crops interaction on unexpected manner, significantly revealing on crop mitigation strategy failure also continuous yield decrease (Cock *et al.*, 2013). For instance, Cagan *et al.* (1998) examined in Slovakia with highly hot and dry climate resulted in decreased populaces of *Trichogramma evanescens* (egg parasitoid) greatly, for which zero parasitism successfulness record on eggs of European corn borer. Higher wind storm could aid in insect dispersal on hibernating places into sites which they reveal more frequent constraints (Rosenzweig *et al.*, 2000). Even though, the extremely event of climate has disturbed the ecosystem which were normally highly vulnerable and susceptible for migration of native and alien pests (Masters and Norgrove, 2010). Although weather modified-stimulated constraints of pests were varying also subjected with many interaction levels, modifying environment with weather with concluding modifications on field utilization were expected into causing world level style of developing altitudinal and latitudinal pest ranges (Barzman *et al.*, 2015). Nearly, the supply of food with global range of about 40 per cent has already diminished due to insect pests (Oerke, 2006) also the weather change substantially and the continuous significant injury to plants or new threatening pest development, this was highly significant as compared with previously as the crop growers around the world started adoption in landscape and field mitigation methods for protecting the livelihoods and production of food. The steps couldn't are prohibited to the farm level only though. The effect of stimulated crop losses and pressure of pests has extended away from the field for food security to regional, international and national level, also with the economic, environmental and social stability. Necessary measures are in need with geographical ranges and various scales, in particular with the loss in yield due to pests related with changes in weather has till dated attained small look over in comparison with animal or human health impact due to weather modification.

Effect of Climate Change on Plant Disease Scenario

Global level examinations and findings on weather modification impact on crop diseases are enormous (Huda *et al.*, 2005). There are very little attempts in India on any crop disease range (Chattopadhyay and Huda, 2009). Even though, at genetic range, modern techniques with greater advanced expression of gene analysis would made this happen for beginning discrimination responding with various abiotic and biotic stress creators also the significant response in trade-offs. At specific crop scale level, more studies have conducted for beginning synthesized impacts of weather factors on infection ranges. In populace range scale, the adopting capacity of disease and plant populaces might be proved as the one such significant indicators of degree of weather modification impacts. Ecologists, at present representing the ecosystem process importance due to disease in crop plants like mass fluxes, energy and so on vis-á-vis the challenge of ranging with infection probability particularly into wider and epidemics constraints (Garrett *et al.*, 2006) revealed that various diseases in particular one crop was highly greater in tropical regions as compared to cold climate regions by indicating that the factor in how temperature increase can affect the outbreak of disease in agricultural crops.

Presently, many studies relating with plant diseases vis-a-vis weather modifications were on-going in wheat (*Puccinia* and *Septoria*), blast and bacterial leaf blight in rice crop and *Meloidogyne* sp. in horticultural crops. This indicating that extremity was resulted in greater along with peak level of CO_2 for major diseases (Chakraborty, 2008), the change in weather off-shoot. It was therefore revealed that weather modifications leading with pathogen changed variants ("Sleeper pathogens are activated by climate change, where remaining might causes as of important one economically" (Bergot *et al.*, 2004). The pathogens which are facultative along with wide range of hosts might have better survival. There is even having the chances of widening the facultative pathogen range of host. The work on those regions in future are needed which would be put forth on existing studies utilizing the double optimum CO_2 levels, as *Colletotrichum gloeosporioides* raised fecundity (Chakraborty *et al.*, 2000) and aggressiveness (Chakraborty and Datta, 2003), the causal organism for legume anthracnose in tropical areas.

The significant one was floods in destructing weather modification impacts due to increase in ocean water level could result in disappearing of lower range fields also to loss of major crop plants. The drought was other one in which the water level are not sufficient in land causing the crops in losing its normal activities also even more becoming highly disease susceptibility. In disease triangle, the weather factors which contributed in involving the characters of pathogen, susceptible host also the environmental conditions suitable for occurrence of infection; also changes in weather affecting the normal conditions of environment either these are on favourable for the pathogen or host plant. For instance in those situations including rain, dew, temperature, relative humidity, soil moisture, wind (aeration) and intensity of light.

In crop type, the extreme temperature impacts are greatly depending with crop reproduction temperature and normal growth. In such areas, temperature raise might proving benefitted into those crop types which normally growing in areas or either by permitting growers to initiate crop planting which growing under high temperature conditions. Even though, these are every time not happens; while the temperature in extreme level reaches maximum as compared with temperature optimum of crop has resulted in worse, appearance, declining crop yield also the pathogen infection may be happens. The economic yield of crop has been impacted mainly with extreme CO_2 levels. The some lab studies revealed that greater CO_2 levels were affecting the crop development positively. Even though, such factors like changes in water level, varying temperatures, low nutrient levels and ozone might reduce or down the expected yield in crop possibly. For example, when the temperature level is extreme as compared with temperature normal requirement of crop, also either not sufficient level of nutrients and water, thereby economically yield increasing might be less. Elevated CO_2 levels were connected with protein content and reduced nitrogen level in alfalfa and soybean plants, resulted with extremely reduced quality level. Decreased quality of grain and forage could decrease the capacity of pasture and rangeland to support livestock which rely on grazing. Although rising CO_2 could enhanced the growth of plant and lessen nutrition benefits values in most of the major growing crop plants. Elevated CO_2 in atmosphere level could mainly impact the essential minerals also the protein concentration through decreasing its quantity on many plant varieties like soybeans, rice also in wheat. Henceforth, the impact of elevated carbon dioxide level in nutrition content of crops was denoted as expected also in human health effect indirectly so on. Even more, because of high utilization also lessen the pesticide efficiency because of pest resistance development; health condition of human beings was affected by pesticide usage additionally also those toxicity residues in their bodies. High precipitation and temperature extremes may reduce some crop development. As already noticed that the event with higher level of flood and drought will damage the crops and reduced the crop yield.

Raised temperature level in summer caused the soil dryness. However, more watering in certain regions might be favourable; also the supply of water might lesser on another area, leading to less water availability for watering while needed in more quantity. Several pests, weeds and fungi were thriving on wetter climates, warmer temperatures also the elevated levels of CO_2. The distribution and ranges of pests and weeds were possible to develop more along with changes in weather condition. This could create further issues on growers crops not exposed prior to those organisms. The parameters of weather modification could impact with the pathogen and host both, for instance, particular temperature range stimulate the growth of pathogen also such temperature range could make greater level of resistant host for infections due to pathogens. The highlighting example in this condition was in oats and wheat was made highly rust disease susceptibility along with temperature increase, where few species of forage crops became highly resistant. Moreover, temperature change was restricted as with changes in CO_2 leads to pests of some category undergone additionally one to five lifecycles in single season, this develops the pest potentiality in overcoming resistance with plants.

Some toxins produced by *Fusarium* spp. (fusarium mycotoxins) are having concentrations at increased level during harvesting period because of higher temperature and relative humidity. Humid situations could further raise the multiplication of weed plants also their biomass was increased along with temperature elevated range. Some aspects could having various impacts depends upon the physiology of plant, for instance, raised level of CO_2 could leads to reduction on rate of decomposition in plants, where also resulted on greater level of fungus inoculum; this quantities might stimulates higher spore production of fungus. On the other hand, high CO_2 concentrations may cause physiological changes to plants, causing them to acquire higher resistance to certain pathogens. In another situation of extremity might like as erosion in soil and less level of water also caused the decreased fertile level of soil and thereby health of plant. The activity of fungicides was even the main predominant level; changes in weather might greatly impact the efficiency of fungicides. More continuous rain has severely affected the fungicide efficiency which is contact in nature, because the rainfall is having the capacity to washout also by eliminating those chemicals on surface of host plant, making those not effective. Even though, the higher rate of metabolism in plants has raised the fungicide intake also were not greatly impacted through this aspect.

IPM and Mitigating Pest Management

According to FAO (Food and Agricultural Organizations), the integrated pest management (IPM) is in approach of ecosystem involves the plant mitigation by combining the various practices and methods in cultivating the crop in healthy manner also by lessening the pesticide utilization for environmental protection. These methods are utilized analytically for agricultural ecosystem detection also its various aspects by managing those optimally in minimizing and mitigating the pests in during the economic health and environmental protection. The possible mitigation strategies (cultural, physical, biological and chemical) are ought to be needed rationally and essentially used by crop growers. However, IPM is more than just a tool and collection of control choices.

The important role in this mitigation method was the making decisions properly due to any of the interference. All planning should be made justified in economically and ecologically both. Further, mitigation methods which involving chemical application sprays systematically might be environment harmful were not accepted in the process of IPM. As an alternative, priority was made with alternation of mitigation methods as compared with the method of prevention. The integrated pest management was implemented on several nations also in regions which differing from its social, natural and economic surroundings by adding with expansion of agricultural range. Even though, modernization in the yield of crop also the safeness might understood with any of the prevailing situations along with IPM implementation. The IPM methods aren't difficult and the simplest mode with submitting the rules and regulations, though rarely involved in action taking along with eco-friendly method with approaches and principles contributing for lessening the chemical utilization also improving the security of food for achieving the sustainability in agriculture. Hence by making this method are highly potential as probable, this must be changed into regional range situations.

Pest Management

Countries that are sufficient in production of food nearly about a decade were presently facing the issues ranged into pollution, pest invasion, high food demand and natural disasters occurrence *viz.*, drought, flooding and wildfire and so on. This condition resulted into wastage, destruction also the production of food in deficient. The glyphoshate herbicide exposure could severely reduce the quality of seeds (Lucke *et al.*, 1995) also the herbicide clopyralid results in potato yield reduction potato (Wasim *et al.*, 2009). The crop plants will further suffering from other side effects applications of pesticide sprays while an impact has happened into beneficent insects and soil microbes. The highest of pesticide applied with 80 – 90 per cent could leads to volatilization in after application in short period of time (Majewski *et al.*, 1998). The drift of pesticides might be resulted on groundwater and soil contamination, affect fertility and non-target vegetation. An important lingering problem is the continuous spray of pesticides. Often, when applying herbicides, farmers tend to apply the highest label-approved rate or sometime exceed it, (López *et al.*, 2012) thus creating concern on ecological impact and biological magnification. However presently, buyers of food were demanding increasingly the produce with quality and natural neither limited nor any treatment with chemicals (Blasco *et al.*, 2002). There is not probability of eliminating the pests entirely. Even though the better mitigation methods for pests are requiring such knowledge or information includes the habit and life cycle of pest.

The most important key component in pest control is correct pest diagnosis also its life cycle (Alston, 1996). Better studying in detailed manner on how species diversity of beneficial insects affecting the insect pest mitigation measures are critical also the management of pests are major significant ecosystem components (Zhang *et al.*, 2007). The mitigation of pests including the selection of intelligent ones also the utilization of pest mitigation measures which would determine suitable ecological, economical also the sociological constraints (Metcalf and Luckmann, 1994). Also similar management measures including increase in pest level monitoring, need based pesticide application or else through proper information so that none of the mitigation measures are essential. This, anyhow, can be combined with the use of multiple actions in a coordinated manner by maintaining the damage due to pests below the economic injury level while providing protection against hazards to plants, humans, animals also the environment. The general practices of pest management have been outlined:

1. Determining how the system of pest life concept needed with change by reducing the population density below economic threshold.
2. Applying modern techniques and biological knowledge for achieving the expected outcomes.
3. Preparing protocols for pest management probable into modern techniques also with the environmental and economic quality compatibility (social and economic approving).

However, management of pest system indicates the tolerating manner for status of pests. Pest mitigation was human community related also the complete achievement of controlling the pests depending greatly on enhancing the community by controlling the insect pests through every management measures (Metcalf and Luckmann, 1994). According to Metcalf and Luckmann (1994), the most important components of pest management in agro-ecosystem include:

1. Identifying the pest to be managed in the crop production system.
2. Defining the management unit.
3. Developing pest management strategy.
4. Development of suitable techniques for monitoring.
5. Establishing economic threshold.
6. Evaluating predictive and descriptive.

a. Insect Traps for Pest Monitoring

Monitoring population density of pest is crucial and very important and is a key towards successful crop protection in agriculture and forestry practices. In earlier, crop growers should normally performed the surveys periodically with dissemination of traps throughout the farm. These are the time and labour consuming activities particularly for large plantations areas. It would be of great advantage to have an affordable system capable of doing task automatically in an accurate and more efficient way (López *et al.*, 2012). The sound IPM depending on population precision and accuracy techniques of monitoring also the management of pests are every time being the challenging and troublesome one. The effective and good technology for mitigation of pests through monitoring was centered initially with the placement of insect-traps. Every trapping system was installed precisely along with the pheromonal chemicals and their related substrates which attracted those pests targeted have been collected (Kamminga *et al.*, 2012). Traps were made in a manner by which none of the pest entered into traps be capacity to come out hence the device for monitoring the pests would be in periodical gathering of information related to surveillance in every trap also by taking the counting in particular pest, respectively. Captured in this way, an efficient pest control management system is realized and achieved (López *et al.*, 2012). The collection of data in traps might requiring the better survey in field also inspection periodically in which the visual recordings can be done by humans by evaluating and recording the captured pest counts (numbers). The range into two frequent next inspections of traps also the surveying would generally in 15-30 days range. This method is labour intensive and costly also the every trap for monitoring could not be coordinated together for populace of pest measurement targeted; thereby offering the worst spatial/temporal measurement and resolution also in consistently the density of population dynamics of pests on farm couldn't monitored accurately (López *et al.*, 2012). To overcome this one by an visualized image sensing technology network probable with routine performing automated monitoring of pest was developed with economically cheap method utilizing the powered battery without wired image-visual sensing which precisely and accurately inspecting the density of pest population by utilizing greater image resolution.

b. Drones in Pest Management

Drones reporting in agro-ecosystem farming have developed during 1998 appearance also gradually raised more in the past ten years. Earlier basic pest infestations scouting in fields were often time-consuming and expensive. This might be challenging practically, like while involved in wider area, that the pests which were very little with naked eye observation or while residing on taller trees or in the soil. In few systems of cropping, better scouting was difficult due to lacking of optimum sampling methods of pests. Henceforth, the one such major operator of implementing the remote sensing techniques with drone related into agro-ecosystem are effective in saving time through monitoring of crop with automation by using the cost-effective techniques for farmers. In comparison with the remote sensing convenient methods, like aerial, ground-based also the drones with orbital sensing are having various benefits which made very effective for utilization on subtlety agriculture. Drones sensing was effectively allowed with broader area coverage as compared with that ground-based and handheld devices. It was flying even at altitude lesser range as compared with orbital systems and manned aircraft, improving the spatial image resolution also by decreasing the pixel mixed numbers. Even, it is cheap in cost for deploying and obtaining as compared with satellites and manned aircrafts also did not having high repeated visiting periods as in case of satellites by permitting the monitoring at greater frequencies.

c. Remote Sensing and Arthropod Pests

The technique of remote sensing has utilized in subtlety agro-ecosystem during past era along with several manners like crop phenology evaluation and yield predictions (Mulla, 2013). This method was utilized in monitoring various stressors of plants (abiotic), like nutritional deficiencies and drought and other plant stressors (biotic), like nematodes, pathogens also the weeds (Peña *et al.*, 2015). Likewise, the technique of remote sensing has greatly implemented for stress detection due to several variety of pests in orchard crops and also on wider field range (Nansen and Elliott, 2016). There are very few examinations regarding about pest stimulated detection of stress utilized remote sensing with drone-based aerially, remote sensing with orbitally or aerial remote sensing with manned aircraft-based, where many examinations utilized based on remote sensing with ground-based manner.

d. Actuation Drones for Precision Application of Pesticides

While sensing drones can used for detection of hotspots of pests, actuation drones could help control the pests at these hotspots. Pest hotspots can effectively controlling by means of variable rate application of pesticides. Aircrafts are utilized with nearly the decades for pesticide sprays, but products are deposited over large areas, and a large amount is lost to drift. This is a concern for neighbouring terrestrial and aquatic ecosystems, also for human health (Damalas, 2015). Major factors determining spray drift are droplet size (influenced by product formulation and nozzle type), weather conditions (direction of wind and its speed) and application method (*e.g.*, spray height above the canopy). The modelling and empirical experiments revealed that drift spray into not targeted regions are to be considered (Al Heidary *et al.*, 2014). Henceforth, the modern application methods of pesticides were in need greatly also the effectiveness for drone utilization on subtlety spraying of acaricides and insecticides. In some manner, where drones gave the more competitiveness as compared with dusters of crop in manned way are those with decrease in pesticide exposure of operator, easy in deployment also the effective drift spray reduction (Faiçal *et al.*, 2014).

Indeed, in Japan, where drone has utilized on agriculture since the 1980s, drones are widely used to spray pesticides on rice, *Oryza sativa* L. These drones are mostly heavier than 25 kg, as they are among the most widely used drones in pest management. The implementation of automated vehicles for aerial dusting of crops was initiated in Japan into Forestry, Agriculture and Fishery. The manufacturing has ended by Japanese multinational corporation, Yamaha during 1986, along with services and broad product ranges; the first globally practically utilized automated helicopter, R-50 came into market on 1987 for application of pesticides by 20 kg spray loading (Xiongkui *et al.*, 2017).

Modern Technologies for Diasease Management Approaches

Modern sensor techniques by way of mechatronics, global positioning systems, geographic information system, machine vision and actuators has again made the development and implementation of intelligent technologies and robotic system were utilized on subtlety agro-ecosystem. Advancement in greenhouse mitigation systems and technologies were improved greatly on earlier years in past, collaborating with wireless sensor network and internet of things both. AI (Artificial Intelligence) and machine learning are applied and used effectively on agriculture for robotic and automated cultivation.

Utilizing learning as Intelligence techniques and machine vision were improved with several factors like irrigation, planting, pruning, weeding, identification or diagnosis of plant disease and harvesting. Instrumental knowledge earning methodologies like transfer and deep knowledge, normally present the result promisingly with identification of plant symptoms and image processing clarity. Moreover, diagnostic specificity may be a challenge for some micro-organism management. Hence, the implementation of mechatronics and robotics methods for mitigation of disease will be followed thereafter the problems associated with diagnosis of plant pathogen will be every time referred with few general issues regarding robots. Recently, the development on many automated diagnosis of diseases and pests are rapidly growing, along with application of sensors in robotics for recognizing each and every plant on several regions and areas also for identifying the presence of diseases and conduct a routine for disease management.

Disease management strategies depend upon climate conditions. Change in climate will cause alterations in the disease geographical and temporal distributions and consequently control methods will have to be adapted to climate change scenarios. High precipitation and temperature changes could modify dynamic residues of fungicide on plant leaves also the product degradation is altered. The plant physiology or morphology changes resulting from high concentration of CO_2 in the atmosphere or from different precipitation and temperature conditions can affect the mode of action, translocation and penetration of systemic fungicides.

Besides that, plant development and growth changes could modify the time of pathogens with greater susceptibility are determined with newer application of fungicides (Pritchard and Amthor, 2005). Therefore, few significant management methods for controlling diseases in plants with related to climate change include:

- ✰ Selection of resistant cultivars/varieties at elevated temperature.
- ✰ New molecules with higher efficacy at increased temperature for disease management.
- ✰ Modernized model for forecasting the predicting disease appearance.
- ✰ Alteration in sowing date which avoids the epidemic cause.
- ✰ Selection of bio agents with broader scale of adoptability to temperature.
- ✰ Disease management through integration of all the existing technologies.
- ✰ Efficient agronomic practices like tillage for mitigation of diseases.

a. Biotechnology in Plant Pathology

Diverse approaches in plant disease management have very much been influenced by the modern advancement in biology at molecular level. Many biotechnological techniques and tools were improved by using different plant pathogens as experimental materials. It has provided way to understand host pathogen correlation under diverse environment to give a novel look to this branch of science parad different aspects are as follows:

(i) Molecular Diagnosis of Plant Pathogens

Conventionally, traditional practices are worked for isolation and identification of pathogens which are effective. This is comparatively the lethargic method, also in need of taxonomists which are reliably skilled for pathogen identification. However, in the past three decades, various technologies has implemented in resulting with diagnosis of pathogen in plant application; this including the utilization of enzyme linked immunosorbant assay and monoclonal antibodies (Clark and Adams, 1977) also the polymerase chain reaction, DNA based technique which amplifies the various folds in millions, thereby improving the detection of pathogen sensitivity. Furthermore, diagnostic PCR has been significantly improved by the introduction of second generation PCR, known as the real time PCR. It is now probable for detection about the absence or presence of the plant pathogen targeted, also in quantification of presence amount on particular samples. Identifying along with detection of plant pathogen is important for potential risk estimation related with the development of diseases also in providing the needful manner with decision making for mitigation of diseases.

DNA micro-array technique was designed originally for studying the expression of genes also in generating the SNP (single nucleotide polymorphism) profiles are at present the emerging and modern diagnostic techniques of pathogens and offering a basement of unlimited multiplexing capability. The rapid database development was made through research in biosystematics and genomics thereby providing the opportunity in unique manner with highly versatile design, throughput highly, specific and sensitive assays in molecular level that will address the high restriction in present techniques and benefit plant pathology. Finally, the so far restricted use of robotics to DNA technology will become economically feasible and thus accessible to farmers and will offer the possibility of using single DNA chip as the diagnostic tool of hundreds of plant pathogens practically (Kumar, 2013).

(ii) Analysis of Molecular Variability in Plant Pathogens

Characterization of genetic diversity in plant pathogens have been made feasible, beyond use of differential hosts, through diverse molecular techniques. Different molecular markers are utilized for characterization of genetic diversity of plant pathogens. In most of the cases, these are RFLP, RAPD, SSR/ISSR, AFLP and ITS (Powell *et al.*, 1996). The markers such as Random Amplified Polymorphic DNA are mostly utilized for fungal pathogens characterization, followed by ITS and AFLP markers. This might be because of their ease and simplicity in use. Guleria *et al.* (2007) collected 19 *Rhizoctonia solani* isolates from paddy cultivating areas in India, used two marker systems that is, ISSR and RAPD for molecular characterization of genetic variability. In this DNA marker types, the Random Amplified Polymorphic DNA markers are having the ability for detecting high gene variation as compared with ISSR markers.

(iii) Mapping of Disease Resistance Genes using DNA Markers

Molecular mapping are utilized with direct selection of disease resistance genes for the use in plant breeding programmes. Commonly used markers are amplified fragment length polymorphism, restricted fragment length polymorphism, single nucleotide polymorphism and simple sequence repeats with predilection of PCR based markers. There are very few reports available from India which is required to be strengthened to support marker assisted selection in plant breeding for disease resistance.

(iv) Transgenics

Development of disease resistance through transgenic research is yet at primitive stage in India. Non availability of resistance to plant pathogens can be overcome by search and the genes which are resistant were transferred from other sexually incompatible species which is possible using genetic engineering approach. It requires to first search for new genes which have broad spectrum resistance to pathogen population present in the region and then following transgenic approach, transferring of resistance to commercial varieties to achieve resistance, working at filled level. Disease resistance transgenic was improved in tobacco and banana through Magainin, a synthetic alternative analogue with small peptide transfer (Chakarbarti *et al.*, 2003). This was such one of anti-microbial peptides reported in earlier through secretions from African clawed frog skin. The peptide is not stable in its native form and therefore, researchers modified it to express in foreign plant systems. The transformed tobacco crop along with the peptide revealed improved resistant level against the *Alternaria alternate, Sclerotinia sclerotium* also in *Botrytis cinerea*. The banana crop in transgenic nature has shown resistant to *Mycosphaerella musicola* and *Fusarium oxysporum* f. sp. *cubense* (Kumar and Gupta, 2012). However, these are remaining needed to see on how these plants perform under natural disease conditions.

(v) Application of Genomics

Genomic has emerged as one of the frontier technologies during this century. Its application in various fields of plant pathology can be enormous in functional, structural or comparative genomics. Using high throughput genome sequencing technologies many plant pathogens are being sequenced world over. A list of pathogens which are at different stages of the genomic sequencing has been given by Jalali (2008). The high sequencing of data genome are being generated on different microorganisms which are utilized with simultaneous detection of multiple plant pathogens. The unique sequence from broader pathogen range could be used to develop micro-arrays with continuous diagnosis of large number of different strains. The primers and probes are developed with differential detection of pathogens also those molecular range of characterization by using the unique.

(vi) Application of RNA Interference

RNAi (RNA interference) are now developed as great method for defending few severely negatively daring diseases with causal organisms like bacteria, viruses and fungi (Wani *et al.*, 2013). RNAi are the regulation for guided RNA mechanism for expression of genes, where the dsRNA (double stranded ribonucleic acid) restricts the gene expression along with additional nucleotide sequencing. The tissue specific inducing silencing of gene application along with combined utilization of promoters which are suitable in silencing various genes continuously would resulting in destructive pathogen mitigation for crop protection. The application of RNAi would also result with achievable management of various significant economic plant pests and diseases. Baum (2007) used RNAi to develop in corn plant in transgenic form which expressed the dsRNA to silence genes of the corn root worm. Similarly, Mao *et al.* (2007) reported that cytochrome p450 cy6AE14 genes of the cotton bollworm were silenced to disable the bollworm from feeding on gossypol in plants. In addition, RNAi approach also has utilized effectively to knockout the expressions of genes and to understand their biological functions (Anandalakshmi, 2013). The RNAi based technologies have tremendous potential in significantly decreasing the utilization or dependence on chemical pesticides and thus pave way for the efficient, cost effective, eco-friendly alternative which will ensure a role model change in pest management in India.

(vii) Silencing of Genes in Post Transcriptional Stage

The RNA silencing mechanism is also a great method to develop crop species resistant to viruses. The expression of virus origin antisense or sense RNA into genetically modified crops which conferred the resistance with RNA mediated virus appeared for inducing the silencing of genes in post transcriptional form including mRNA degradation, RNA silencing, an evolutionary mechanism protecting cells from pathogenic DNA and RNA, is viewed as an adaptive immune system of plants against viruses (Krishnaraj, 2013). Several lines of research indicate that silencing of RNA are locally stimulated then after distributed into the entire organisms, also the process of this aspect are reflected their role likely defense virally.

b. Role of Nanotechnology

Nanotechnology offers an imperative role in improving the existing crop management techniques. Generally only very low concentrations of agrochemicals have attained the sites which are target to crop plants because of photolysis degradation, chemical leaching, microbial degradation and hydrolysis. Hence, repetitive application is needed for effective control causing the effects which are not favourable includes water and soil pollution and rapid occurrence of resistance. The agrochemicals with Nano-formulations are developed through the means of holding every essential properties like timely managed release in responding to certain stimuli, effective concentration (with high stability and solubility), improved target activity and less eco-toxicity with easy and safe delivery mode. Therefore, an urgent need is to evaluate and develop natural, biodegradable, and environment safe Nano-formulated compounds.

Conclusion

The damage in India due to pests varied from several regions of agricultural climate around throughout the nation because of various significant constraint factors abiotic nature includes rainfall, temperature and relative humidity. This unexpected significant loss in yield has changed greatly due to elevated insect-pest damage and crop diversity because of climate change. The climate change impacts on agriculture farming intermediated with populace modifications of significant major pests are needed to give more diligence for devising and planning management and adaptation methods for upcoming programmes of pest management. Current management of data modernization were made the farming with smartly growing as exponential data has becoming the major factor in modernized farming by helping growers through making decision critically. Worthy benefits are appeared through the aspect data from sensor collection by aiming with increasing sustainability and productivity. The robotics with data based farming solutions collaborating AI (artificial intelligence) technologies, made the basement for future agriculture farming in sustainable manner. Changing disease scenarios due to global climate change have considered as important for good farming methods also the utilization of environment fairly strategies in disease management for sustainable crop production. In such situations, weather based disease monitoring, inoculum monitoring, especially for soil borne diseases and rapid diagnostics would play an important role. Integrated mitigation of disease methods has been improved in decreasing the fungicide chemical dependence level. Another multiple stable approach methods include seeds with good and instinctive manner of durable and broadly resistant to diseases also the inter-crop methods for natural biological mitigation organisms by growing the refuge crop system. Again, early warning and monitoring methods for epidemic disease forecasting has improved for significant pathogen-hosts are having the main adaptation on food security and farmer earnings at greatly. However, few evaluations were at present on some regions, nations, particular pathogens and their hosts in concerning along with the security of food. Now, the suggestions should be shifting with assessment of impacts into improving management and adaptation options and methods. Primarily in want of evaluating the weather modification, effectiveness of present cultural, physical and biological mitigation strategies including disease resistant varieties and chemical methods; then secondarily, by including the scenario of weather in future on every aimed research for development of new methods and tools.

REFERENCES

Al Heidary, M., Douzals, J.P., Sinfort, C. and Vallet, A. (2014). Influence of spray characteristics on potential spray drift of field crop sprayers: a literature review, Crop Protection, 63: 120-130.

Alston, D.G. (1996). Important component of a successful pest management program, integrated pest management fact sheet, USU extension publication, Uta state University, Logan, 2-5.

Anandalakshmi, R. (2013). Application of RNAi for engineering disease tolerance in crops with special reference to horticultural crops, Journal of Mycology and Plant Pathology, 43(1): 111.

Barzman, M., Lamichhane, J.R., Booij, K., Boonekamp, P., Desneux, N., Huber, L., Kudsk, P., Langrell, S.R.H., Ratnadass, A., Ricci, P., Sarah, J.L. and Messean, A. (2015). Research and development priorities in the face of climate change and rapidly evolving pests. In: Sustainable agriculture reviews (Lichtfouse, E., Eds.) Springer, Cham. (17): pp 1–27.

Baum, J.A., Bogaert, T., Clinton, W., Heck, G.R., Feldmann, P., Ilagan, O., Johnson, S., Plaetinck, G., Munyikwa, T., Pleau, M., Vaughn, T. and Roberts, J. (2007). Control of coleopteran insect pests through RNA interference, Nature Biotechnology, 25: 1322-1326.

Bergot, M., Cloppet, E., Pérarnaud, V., Déqué, M., Marçais, B. and Desprez-Loustau, M.L. (2004). Simulation of potential range expansion of oak disease caused by *Phytophthora cinnamomi* under climate change, Global Change Biology, 10: 1-14.

Blasco, J., Aleixos, N., Roger, J., Rabatel, G. and Molto, E. (2002). AE-Automation and emerging technologies: Robotic weed control using machine vision, Biosystems Engineering, 83(2): 149-157.

Cagan, L., Tancik, J. and Hassan, S. (1998). Natural parasitism of the European corn borer eggs *Ostrinia nubilalis* Hbn. (Lep., Pyralidae) by *Trichogramma* in Slovakia-need for field releases of the natural enemy, Journal of Applied Entomology, 122: 315-318.

Chakarbarti, A., Ganapathi, T.R., Mukherjee, P.K. and Bapat, V.A. (2003). MSI-99, a magainin analogue, imparts enhanced disease resistance in transgenic tobacco and banana, Planta, 216: 587-596.

Chakraborty, S. (2008). Meeting Sukuma Chakraborty, ICPP Newsletter, 4: 2.

Chakraborty, S. and Datta, S. (2003). How will plant pathogens adapt to host plant resistance at elevated CO_2 under a changing climate, New Pathologist, 159: 733-742.

Chakraborty, S., Pangga, I.B, Lupton, J., Hart, L., Room, P.M. and Yates, D. (2000). Production and dispersal of *Colletotrichum gloeosporiodies* spores on *Stylosanthes scabra* under elevated CO_2, Environmental Pollution, 108: 381-387.

Chattopadhyay, C. and Huda, A.K.S. (2009). Changing climate forcing alteration in cropping pattern to trigger new disease scenario in oilseeds and pulses in Indian sub-continent. In: Proceeding: National Symposium on Climate Change, Plant Protection, Food Security Interface, 17-19 December 2009, Kalyani, pp.70.

Clark, M.F. and Adams, A.N. (1977). Characteristics of the microplate method of enzyme linked immunosorbent assay for the detection of plant viruses, Journal of General Virology, 34: 475-483.

Coakley, S.M., Scherm, H. and Chakraborty, S. (1999). Climate change and plant disease management, Annual Review of Phytopathology, 37: 399-426.

Cock, M.J.W., Biesmeijer, J.C., Cannon, R.J.C., Gerard, P.J., Gillespie, D., Jimé-nez, J.J., Lavelle, P.M. and Raina, S.K. (2013). The implications of climate change for positive contributions of invertebrates to world agriculture, CAB Reviews Perspectives in Agriculture Veterinary Science Nutrition and Natural Resources, 8: 28.

Damalas, C. A. (2015). Pesticide drift: seeking reliable environmental indicators of exposure assessment. In: Environmental indicators (R.H. Armon and O. Hönninen, Eds.), Springer, Dordrecht, Netherlands.

De Sassi, C. and Tylianakis, J.M. (2012). Climate change disproportionately increases herbivore over plant or parasitoid biomass, PLoS ONE, 7: e40557.

Faiçal, B.S., Pessin, G., Filho, G.P.R., Carvalho, A.C.P.L.F., Furquim, G. and Ueyama, J. (2014). Fine-tuning of UAV control rules for spraying pesticides on crop fields. In: IEEE International Conference on Tools with Artificial Intelligence (ICTAI), Limassol, Cyprus, pp. 527-533.

Garrett, K.A., Dendy, S.P., Frank, E.E., Rouse, M.N. and Travers, S.E. (2006). Climate change effects on plant disease: Genomes to ecosystems, Annual Review of Phytopathology, 44: 489-509.

Guleri, S., Aggarwal, R., Thind, T.S. and Sharma, T.R. (2007). Morphological and pathological variability in rice isolates of *Rhizoctonia solani* and molecular analysis of their genetic variability, Journal of Phytopathology, 155: 654-661.

Hatfield, J.L., Boote, K.J., Kimball, B.A., Ziska, L.H., Izaurralde, R.C., Ort, D.R., Thomson A.M. and Wolfe, D. (2011). Climate impacts on agriculture: implications for crop production, Agronomy Journal, 103: 351-370.

Huda, A.K.S., Spooner-Hart, R., Murray, G., Hind- Lanoiselet, T., Ramakrishna, Y.S., Desai, S., Thakur, R.P., Chattopadhyay, C., Jagannathan, R., Khan, S.A. and Rathore, L.S. (2005). Climate related agricultural decision making with particular reference to plant protection, Journal of Mycology and Plant Pathology, 35: 513.

Iltis, C., Martel, G., Thiéry, D., Moreau, J. and Louâpre, P. (2018). When warmer means weaker: high temperatures reduce behavioural and immune defences of the larvae of a major grapevine pest, Journal of Pest Science, 91: 1315-1326.

Jalali, B.L. (2008). Molecular plant pathology: Where do we stand, Journal of Mycology and Plant Pathology, 38(3): 419-429.

Kamminga, K.L., Koppel, A.L., Herbert, D.A. and Kuhar, T.P. (2012). Biology and management of the green stink bug, Journal of Integrated Pest Management, 3(3): 1-8.

Krishnaraj, P.U. (2013). Post transcriptional gene silencing: A tool to develop virus disease resistance, Journal of Mycology and Plant Pathology, 43(1): 133.

Kumar, S. (2013). Molecular diagnostics in plant pathogens: Recent advances, Journal of Mycology and Plant Pathology, 43(1): 135.

Kumar, S. and Gupta, O. (2012). Expanding dimensions of plant pathology, JNKVV Research Journal, 46(3): 286-293.

Larsen, A.E., Gaines, S. and Deschênes, O. (2017). Agricultural pesticide use and adverse birth outcomes in the San Joaquin Valley of California. Nature Communications, 8(1): 302.

López, O., Rach, M.M., Migallon, H., Malumbres, M.P., Bonastre, A. and Serrano, J.J. (2012). Monitoring pest insect traps by means of low-power image sensor technologies, Sensors, 12(11): 15801-15819.

Lucke, D., Landiver, J.A. and Moseley, D. (1995). The effect of rate and timing of glyphosate applications of defoliation deficiency, regrowth inhibition, lint yield, fiber quality and seed quality. In: Proceeding of Beltwide cotton conference, National cotton council of America, pp. 1088-1090.

Majewski, M.S., Foreman, W.T., Goolsby, D.A. and Nakagaki, N. (1998). Airborne pesticide residues along Mississipi river, Environmental Science Technology, 32(23): 3689-3698.

Mao, Y.B., Cai, W.J., Wang, J.W., Hong, G.J., Tao, X.Y., Wang, L.J., Huang, Y.P. and Chen, X. Y. (2007). Silencing a cotton bollworm P_{450} mono-oxygenase gene by plant mediated RNAi impairs larval tolerance of gossypol, Nature Biotechnology, 25: 1307-1313.

Masters, G. and Norgrove, L. (2010). Climate change and invasive alien species, CABI Work Paper, 1: 1-30.

Metcalf, R.L. and Luckmann, W.H. (1994). Introduction to insect pest management, John Wiley and Sons.

Mulla, D.J. (2013). Twenty five years of remote sensing in precision agriculture: key advances and remaining knowledge gaps, Biosystems Engineering, 114: 358-371.

Musser, F.R. and Shelton, A.M. (2005). The influence of post-exposure temperature on the toxicity of insecticides to *Ostrinia nubilalis* (Lepidoptera: Crambidae). Pest Management Science, 61: 508-510.

Nansen, C. and Elliott, N. (2016). Remote sensing and reflectance profiling in entomology, Annual Review of Entomology, 61: 139-158.

Oerke, E.C. (2006). Crop losses to pests, Journal of Agricultural Science, 144: 31-43.

Peña, J.M., Torres-Sánchez, J., Serrano-Pérez, A., de Castro, A.I. and LópezGranados, F. (2015). Quantifying efficacy and limits of unmanned aerial vehicle (UAV) technology for weed seedling detection as affected by sensor resolution, Sensors, 15: 5609-5626.

Powell, W., Machray, G.C. and Provan, J. (1996). Polymorphism revealed by simple sequence repeats. Trends in Plant Science, 1: 215-222.

Pritchard, S.G. and Amthor, J.S. (2005). *Crops and Environmental Change*, Food Products Press, Binghamton, pp. 421.

Rosenzweig, C., Iglesias, A., Yang, X.B., Epstein, P.R. and Chivian, E. (2000). Implications of climate change for U.S. agriculture: Extreme weather events, plant diseases and pests. Centre for Health and the Global Environment, Harvard Medical School, Cambridge, MA.

Rosenzweig, C., Iglesias, A., Yang, X.B., Epstein, P.R. and Chivian, E. (2001). Climate change and extreme weather events: implications for food production, plant diseases, and pests, Global Change and Human Health, 2: 90-104.

Trumble, J.T. and Butler, C.D. (2009). Climate change will exacerbate California's insect pest problems, California Agriculture, 63: 73-78.

Wani, S.H., Sanghera, G.S. and Singh, N.B. (2013). Biotechnology and plant disease control-Role of RNA interference, American Journal of Plant Sciences, 1: 55-68.

Wasim, A.M., Sengupta, D. and Ashim, C. (2009). Impact of pesticides use in agriculture: Their benefit and hazard, Interdisciplinary Toxicology, 2(1): 1-12.

Xiongkui, H., Bonds, J., Herbst, A. and Langenakens, J. (2017). Recent development of unmanned aerial vehicle for plant protection in East Asia, International Journal of Agricultural and Biological Engineering, 10: 18-30.

Zhang, W., Taylor, P., Kremer, C., Carney, K. and Sott, S.M. (2007). Ecosystem services and dis-services to agriculture, Ecological economics, 64(2): 253-260.

Chapter 13

Application of Climate Smart Agriculture to Crop Production in India

Anmol Gupta[1], *Priya Dubey*[2]*, *Mohit Pandey*[3] *and Rajneesh Singh*[4]

[1]*IIRC-3, Plant-Microbe Interaction and Molecular Immunology Laboratory, Department of Biosciences, Faculty of Science, Integral University, Lucknow, Uttar Pradesh*
[2]*Department of Environmental Technologies and Envis,*
[4]*Molecular Biology and Biotechnology Division,*
CSIR-National Botanical Research Institute, Uttar Pradesh
[3]*Department of Geology, University of Lucknow, Lucknow, Uttar Pradesh*

India is the seventh largest country by area and the second most populated country in the world. Most scenarios for farming under changing climate is that the agriculture sector will be significantly impacted from increased temperatures and Shifting patterns of precipitation. Increasing variability of rainfall activity and other meteorological parameters further increases the region's challenges. Some agricultural technologies and practices change for the production of agriculture as it may allow 30-50 per cent of land use for forestation that leads to the increased terrestrial stock of carbon. This chapter, aimed to assess the impact of climate change on agriculture, and the role of Climate Smart Agriculture (CSA) with sustainable agricultural practices like low-input sustainable agriculture (LISA) for ensuring food security while safeguarding the critical natural resources for human-wellbeing and also for attaining UN-Sustainable Development Goals. Climate Smart Agriculture (CSA) has attracted interest given its promise to increase agricultural productivity under a changing climate while reducing its emissions and of greenhouse gases (GHG). Here we are trying to show that under unfavorable climate regimes, a subset of agronomic practices that are often included under the CSA rubric may contribute to increasing agricultural production while contributing to reducing GHG emissions. We conclude that the large-scale implementation of CSA will facilitate agricultural sustainability, and therefore, suitable policy frameworks are imperative for its worldwide adoption. However, essential investments and cooperation are needed for CSA to have a meaningful effect, and its principles need to be broadly applied across the whole sector.

Keywords: *Agriculture, Climate change, Climate smart agriculture, Low-input sustainable agriculture.*

Introduction

The agriculture division plays an essential role within the growth of the Indian financial system. It not only provides employment to the mainstream population of the country but also serves as a raw material source for a huge number of industries. It contributes around 12.55 per cent of its overall exports and 14 per cent of India's GDP (Government of India, 2017). Agriculture is littered with both natural disasters (caused by epidemics outbreak) and man-made disasters which dangerously affect farm manufacturing and its income. In addition, growing fragmentation of holdings, frequent climatically variations (Boomiraj *et al.*, 2010; Rao *et al.*, 2013), and rising input prices and postharvest losses cause a huge venture challenge to sustaining agricultural growth. Numerous marketplace and policy instruments are evolved and established to handle combined risks within the agriculture sector. Climate change and related environmental changes are the essential sustainability challenges for humanity during this 21st century. Whereas mankind has witnessed climate change throughout the historical centuries, its real-time variability is immediate and the outcomes are beyond human control. These changes are deleteriously affecting the natural biomes of the earth. Among the numerous systems, agro-ecosystems are more susceptible to global climate change (Xiao *et al.*, *et al.*, 2007; Singh *et al.*, *et al.*, 2013; Dubey *et al.*, *et al.*, 2020a). Approximately 54.6 per cent Indian population still relied on agricultural activities as it is a principal source of financial assistance thereby functions as a supply of raw material for an oversized variety of industries. India accounts for approximately 2.4 per cent of the global geographic region and 4 per cent of its water resource; however, it supports approximately 15 per cent of the livestock and 17 per cent of the world's human population. Augmented growth of agriculture production is so important, to attain higher contribution towards the gross domestic product (GDP) and meet the expanding demand for food, but also to boom farmers' earnings to safeguard their inclusiveness. It has been predictable that the crop yield might be significantly decreased due to the deterioration of freshwater availability. Furthermore, the per capita accessibility of cultivable land is constantly reducing over the decades (*i.e.*, 0.41 hectare in 1961 to 0.25 hectare in 2015, and by 2050 it is calculable to be 0.20 hectare) (Hens and Quynh, 2016; Dubey *et al.*, *et al.*, 2020). The un-systematic usage of chemical fertilizers has previously crossed the planetary boundaries. Moreover, several anthropogenic activities which include the burning of fossil fuels and industrialization have additionally resulted in augmented greenhouse gases (GHGs) emission at the local and global levels. Deforestation gives rise to the alteration of micro-climate by an upsurge in the mean temperature and reduces within the mean downfall of a specific region (Lawrence and Vandecar, 2015). Climate change and changeability could also lead to critical consequences for cereal vegetation production by 2050 in developing nations. The higher exhaustion of fossil fuels alone contributes to around 56.6 per cent of carbon dioxide (CO_2) emissions (Murray and Ebi, 2012). Apart from these, many unscientific agricultural practices forestry and land uses are also the promising sources of increased carbon dioxide (CO_2), nitrous oxide (N_2O), and methane (CH_4), concentration in the atmosphere. The involvement of these gases is accounted to be about 22 per cent of the man-made GHG emissions (IPCC, 2019). Recent Inter-governmental Panel on Climate Change (IPCC) special report and progression in the United Nations Framework Convention on Climate Change (UNFCCC) negotiations on agriculture (UNFCCC, 2017) have strengthened to reduce the greenhouse gas emissions, including the pricing of carbon consequently levy a carbon tax (Masson *et al.*, 2018). Nevertheless, the latest investigation on the subject (Frank *et al.*, 2017; Hasegawa *et al.*, 2018) indicates that a tax on GHG emissions may lead to stunning tradeoffs between food security and emissions abatement. Climate smart agriculture is encompassed of agricultural practices that support farmers to adapt to climatic specific factors such as temperature and rainfall. It is in this ecosystem that the concept of Climate Smart Agriculture (CSA) has become progressively important. CSA recommends a framework that supports policymaking in the agriculture division by considering three foundational results and by completely accounting for the trade-offs and synergies among them. It is made up of agricultural systems that impart sustainable and justifiable increases in incomes and agricultural productivity; better adaptation and resilience of food systems towards climate change from the farm- to the nationwide-level; and reduction of GHG emissions, where possible (Lipper *et al.*, 2014). Therefore, Farmers should determine what can be thought-about climate smart given their agricultural, biophysical, and socio-economic context. As an outcome, the utilization of the CSA approach is knowledge-intensive and might need substantial institutional support (Neufeldt *et al.*, 2013; McCarthy *et al.*, 2011). Many fields and farm-based sustainable agricultural operations like land management practices that are already in use are-agroforestry conservation tillage, residue management, intercropping, crop shifting, multipurpose crops, alley farming, drip irrigation for water, water harvesting, and nutrient management depending on the water supply, and irrigated and rain-fed lands (Table 13.1). Improvement of these activities at native and regional levels in the climate change scenario would boost food security and livelihood opportunity. Food and Agriculture Organization, UN definition of climate smart agriculture highlights biodiversity conservation, forest lands management, and abridged inter-sectoral competition for water (FAO, 2010). The World Bank (2011), category of climate smart agriculture covers combined planning of agricultural land with household requirement and strengthened the production from small landholdings, climate resilience, multipurpose water harvesting, and environmental sustainability.

Agro-ecological resource management and Food production is additionally a possible beneficiary of such developments (Gangwar *et al.*, 2019). Farmers are forced to scale back the ecological and economic cost, maximize their physical yield with the advanced quality of their food items. These necessities go hand in hand with adherence to strict social, health, environmental, and safety guidelines (Yick *et al.*, 2008). The utilization of ICT-enabled farm management information system (**FMIS**) has shown great promises for attaining these goals (Sørensen *et al.*, 2011), particularly within the context of Precision Farming (Blackmore 1994). The first adopters prevail with the flexible rate application of pesticides and fertilizers (Lawson *et al.*, 2011). This chapter aimed to look at the impact of climate change on agriculture and various advanced/futuristic steps taken to reduce the negative effects together with the role of LISA and alternative smart agriculture techniques to safeguard food security in an exceedingly warming condition.

Table 13.1: Effect of CSA Technology on Crops

Sl. No.	*CSA Technology*	*Description*	*Crops*	*Possible Effects on Crop Yields and GHGs Emissions*
1.	No tillage	Minimum soil disturbance, often in amalgamation with crop rotation, and usage of cover crops Unresolved effect on GHG emission	Maize and Wheat	✰ Positive or impartial effects on crop yields
2.	Integrated soil fertility management	Combination of chemical fertilizers, crop residues, and manure/compost	Maize and wheat	✰ Beneficial effects on yields ✰ Irregular effects on GHG emissions
3.	Nitrogen use efficiency	Practices and technologies to enhance the uptake of inorganic soil nitrogen by crops	Rice	✰ Progressive results on yields ✰ Decline of GHG emissions

Source: Pinto *et al.*, 2020.

Increasing Affirmation of Climate Change

The Intergovernmental Panel on Climate Change (IPCC) recommends that the present scenario of climate change has affected many parts of Asian countries due to an increase in rainfall intensity and temperature. This climate change is led to hampering the agricultural produce due to various stresses, worsening the marine, forest, and coastal ecosystems; thereby affecting the well-being of the human and causing climate induced water-borne and vector-borne diseases (Cruz *et al.*, 2007). The increase in earth temperature causes numerous havoc effects on the ecosystem such as the melting of glaciers to cause a rise in sea level, which ultimately cause floods in the river valleys, subsequently diminishing the flow of freshwater resulting insufficient supply of water for agricultural and domestic activities (Haberli and Hoelzle, 2001; Rai and Gurung 2005). Many anthropogenic activities cause an upsurge in GHGs concentration like CH_4, CO_2, and N_2O causing serious concern about climatic changes thereby affecting the agro-ecosystems (Aggarwal, 2003).

In this viewpoint, Cruz *et al.* (2007) has been trying to confirm fact that these abrupt climate changes are estimated to reduce the number of monsoon seasons, while the extreme rainfall events will cause a severe disaster like flood conditions. Some scholars predicted that it occurs due to elevated concentration of aerosol in the air thereby causing warming of the land surface as well as the atmosphere which further disturb the patterns of local circulation (Kumar and Ashrit, 2001; Sarkar *et al.*, 2020). Recently, due to increased air pollution, cloud cover, and atmospheric aerosols a reduction in evapotranspiration is also witnessed at several places (Sarkar *et al.*, 2020). Currently, 15 per cent of aquifers are at extreme conditions and are expected to be aggravated to 60 per cent by the end of the year 2030 (Briscoe and Malik, 2006).

Impact of Climate Change on Agriculture and Crop Production

The combined result of climate change and rising atmospheric CO_2 concentration causes global warming that hampers agricultural crop production harshly across the country. (Rötter and Van de Geijn, 1999; Ruchita and Rohit, 2017; Dubey *et al.*, 2020a) (Figure 13.1). Considering the Indian scenarios it is predicted that the rise of 2 °C temperature and +7 per cent precipitation change will led to the net revenue loss by ~9 per cent while 15 per cent to 25 per cent on increasing temperature and precipitation (Kumar and Parikh, 2001). In Rabi season, (Mall *et al.*, 2006). Birthal *et al.* (2014) estimated that the erratic rainfall and elevated temperature led to the yield reduction of 6 per cent by 2035 and will additionally expand to 12 per cent and 16 per cent by the year 2065 and 2100in arid and semi-arid tropics. Although the share of agriculture in national income has dropped to <15 per cent in 2012–13, but it keeps its significance from the viewpoints of employment generation, food security, and poverty reduction (Birthal *et al.*, 2014).

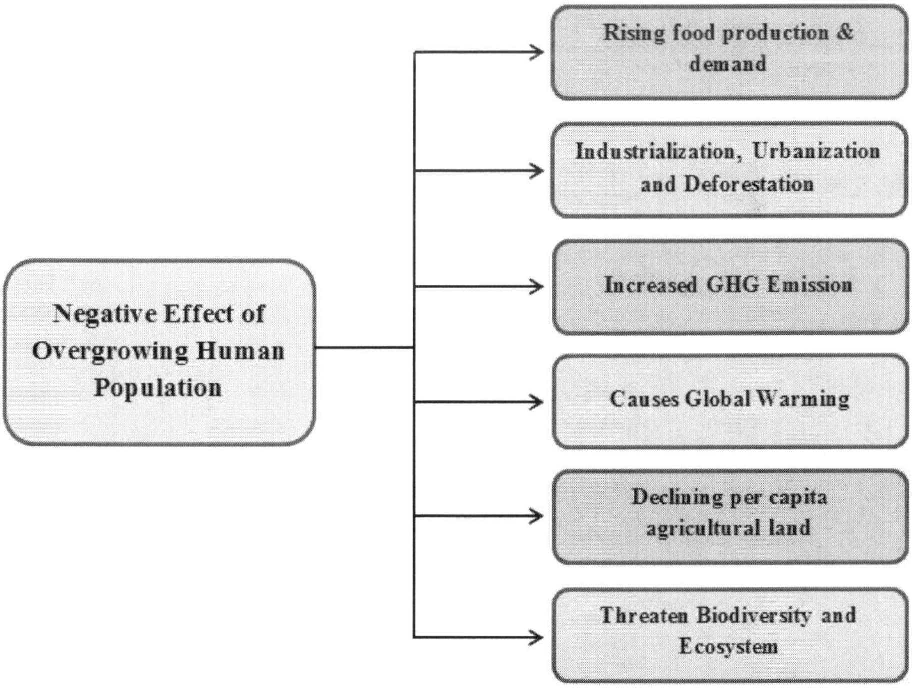

Figure 13.1: Impact of Various Anthropogenic Activities on Climate Change Affecting Agricultural Production and Threatening Food Security.

An Alternative to Reduce the Impact of Climate Change and Enhance the Crop Production

Climate Smart Agriculture (CSA)

The perception of climate smart agriculture (CSA) is acquiring popularity as a combining concept on agriculture and climate change. CSA is defined as "agriculture that sustainably increases the resilience, crop productivity, reduces the GHGs emission and augments the accomplishment of national food security and development goals in a sustainable manner is known to as Triple Wins" (FAO 2010). The forestry, climate smart agriculture, and fisheries are a substantial pathway to the achievement of SDG (FAO, 2010). It is an amalgamation of the three dimensions (3D's) of sustainable development *viz.* social, economic, and environmental which help in addressing the climate challenges and food security. It is made up of three main pillars: (1) Increasing agricultural produce and income viability. (2) Building resilience and adaptive towards climate change. (3) Reducing emissions of GHGs (FAO 2010). Climate smart agriculture has mainly 3 goals in the diverse area for crop production (Figure 13.2).

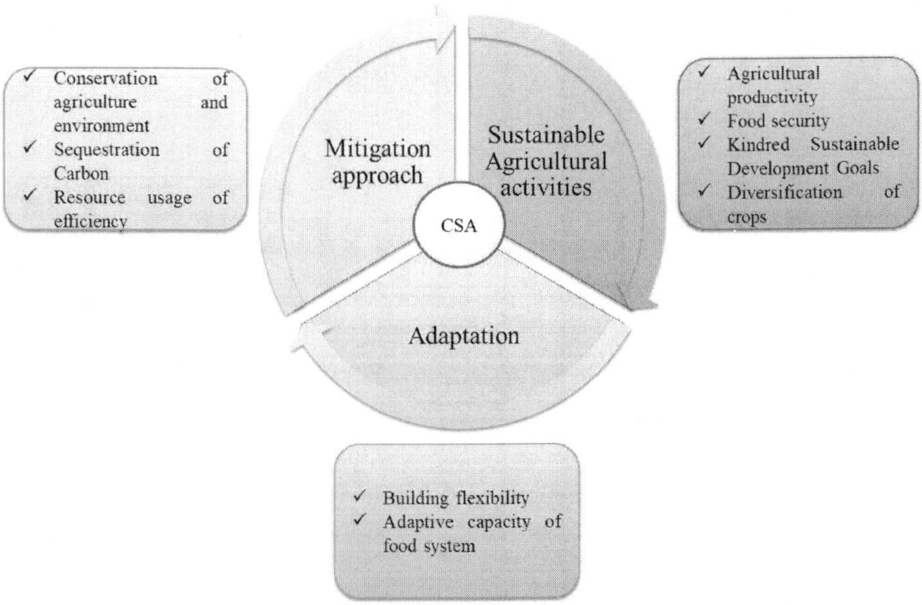

Figure 13.2: The Goal of Climate smart Agriculture (*Source*: Venkatramanan and Shah 2019).

Climate Smart Crops and Cultivars

The climatic irregularities change the growing period of the crops. To serve the requirements of changing climatic situations, various research institutes and universities are engaged in evolving climate smart crop cultivars that are cost and time-consuming. The farmers have been recognized the advantage of improved varieties under rain-fed systems across India. For instance, Bhoosamrudhi sites have shown the lead in crop yield varies from 12 to 55 per cent in chickpea, pigeon pea, and pearl millet crops.

Low Input Sustainable Agriculture (LISA)

Developing countries are mostly dependent on agriculture for rural livelihoods and progress in their lifestyle. Due to the presence of several restrictions such as costly inputs and land pressure, the profitability of the farming process is affected thereby threatening the inclusive sustainability of agricultural production. However, the agriculture sector exclusively accounts for 13.5 per cent of greenhouse gas emissions (IPCC, 2007). So there is an urgent need for shifting of our traditional agricultural practices to climate resilient smart sustainable agriculture. Hence, LISA is a sort of farming system that can produce fiber and food profitably without damaging the quality of natural resources, guaranteeing to provide nutritious foods and overcome the health hazards by reducing its reliance on external inputs (Najafabadi *et al.*, 2012). Mainly, the concept of LISA was developed to make awareness among farmers about the adverse effect of highly capital-intensive, chemicals such as fertilizers, and pesticides, and intensive farming systems on the ecosystem (Schaller, 1991; Dubey *et al.*, 2020a). It is an amalgamation of the locally accessible natural resources (water, soil, local plants, vegetation, and animals) and human resources such as knowledge, skills, and human labor to maintain a cultural, social, ecological, and economical production system sustainably. LISA also advocates the judicious and minimal usage of external resources used for sustainable farming. This farming system can undergo the adverse effects of climate change whereas retaining agricultural productivity through the usage of various technologies.

Agroforestry

The incorporation of trees along with other agricultural crops in a land usage system is called as "Agroforestry". It provides many benefits such as providing fuel, fiber, timber, food (fruits, honey, and oils), and pharmaceuticals that play an important role in carbon sequestration, biodiversity management, and soil restoration (Dagar and Minhas, 2016). Agroforestry helps in improving the soil organic matter and soil fertility which helps many leguminous crops to grow, and thus, smoothen the nutrient cycling and recover the soil conditions (Mbow *et al.*, 2014). Additionally, agroforestry practices also increase water productivity by maintaining the hydrological cycle (Ong and Swallow, 2003; van Noordwijk *et al.*, 2004). The agroforestry method such as kallar grass (*Leptochloa fusca*) and Vilayati babul (*Prosopis juliflora*) grown in Lucknow, Uttar Pradesh, India, is used to ameliorate the alkalinity of the soils (Singh *et al.*, 2017).

Organic Farming

The basic concept of organic farming includes inter-cropping, crop rotations, mulching, ploughing, crop diversification, green maturing, composting, pest control techniques, crop residual management, *etc.* (Rakshit *et al.*, 2010; Dhiman 2020). These actions help in lowering GHG emissions and ensured balanced carbon sequestration.

It reduces the nitrogen dependency on fertilizers and pesticides thereby lowering the CO_2 emissions (Müller, 2009). Furthermore, with the usage of organic farming approach *viz.* the animal and green manures, composting techniques, intercropping, and cover cropping led to increases the organic matter content in soil (Niggli *et al.*, 2008; Dubey *et al.*, 2016a, 2020b). Such practices also recover the nutritional quality of agricultural crops and have many advantages (Table 13.2).

Table 13.2: The Beneficial Effect of Organic Farming on Crop Residue

Crop	*Compost*	*Advantage*	*References*
Olive	Prepared from the olive pomace, olive mill wastewater, sheep manure, and olive tree pruning, and legume cover crop as fava bean.	Increases the organic matter content and availability of nutrients (N, P, K) into the soil for crop development and lowering the GHG emissions.	Chehab *et al.*, 2019

Species richness of micro and macrofauna and flora is the main indicator for the steadiness of biological farming (Table 13.3). Recently, Rollan *et al.* (2019) documented that the bird species in Europe near the Mediterranean vine-yards of Catalonia (in Spain) is positively affected due to organic practices. Lori *et al.* (2017) documented that organic farming led to an increase in the microbial presence and its activity in agricultural soils universally by *metagenomic*-analysis.

Table 13.3: An Indicative List of Various Sustainability Indicators of LISA Practices for Natural Food Production

Sl.No.	*Indicator*	*Sustainability Component*	*Reference*
1.	Soil organic carbon	Soil fertility and health	Dubey *et al.* (2019), Das *et al.* (2014), Kumar *et al.* (2018), Abo-baker (2017), Mariotte *et al.* (2018).
2.	Soil enzymes such as urease, dehydrogenase, β-glucosidase, *etc.*	Soil functioning	Dubey *et al.* (2019), Zhang *et al.* (2017), Sarkar *et al.* (2017b).
3.	Soil N, P, K content	Soil nutrient dynamics	Biratu *et al.* (2018), Kumar *et al.*, 2018, Abo-baker (2017).
4.	Soil microbial diversity	Soil biodiversity	Lori *et al.* (2017), Zhang *et al.* (2017), Dubey *et al.* (2016b, 2019).
5.	Bees (bumble bees/ honey bees)	Biodiversity conservation, crop pollination, ecosystem services	Giacomini *et al.* (2018), Vanbergen and the Insect Pollinators Initiative (2013).

Livestock Production

Livestock production plays a predominant role in food security but climate change further interrupted livestock production, affecting the availability of water, animal production, and health. The collective effect of elevated CO_2, temperature, and precipitation will further reduce the quality of crop and fodder thereby enhances livestock diseases (Rojas-Downing *et al.*, 2017). Therefore, we need to adopt some suitable measures, or climate change may exterminate almost 15 to 37 per cent of total species in the biosphere (Thomas *et al.*, 2004). Integrated livestock production is considered to be a suitable way to upsurge the *in-situ* biodiversity in the agri-environment (Murgueitio *et al.*, 2011; Broom *et al.*, 2013).

Aquaculture

The incorporation of aquaculture into the agricultural systems is another promising feature that provides a diversity of benefits such as diversification, increased productivity, and sustainability of crops. Land scarcity has often restricted the construction of fish ponds hence, aquaculture can be achieved with assimilated farming practices such as agroforestry-fish system, and rice-fish system to improve the farmers' livelihood through aquaculture and fishing. Aquatic foods are an essential source of amino acids, protein, vitamins, fatty acids, and minerals (Tacon and Metian, 2013; Huong *et al.*, 2017). Furthermore, the risk of production failure is also curtailed because of the prevalence of multiple production activities. The Rice-fish system requires less pesticide (68 per cent) and fertilizer (24 per cent) consumption as compared to the rice monoculture (Lansing and Kremer, 2011). Above all, the rice yield is thought to be improved by 15 per cent with the usage of organic fertilization (aquatic excreta) (Reddy and Kishori, 2018). The world-wide freshwater scarcity has led to boosting the brackish-water aquaculture.

Drip Irrigation (DI): Great Promise, Slow Progress

The motivation to implement an innovative technology is clearly depending on the quantum of advantage that it provides. It helps in delivering the water directly to the plant's root and opposed the flooding in the fields. Various technical studies have mentioned the potential advantages of DI like water and energy savings (Table 13.4)

Table 13.4: Presumed Benefits of Drip Irrigation

Sl.No.	*Benefits*	*Remarks*
1	Water use	Research analysis specifies that sugarcane cultivation uses around 44 to 55 per cent less water when compared to flood irrigation.
2	Electricity use	Drip irrigation saves the electricity consumption of around 25-50 per cent.
3	Fertilizer use	Drip irrigation prevents excessive washing of fertilizers that are applied to various crops.
4	Drought resilience	Drip irrigation managed the water usage sustainably thereby provide water to crops in a systematic manner under drought conditions.
5	Soil maintenance	Flood irrigation caused salinity thereby affecting sugarcane farming areas of Maharashtra.
6	Crop yield	DI can increase crop productivity. For example, that it can increase sugarcane farming by around 23 per cent in Maharashtra state itself.

Source: Surendran *et al.* (2016) and Tankha *et al.* (2020).

Building Strength Towards Climate Change in Agricultural Sector

Agricultural Conservation

Mixed cropping or intercropping systems are flexible than the mono-cropping system in cropland areas because they help in the efficient utilization of soil nutrients. This method is also stable under various adverse conditions like pests and weather situations. Water and land management are the basic components for the safe exclusion of excess water in a proper manner. The broad bed and furrow (BBF) system is better-quality in-situ soil conservation and drainage technology for clayey soils during the rainy season (El-Swaify *et al.,* 1985). The improved management led to a change in biophysical properties thereby changes the hydrological cycle as runoff was decreased in BBF fields.

Soil and Water-Conserving Practices

Since the water and soil resources are depleting due to variability in climate there is a need to adopt conservation practices. Climate change, mainly the variation in rainfall will enhance the risk of runoff, soil erosion, and other associated environmental damages (SWCS, 2003). Therefore, we need to introduce the resource-conserving techniques (RCTs) in the prevailing cropping system that will improve resource efficiency by saving the fuel, water, and labor requirements, reducing the cost of production, improving yields, and mitigating the greenhouse gas emissions (Pathak *et al.,* 2012). The notable significance of the tillage conservation practice includes preservation of soil structure, improving the soil organic carbon, conservation of soil-water, reducing the erosion risks, and enhancing soil quality (Busari *et al.,* 2015). Many bio-engineering measures are also used in the conservation of water and soil such as usage of vegetative barriers, intercropping, mulching, mixed cropping, *etc.* Contour bunding in West Africa (Southern Mali) has improved the rainwater infiltration and reduced soil erosion thereby eventually increased vegetation productivity (Tabo, 2020). The application of various conservation technologies has reduces the runoff by 8 to 40 per cent and soil loss by 6 to 35 per cent in sloping areas of India thereby leading to improve crop productivity by 3 to 28 per cent (CSWCRTI, 2011).

Farm Management Information Systems (FMIS)

The primary objective of FMIS is to create models for crop production similar to industrial production. The model was proposed to support precision farming (Zaks and Kucharik, 2013). For effective administration of farm production, temporal and spatial crop monitoring is required (Blackmore, 2002). These systems are essential to support the documentation of data and its collection, and utilize this information for strategic farm management decisions through communication technologies and autonomous sensing. The idea of FMIS includes the participation of farmers in various agricultural practices that are site-specific. Information integration is required for effective management of agricultural ecological resources and improved the economic possibility of farm products (Wang *et al.,* 2006). Climate smart landscape management contains the sustainable usage of natural resources like forest, land, grasslands, water, and ecological services (FAO, 2010; Scherr and Sthapit, 2009; Branca *et al.*, 2011). The living conditions of several insects, birds, and animals have been improved through rainwater harvesting (RWH) in the forest and provide ecological connectivity. Expansion of forests, grasslands, and wetland offers biodiversity conservation, local livelihood, and tourism. Climate smart agricultural landscape operates on planting a different variety of dates various livestock species, rainwater harvesting, and sustainable practices in the field. Therefore, we need new methods towards sustainable agriculture that strengthens ecosystem health and its adaptability.

Land Usage and Agricultural Management Practices

Agriculture land usage covers around 38.5 per cent of the worldwide total land area, containing 28.4 per cent arable land area (DSI MSU, 2015). Change in agricultural land usage is associated with 2 concepts: (1) per unit profit and (2) opportunity cost of the land. Profit mentions the farmer's willingness to manage the specific usage of land which will differ depending on the biophysical features of land, inputs, price, and access to service and markets. Land degradation is caused due to conversion of land from forest land to grasslands or cropland. The farmers consume their own land resources and explore another suitable area for their cultivation. Therefore, agro-ecological areas are brought under grasslands threatening food security (Asner *et al.*, 2004). According to the IPCC, the agricultural

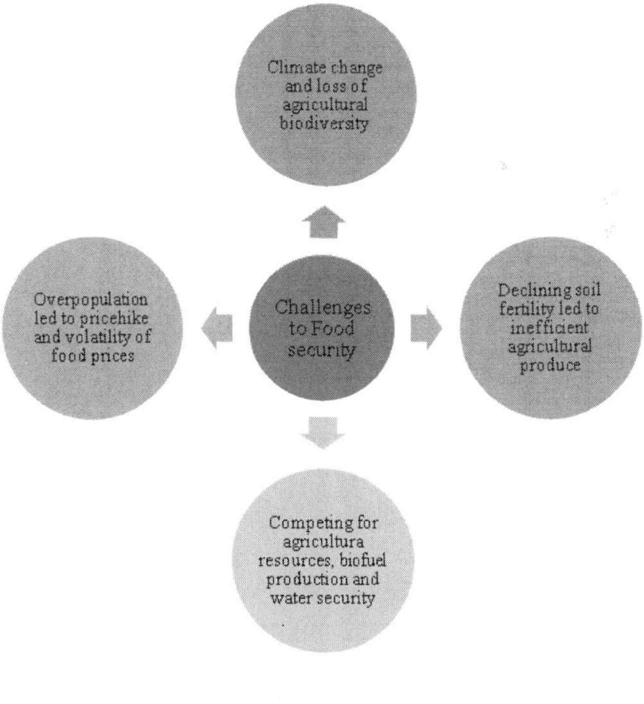

Figure 13.3: Factors that Causes Challenges to Food Security (Venkatramanan and Shah, 2020).

cropland sequesters 6 per cent of global carbon while pastureland tropical savannas temperate grasslands were collectively sequestered 27 per cent of carbon (IPCC, 2000). Various factors cater to the needs of the overgrowing population and challenging food security (Figure 13.3). The remarkable technological changes, mitigation, and adoption are done improving the farm practices which may allow 30-50 per cent land usage for forestation (Boumer *et al.*, 1998; Shvidenko *et al.*, 2002). As a result, Carbon sequestration in forest soils reduces the CO_2 level in the atmosphere, making the relaxing conditions for flora and fauna, and projected climate changes (Carnaval *et al.*, 2009).

Conclusion

The previous discussion has discovered that despite the impact of climate change, a variety of smart techniques has been existed for crop intensification by acquiring various interventions of natural resource management (NRM) and its mechanization, and suitable crop cultivars. The current chapter focuses to tackle the change in climate and the use of various climate smart approaches to meet the sustainable goal. In-situ interventions must be encouraged on a large scale to promote crop development in dryland agriculture. Therefore, a large-scale awareness is required which includes researchers, policy-makers, development agencies together with the farming community. To overcome the climate change risks there is a need for the participation of large-scale farmers that are helpful in (i) suitable cultivar selection, and (ii) enhancing awareness among stakeholders about climate smart agriculture. Climate smart agriculture (CSA) is a unified approach that can address many challenges that are inter-linked with a change in the climate. Additionally, there are some water and land management innovative methods that are used for enhancing water and land usage efficiency. These interventions enhance crop production and thereby significantly reducing the cost of cultivation. To strengthen the present extension system, public-private participation, development organizations, private institutes, and research organizations should work in a tandem manner to achieve the stem-level outcomes. Through this chapter, we have examined the positive effects of smart agricultural technology and, various management practices, and an approach towards sustainable agriculture (LISA) in the domain of climate smart agriculture.

REFERENCES

Aggarwal, P. K. (2003). Impact of climate change on Indian agriculture. *Journal of Plant Biology-new Delhi*, *30*(2), 189-198.

Birthal, P. S., Negi, D. S., Kumar, S., Aggarwal, S., Suresh, A., and Khan, M. (2014). How sensitive is Indian agriculture to climate change?. *Indian Journal of Agricultural Economics*, *69*(902-2016-68357), 474-487.

Blackmore, S. (1994). Precision farming: an introduction. *Outlook on agriculture*, *23*(4), 275-280.

Blackmore, S. (2002). Developing the principles of precision farming. *The Centre ofPrecision Farming, Dinamarca.*

Boomiraj, K., Wani, S. P., Garg, K. K., Aggarwal, P. K., and Palanisami, K. (2010). Climate change adaptation strategies for agro-ecosystem–a review. *Journal of Agrometeorology, 12*(2), 145-160.

Briscoe, J., and Malik, R. P. (2006). *India's water economy: bracing for a turbulent future.* New Delhi: Oxford University Press.

Broom, D. M., Galindo, F. A., and Murgueitio, E. (2013). Sustainable, efficient livestock production with high biodiversity and good welfare for animals. *Proceedings of the Royal Society B: Biological Sciences, 280*(1771), 20132025.

Bulluck Iii, L. R., Brosius, M., Evanylo, G. K., and Ristaino, J. B. (2002). Organic and synthetic fertility amendments influence soil microbial, physical and chemical properties on organic and conventional farms. *Applied Soil Ecology, 19*(2), 147-160.

Busari, M. A., Kukal, S. S., Kaur, A., Bhatt, R., and Dulazi, A. A. (2015). Conservation tillage impacts on soil, crop and the environment. *International Soil and Water Conservation Research, 3*(2), 119-129.

CSWCRTI, 2011. CSWCRTI Vision 2030. Central Soil and Water Conservation Research and Training Institute (CSWCRTI), Dehradun, Uttarakhand.

Dagar, J. C., and Minhas, P. S. (2016). Global perspectives on agroforestry for the management of salt-affected soils. In *Agroforestry for the management of waterlogged saline soils and poor-quality waters* (pp. 5-32). Springer, New Delhi.

De Pinto, A., Cenacchi, N., Kwon, H. Y., Koo, J., and Dunston, S. (2020). Climate smart agriculture and global food-crop production. *PloS one, 15*(4), e0231764.

Dhiman, V. (2020). Organic Farming for Sustainable Environment: Review of Existed Policies and Suggestions for Improvement. *International Journal of Research and Review, 7*(2), 22-31.

Dubey, P. K., Singh, G. S., and Abhilash, P. C. (2016). Agriculture in a changing climate.

Dubey, P. K., Singh, G. S., and Abhilash, P. C. (2019). *Adaptive agricultural practices: Building resilience in a changing climate.* Springer.

Dubey, P. K., Singh, G. S., and Abhilash, P. C. (2020). Adaptive Agricultural Practices Employed in Eastern Uttar Pradesh, India. In *Adaptive Agricultural Practices* (pp. 93-122). Springer, Cham.

FAO (Food and Agriculture Organization). 2010. "ClimateSmart" Agriculture: Policies, Practices and Financing for Food Security, Adaptation and Mitigation. Rome, Italy: FAO.

Frank, S., Havlík, P., Soussana, J. F., Levesque, A., Valin, H., Wollenberg, E. and Obersteiner, M. (2017). Reducing greenhouse gas emissions in agriculture without compromising food security?. *Environmental Research Letters, 12*(10), 105004.

Hasegawa, T., Fujimori, S., Havlík, P., Valin, H., Bodirsky, B. L., Doelman, J. C. and Witzke, P. (2018). Risk of increased food insecurity under stringent global climate change mitigation policy. *Nature Climate Change, 8*(8), 699-703.

Hens, L., and Quynh, L. X. (2016). Environmental space.

Kumar, K. R., and Ashrit, R. G. (2001). Regional aspects of global climate change simulations: Validation and assessment of climate response over Indian monsoon region to transient increase of greenhouse gases and sulphate aerosols. *Mausam, 52.*

Kumar, K., and Parikh, J. (2001). Socio-economic impacts of climate change on Indian agriculture. *International Review for Environmental Strategies, 2*(2).

Lansing, J. S., and Kremer, J. N. (2011). Rice, fish, and the planet. *Proceedings of the National Academy of Sciences, 108*(50), 19841-19842.

Lawrence, D., and Vandecar, K. (2015). Effects of tropical deforestation on climate and agriculture. *Nature climate change, 5*(1), 27-36.

Lawson, L. G., Pedersen, S. M., Sørensen, C. G., Pesonen, L., Fountas, S., Werner, A. and Blackmore, S. (2011). A four nation survey of farm information management and advanced farming systems: a descriptive analysis of survey responses. *Computers and Electronics in Agriculture, 77*(1), 7-20.

Lipper, L., Thornton, P., Campbell, B. M., Baedeker, T., Braimoh, A., Bwalya, M. and Torquebiau, E. F. (2014). Climate smart agriculture for food security. *Nature climate change, 4*(12), 1068-1072.

Mall, R. K., Singh, R., Gupta, A., Srinivasan, G., and Rathore, L. S. (2006). Impact of climate change on Indian agriculture: a review. *Climatic change, 78*(2), 445-478.

Masson-Delmotte, V., Zhai, P., Pörtner, H. O., Roberts, D., Skea, J., Shukla, P. R. and Waterfield, T. (2018). Global warming of 1.5 C. *An IPCC Special Report on the impacts of global warming of, 1*, 1-9.

Mbow, C., Smith, P., Skole, D., Duguma, L., and Bustamante, M. (2014). Achieving mitigation and adaptation to climate change through sustainable agroforestry practices in Africa. *Current Opinion in Environmental Sustainability, 6*, 8-14.

McCarthy, N., Lipper, L., and Branca, G. (2011). Climate smart agriculture: smallholder adoption and implications for climate change adaptation and mitigation. *Mitigation of Climate Change in Agriculture Working Paper, 3*(1), 1-37.

Muller, A. (2009). Benefits of organic agriculture as a climate change adaptation and mitigation strategy in developing countries. *rapport nr.: Working Papers in Economics 343.*

Murgueitio, E., Calle, Z., Uribe, F., Calle, A., and Solorio, B. (2011). Native trees and shrubs for the productive rehabilitation of tropical cattle ranching lands. *Forest Ecology and Management, 261*(10), 1654-1663.

Murray, V., and Ebi, K. L. (2012). IPCC special report on managing the risks of extreme events and disasters to advance climate change adaptation (SREX).

Najafabadi, M. O., Khedri, K., and Lashgarara, F. (2006). Requirement of low input sustainable agricultural implementation: A factor analysi of experts perspective.

Neufeldt, H., Jahn, M., Campbell, B. M., Beddington, J. R., DeClerck, F., De Pinto, A. and Zougmoré, R. (2013). Beyond climate smart agriculture: toward safe operating spaces for global food systems. *Agriculture and Food Security*, 2(1), 1-6.

Niggli, U., Schmid, H., and Fliessbach, A. (2008). *Organic farming and climate change*. International Trade Centre (ITC), Geneva.

Ong, C. K., and Swallow, B. M. (2003). 13 Water Productivity in Forestry and Agroforestry. *Water productivity in agriculture: limits and opportunities for improvement*, 1, 217.

Parry, M., Parry, M. L., Canziani, O., Palutikof, J., Van der Linden, P., and Hanson, C. (Eds.). (2007). *Climate change 2007-impacts, adaptation and vulnerability: Working group II contribution to the fourth assessment report of the IPCC* (Vol. 4). Cambridge University Press.

Pathak, H., Aggarwal, P. K., and Singh, S. D. (2012). Climate change impact, adaptation and mitigation in agriculture: methodology for assessment and applications. *Indian Agricultural Research Institute, New Delhi*, 1-302.

Prein, M. (2002). Integration of aquaculture into crop–animal systems in Asia. *Agricultural systems*, *71*(1-2), 127-146.

Rai, S. C., and Gurung, T. (2005). An overview of glaciers, glacier retreat, and subsequent impacts in Nepal, India and China.

Rakshit, A., Sarkar, N. C., Sen, D., and Maiti, R. K. (2010). Basics of conversion to organic farming. *International Journal of Agriculture, Enviornment and Biotechnology*, *3*(2), 253-256.

Rao, A. K., Wani, S. P., Singh, K. K., Ahmed, M. I., Srinivas, K., Bairagi, S. D., and Ramadevi, O. (2013). Increased arid and semi-arid areas in India with associated shifts during 1971-2004. *Journal of Agrometeorology*, *15*(1), 11-18.

Reddy, P. R., and Kishori, B. (2018). Integrated Rice and Aquaculture Farming. In *Aquaculture-Plants and Invertebrates*. IntechOpen.

Rojas-Downing, M. M., Nejadhashemi, A. P., Harrigan, T., and Woznicki, S. A. (2017). Climate change and livestock: Impacts, adaptation, and mitigation. *Climate Risk Management*, *16*, 145-163.

Rollan, À., Hernández-Matías, A., and Real, J. (2019). Organic farming favours bird communities and their resilience to climate change in Mediterranean vineyards. *Agriculture, ecosystems and environment*, *269*, 107-115.

Rötter, R., and Van de Geijn, S. C. (1999). Climate change effects on plant growth, crop yield and livestock. *Climatic change*, *43*(4), 651-681.

Ruchita, S., and Rohit, S. (2017). Effect of global warming on Indian agriculture. *Journal of Climatology and Weather Forecasting*, *5*(1).

Sarkar, D., Kar, S. K., Chattopadhyay, A., Rakshit, A., Tripathi, V. K., Dubey, P. K., and Abhilash, P. C. (2020). Low input sustainable agriculture: a viable climate smart option for boosting food production in a warming world. *Ecological Indicators, 115,* 106412.

Schaller, N. (1991). Background and status of the low-input sustainable agriculture program. *Sustainable Agriculture Research and Education (SARE) Program.*

Singh, S. P., Saralch, H. S., and Kaur, A. (2013). Impact of climate change on biodiversity and agriculture. *International Journal of Farm Sciences, 3*(2), 113-116.

Singh, Y. P., Singh, G. B., Sharma, D. K., and Arora, S. (2017). Evaluation of different agro-forestry systems for their effect on alkali soils.

Smith, P., and Wollenberg, E. (2012). Achieving mitigation through synergies with adaptation. *Climate Change Mitigation and Agriculture,* 50-57.

Sørensen, C. G., Pesonen, L., Bochtis, D. D., Vougioukas, S. G., and Suomi, P. (2011). Functional requirements for a future farm management information system. *Computers and Electronics in Agriculture, 76*(2), 266-276.

SWCS (Soil and Water Conservation Society), 2003. Conservation implications of climate change: Soil erosion and runoff from cropland. A Report from the Soil and Water Conservation Society, Ankeny, Iowa.

Tabo, R. (2020). Regional Science Review and Planning Meeting 2020.

Tacon, A. G., and Metian, M. (2013). Fish matters: importance of aquatic foods in human nutrition and global food supply. *Reviews in fisheries Science, 21*(1), 22-38.

Tankha, S., Fernandes, D., and Narayanan, N. C. (2020). Overcoming barriers to climate smart agriculture in India. *International Journal of Climate Change Strategies and Management.*

Thomas, C. D., Cameron, A., Green, R. E., Bakkenes, M., Beaumont, L. J., Collingham, Y. C. and Williams, S. E. (2004). Extinction risk from climate change. *Nature, 427*(6970), 145-148.

Tunde, A. B., Kuton, M. P., Oladipo, A. A., and Olasunkanmi, L. H. (2015). Economic analyze of costs and return of fish farming in Saki-East Local Government Area of Oyo State, Nigeria. *Journal of Aquaculture Research and Development, 6*(2), 1.

UNFCCC Decision 4/CP.23. Report of the Conference of the Parties on its twenty-third session, held in Bonn from 6 to 18 November 2017. Bonn: United Nations Framework Convention on Climate Change (UNFCCC); 2018.

UNFCCC. Paris Agreement. United Nations Treaty Collection 2015.

Van Huong, N., Cuong, T. H., Thu, T. T. N., and Lebailly, P. (2017). Efficiency of Different Integrated Agriculture Aquaculture Systems in the Red River Delta of Vietnam. *Fisheries and Aquaculture Journal, 8*(4), 1G-1G.

Van Noordwijk, M., Agus, F., Didik Suprayogo, D., Hairiah, K., Pasya, G., Verbist, B., and Farida, A. (2004). Role of agroforestry in maintenance of hydrological functions in water catchment areas. *Proceedings: Hydrological Impact of Forest, Agroforestry and Upland Cropping as a Basis for Rewarding Environmental Service Providers in Indonesia,* 20-34.

Venkatramanan, V., and Shah, S. (2019). Climate smart agriculture technologies for environmental management: the intersection of sustainability, resilience, wellbeing and development. In *Sustainable green technologies for environmental management* (pp. 29-51). Springer, Singapore.

Wang, N., Zhang, N., and Wang, M. (2006). Wireless sensors in agriculture and food industry—Recent development and future perspective. *Computers and electronics in agriculture, 50*(1), 1-14.

Xiao, G. J., Zhang, Q., and Wang, J. (2007). Impact of global climate change on agro-ecosystem: A review. *Ying yong sheng tai xue bao= The journal of applied ecology/ Zhongguo sheng tai xue xue hui, Zhongguo ke xue yuan Shenyang ying yong sheng tai yan jiu suo zhu ban, 18*(8), 1877-1885.

Yick, J., Mukherjee, B., and Ghosal, D. (2008). Wireless sensor network survey. *Computer networks, 52*(12), 2292-2330.

Zaks, D. P., and Kucharik, C. J. (2011). Data and monitoring needs for a more ecological agriculture. *Environmental Research Letters, 6*(1), 014017.

Chapter 14

Role of Agriculture Extension for Climate Smart Agriculture

Shivani Jha[1]* and Smriti Singh[2]

[1]PHD, Punjab Agricultural University, Ludhiana, Punjab
[2]PHD, Govind Ballabh Pant University of Agriculture and Technology, Pantnagar, Uttarakhand

Climate change is probably the foremost serious environmental threat to fight against hunger, malnutrition, disease, and poverty in the world. Agriculture the most vulnerable and sensitive sector suffering from global climate change due to its dependency on local climate parameters like rainfall, temperature, soil health, etc. Agriculture has to become "climate smart" to alleviate the challenges posed by global climate change, that is, sustainably increase agricultural productivity and incomes. The adaption of climate related knowledge, technologies, and practices to local conditions promoting joint learning by farmers, researchers, extension workers, and widely disseminating CSA practices, is critical. There is a requirement for site-specific assessments to spot suitable agricultural technologies and practices needed for CSA. So extension can play a crucial role in helping the farmers to deal with the various impacts of global climate change by using the appropriate approach to make awareness and make them aware of the various adaptation and mitigation strategies.

Keywords: *Climate change, Climate smart agriculture, Climate smart, Extension, Challenges.*

Introduction

Climate change is one of the most emerging issues within the world nowadays. The climate is changing continuously no matter what investments in mitigation are made. It is a significant variation in weather patterns occurring over periods starting from decades to many years. Changing climatic conditions most severely affect agriculture because it depends on local climatic parameters like temperature, rainfall, humidity, *etc.* Climate change directly affects agricultural production as the sector is dependent on climatic conditions and is one of the most vulnerable sectors to risk and impact of global climate change. Climatic change could affect agriculture as follows: quantity and quality of crops, change of agricultural input doses such as herbicides, insecticides, and fertilizers, the intensity of soil drainage (leading to nitrogen leaching), erosion, loss, and gain of cultivated lands and organisms may become more or less competitive, like flood-resistant or salt-resistant. Overall, agricultural productivity for the whole world is projected to be between 3 and 16 per cent by 2080. To alleviate challenges posed by climate change, agriculture has got to become "climate smart", that is, sustainably increase agricultural productivity and incomes, adapt and build resilience to climate change. Climate Smart Agriculture (CSA) integrates economic, social, and environmental development by jointly addressing food security and climate challenges. Extension providers can play a major role through technology development and information dissemination, strengthening farmers' capacity, facilitation and brokering, and advocacy and policy support. Rural Advisory Services (RAS) can contribute by disseminating climate information and technologies on production practices for climate adaption through innovative approaches, like plant clinics, participatory video, climate smart villages, climate training or workshops, *etc.* So it's necessary to understand the role of extension in CSA and what are the various extension methods utilized in CSA to assist the farmers.

Agriculture in developing countries must undergo a significant transformation if it's to satisfy the growing and interconnected challenges of food insecurity and global climate change (FAO, 2010). Climate change is the major challenge that affects development in the 21st century. It is one of the major threats to humankind and affects many sectors like forestry, agriculture, the environment, and human lives as well. The croplands, pastures, and forests that occupy approximately 60 per cent of the earth's surface are progressively being exposed to threats from increased climatic variability. The challenge of rapidly boosting productivity is compounded by the present and expected impacts of global climate change Changes to precipitation and temperature, especially in marginal areas, are expected to scale back productivity and make production more erratic (Cline, 2008; Lobell *et al.*, 2008; Boko *et al.*, 2007). Ensuring that agriculture becomes climate smart may be a priority for addressing the necessity for adequate, nutritionally balanced food for a growing and more demanding population during a situation of resource limitations, and climate change and variability. Consequently, there's a requirement to simultaneously improve agricultural productivity and reduce yield variability overtime under adverse climate. A proposed means to realize this is often increased adoption of a 'Climate Smart Agriculture' (CSA) approach (FAO, 2010). CSA, which is defined by its intended outcomes, instead of specific farming practices, is composed of three main pillars: sustainably increasing agricultural productivity and incomes; adapting and building resilience to global climate change, and reducing and/or removing greenhouse emission emissions relative to standard practices (FAO, 2013a). So CSA is an important approach in agriculture to deal with the most challenging issue of the world. Much research has been conducted on the biophysical aspect of climate change but socioeconomic research regarding the impact of these CSA practices is particularly lacking. So there is a need for a strong extension network in climate smart agriculture to change the behavior of farmers or to provide them different location-specific adaptation and mitigation strategies.

Climate Change and Agriculture

Agriculture is the backbone of the financial system of most countries. Additionally to food, agriculture additionally provides employment opportunities to an oversized population. Agricultural production has continually been closely connected with variations in weather. Global climate change is projected to own vital impacts on conditions touching agriculture, as well as temperature, greenhouse emission, glacial run-off, precipitation. These conditions confirm the carrying capability of the region to supply enough food for the human population and domesticated animals. The general impact of global climate change on agriculture depends on the varied measures adopted to balance these effects. In general, global climate change may have an effect on agriculture in many ways: productivity, in terms of amount and quality of crops; agricultural practices, through changes of irrigation and agricultural inputs like herbicides, pesticides, and fertilizers; environmental effects, specifically during a relevancy frequency and intensity of soil evacuation (leading to chemical element leaching), reduction of crop diversity; rural area, through the loss and gain of cultivated lands, land speculation, land renunciation, and hydraulic amenities; and adaptation, (organisms might become a lot of or less competitive, like flood-resistant or salt-resistant forms of rice). In Africa and geographical area, several rainfed crops are close to their most temperature tolerance, thus their yields are probably to fall sharply for even tiny climate changes. In general, global climate change may have an effect on agriculture in many ways:

- ☆ Productivity, in terms of amount and quality of crops
- ☆ Agricultural practices
- ☆ Environmental effects
- ☆ Rural area, through the loss and gain of cultivated lands, land speculation, land renunciation, and hydraulic amenities
- ☆ Adaptation

Impact of Temperature Change on World Agriculture

Climate change incorporates a direct impact on food production across the globe. A rise in mean seasonal temperature will cut back the length of the various crops and hence cut back final yield. World agriculture faces a big decline inside this century due to heating. Temperature change is anticipated to extend the gap between developed and developing countries as a results of additional severe impacts in already vulnerable developing regions, exacerbated by their comparatively lower capability to new threats (Padgham, 2009).

Agriculture and Temperature Change in India

India could also be an oversized country with varied climates. Various seasons mean various crops and farming systems. There's a high dependency of agriculture on the monsoon rains and an in-depth link exists between climate and water resources. Poor rural households, whose livelihoods rely on agriculture and natural resources, can bear a disproportionate burden of adverse impacts of temperature change (Kates, 2000; architect *et al.*, 2007; Satapathy *et al.*, 2011).

Role of Rural Advisory Services

Climate change is already touching agriculture and food security. The expected effects of world temperature change, that embrace higher temperatures, additional frequent extreme weather events, water shortages, rising ocean levels, ocean natural process, land degradation, the disruption of ecosystems, and therefore the loss of diversity, may seriously compromise agriculture's ability to feed the foremost vulnerable populations and impede progress towards the reduction of hunger, deficiency disease, and poorness (FAO, 2016a). Enhancing the capacities of farmers to manage risks and adopt effective temperature change adaptation and mitigation ways deserves special attention.

Promoting climate smart agriculture involves strategies and agricultural practices supporting climate smart agriculture. RAS serves as a link between farmers and new sources of information and tools, and promoting the activity changes which will lead agricultural communities to adopt climate smart agricultural practices (Simpson and Burpee, 2014).

RAS is prepared to effectively support the promotion of climate smart agriculture among farmers for a range of reasons.

1. RAS employees have shut operating relationships with rural communities, particularly at the sector level. Solely RAS have personnel accessible on all-time to directly work with rural communities.
2. RAS understanding extends to the assets of farm households, the social dynamics inside and between communities, and farmers' adaptive capacities.
3. RAS providers can play a needed mediation role for the extensive scaling up of climate smart agriculture. This requires broader expertise and new levels of innovative thinking to harness appropriate technologies, practices, and agricultural products. It will therefore in many cases involve promoting interactions and knowledge flows among a broader range of stakeholders performing a good range of roles within the policy and practice spectrum.
4. RAS have traditionally acted as intermediaries between research and farmers.
5 During and post-natural disasters and extreme weather events, RAS often assist rural communities in coping with the crises by providing relief and engaging in rebuilding rural livelihoods post-disaster (Shepherd *et al.*, 2013). RAS contribution in promoting climate smart agriculture practices.

RAS use a variety of approaches and tools to succeed in farmers with new knowledge, including:

- Demonstrations
- Training activities
- Individual farm visits
- The training of lead farmers or farmer trainers to train others,
- Training of input and service providers,
- Group discussions,
- Exposure visits to innovative farmers
- Farmer Field Schools
- Plant clinics
- Field days
- Messages delivered through various media (*e.g.* mobile phone messaging, farm radio, participatory videos, television)

Some of innovative extension approaches used worldwide to deal with the adverse impacts of climate change are given below:

Climate Awareness Mass Media Campaigns

The high reach of mass extension campaign approaches are very attractive. As we all know, the main constraints of national extension systems are a shortage of field extension personnel and limited resources to succeed in large numbers of farmers spread widely across geographical areas. To tackle these constraints, the extension is often more efficiently performed using mass media.

Climate Training

Education/training of intermediaries/extension personnel is vital to update their knowledge associated with global climate change, their impacts, and consequences or on different adaptation and mitigation strategies. As we all know, climate smart agriculture may be a new and emerging issue so extension providers should be trained on this. University education is that the fundamental basis for any scientific education that supports the training in checking out solutions at the extension level, that Science Field Shops were established (Stigter and Winarto, 2013). The extension intermediaries trained at that level, for instance in Climate Field Schools, train the farmers in Farmer Field Schools (Integrated Pest Management Farmer Field Schools as an existing but deteriorated example), and establish climate services for agriculture with the farmers in their fields (Sala *et al.*, 2016).

Plant Clinics

The plant clinic approach is analogous to human health clinics; they're the frontline contact point of the national extension system and permit direct information exchange between extension workers and farmers on "any problem and any crop". Plant clinics are a channel for facilitating face-to-face exchange and a two-way flow of data and knowledge between extension workers and farmers and link to other components of a plant health system (Boa *et al.*, 2015). The varied crop problems delivered to plant clinics are often associated with either abiotic factors (*e.g.* nutrient deficiency, waterlogging, chemical misuse, *etc.*) or biotic factors (*e.g.* pathogens, insects, rats, *etc.*). Plant doctors should be intimate with farmers and farming conditions, speak the local language and know what inputs are available. Plant doctors are usually trained extension workers and remain employed by their current organizations.

Farmers Field Schools (FFS)

The Farmer Field School (FFS) may be a participatory, non-formal extension approach supported experiential learning that puts farmers and their demands at the center (FAO, 2002). It provides farmers with a low-risk setting to experiment with new agricultural management practices, discuss and learn from their observations, which allows them to develop new practical knowledge and skills (Settle *et al.*, 2014). FFSs have also integrated elements of global climate change adaption, like the FAO FFS program on Integrated Plant and Pest Management (IPPM) that promoted improved and adapted varieties and agroforestry practices in Mali and Niger (FAO, 2015). Climate Field Schools in Indonesia raised awareness of global climate change and promoted solutions to deal with changing rainfall patterns, like recording and interpretation of on-farm rainfall measurements and in-field water harvesting (Winarto *et al.*, 2008).

Plant Health Rally Approach

it's an extension method for quickly raising awareness about major agricultural risks or threats on important crops, marketing the utilization of improved agricultural practices, and gathering feedback from farmers on major issues that affect production. A plant health rally could also be spontaneous, attracting people with a banner and other announcements, or may target farmers who are specifically mobilized for the event. The approach also can contribute to global climate change mitigation; for instance, plant health rallies would find a perfect fit as a vehicle for putting mitigation research (such as urea deep placement technology in rice production) into use and thus reducing greenhouse emission emissions from paddy rice.

Contingency Crop Planning

It's a document that has recommendations across the key aspects of crop management and cultural practices. This type of calendar is extremely useful in terms of crop planting, irrigation scheduling, and plant protection measures for farmers. It's done on the basis of local weather for the local crops grown therein particular area. It's participatory because the local knowledge of farmers about the crops or various agricultural practices and knowledge domain of scientists both plays important role in crop planning. It's prepared with the assistance of the agro-meteorological department.

ICT Supported Network

The term Information and Communication Technologies (ICTs) played a crucial role as a medium of data and communication in global climate change awareness, adaptation and mitigation strategies. However, the supply and adoption of ICTs is varied between areas, developed and developing countries, urban and rural areas, and within rural areas themselves. The use of mobile phones, videos, radios, *etc.* was done to deal with the difficulty of global climate change by creating awareness among the farmers about the supply of various adaptation and mitigation strategies.

Farmer-to-Farmer Extension (F2FE)

F2FE is that the provision of coaching by farmers to farmers, often through the creation of a structure of farmer-promoters and farmer-trainers. F2FE offers great promise for effectively scaling up climate smart agriculture (CSA). The approach empowers farmers as change agents and helps to extend adoption because farmers are more willing to find out from their colleagues than from extension staff (Franzel *et al.*, 2015). F2FE programs contribute to all or any CSA, that is, they assist to improve productivity, build resilience and reduce greenhouse emission emissions.

Conclusion

Extension services play a key role in CSA due to their role in knowledge transfer to farmers' fields. the right mixture of different extension approaches utilized in CSA will largely depend upon factors such as the complexity of extension messages, the target population and its geographical spread, the available technology, the sort and sort of data to be collected from farmers, and lastly, on the financial means available for extension. On-the-ground implementation of extension also must go hand-in-hand with advocacy and awareness-raising of decision-makers on the approaching threat of global climate change" global climate change for agriculture to form extension more aware of climate change and contribute to deal with the triple challenge of food security, adaptation, and mitigation.

REFERENCES

Bentley, J.W., Boa, E., Van Mele, P., Almanza, J., Vasquez, D. and Eguino, S. (2003). Going Public: A New Extension Method. International Journal of Agricultural Sustainability, 1(2): 108-123.

Boa, E., Danielsen, S. and Haesen, S. (2015). Better Together: Identifying the Benefit of a Closer Integration Between Plant Health, Agriculture and One Health. In One Health: The Theory and Practice of Integrated Health Approaches, Chapter 22: 258-259.

Boko, M., Niang, I., Nyong, A., Vogel, A., Githeko, A., Medany, M., and Yanda, P. Z. (2018). Africa Climate Change 2007: Impacts, Adaptation, and Vulnerability: Contribution of Working Group II to the Fourth Assessment Report of the Intergovernmental Panel on Climate Change.

Cline, W. (2008) 'Global warming and agriculture'. Finance and Development, 45(1): 23-7.

FAO. (2002). From farmer field school to community IPM – Ten years of IPM training in Asia. Bangkok, Thailand. Food and Agriculture Organisation of the United Nations (FAO), Rome, Italy.

FAO. (2010) "Climate Smart" Agriculture: Policies, Practices and Financing for Food Security, Adaptation and Mitigation', Food and Agriculture Organisation of the United Nations, Rome.

FAO. (2013a). Sourcebook on Climate Smart Agriculture, Forestry and Fisheries, Food and Agriculture Organisation of the United Nations (FAO), Rome, Italy. Retrieved from http: //www.fao.org/climatechange/37491-0c425f2caa2f5e6 f3b9162d39c8507fa3.pdf.

FAO. (2013b). Climate Smart Agriculture: Sourcebook. Food and Agriculture Organisation of The United Nations, Rome, Italy.

FAO. (2015). Building resilient agricultural systems through farmer field schools – Integrated Production and Pest Management Programme (IPPM). Food and Agriculture Organisation of The United Nations, Rome, Italy.

IPCC. (2001) Climate change 2001: The scientific basis. Contribution of Working Group I to the Third Assessment Report of the Intergovernmental Panel on Climate Change (Houghton, J. T., Ding, Y., Griggs, D. J., Noguer, M., Van der Linden, P. J., Dai, X., Maskell, K. and Johnson, C. A. (Eds.)) Cambridge University Press, Cambridge and New York.

IPCC. (2007). Climate Impact s, Adaptation and Vulnerability. Working Group II to the 41 Intergovernmental Panel on Climate Change Fourth Assessment Report, DRAFT technical summary 2006, Intergovernmental Panel on Climate Change, Geneva.

K Mendelsohn, R., A. Basist, F. Kogan, and P. Kurukulasuriya. (2007). Climate analysis with satellite versus weather station data. Climate Change, 81: 71-83.

Kates, R. (2000). Cautionary tales: Adaptation and the global poor. Climatic Change. 45(1): 5–17.

Lobell, D. B., Schlenker, W., and Costa-Roberts, J. (2011). Climate trends and global crop production since 1980. Science, 333(6042), 616-620.

Mahato, A. (2014).Climate Change and its Impact on Agriculture International. Journal of Scientific and Research, 4(4): 3.

Nelson, G, Rosegrant, M, Palazzo, A, Gray, I, Ingersoll, C, Robertson, R, Tokgoz, S, Zhu, T, Sulser, T, Ringler, C, Msangi, S, and L You. (2010). Food Security, Farming, and Climate Change to 2050. International Food Policy Research Institute, Washington DC.

Padgham, J. (2009) Agricultural development under a changing climate: Opportunities and challenges 42 for adaptation, Joint Departmental Discussion Paper-Issue 1, Agriculture and Rural Development and Environment Departments. The World Bank, Washington DC.

Rao, V. U. M., Rao, B. B. and Khandgonda, I. R. (2011). Perception of Indian farmers on climate change. Central Research Institute for Dryland Agriculture, Hyderabad. P-27.

Sala S, Rossi F, David S. (2016) Supporting agricultural extension towards Climate Smart Agriculture: An overview of existing tools. Global Alliance for Climate Smart Agriculture (GASCA)/FAO, Italy, P- 28-30.

Satapathy, S. (2011) 'Adaptation to Climate Change with a Focus on Rural Areas and India' Deutsche Gesellschaft für Internationale Zusammenarbeit (GIZ) GmbH, India Project on Climate Change Adaptation in Rural Areas of India.

Settle, W., Soumaré, M., Sarr, M., Garba, H. and Poisot, A.S. (2014). Reducing pesticide risks to farming communities: cotton farmer field schools in Mali. Philosophical Transactions of the Rural Society B. Retrieved from http: // dx.doi.org/10.1098/rstb.2012.0277.

Stigter C. J. and Winarto, Y.T. (2013). Science Field Shops in Indonesia. A start of improved agricultural extension that fits a rural response to climate change. Journal of Agri-Food and Applied Sciences, 2(2): 112-123.

Winarto, Y.T., Stitger, K., Anantasari, E., And Hidayah, S.N. (2008). Climate Field Schools in Indonesia: Improving "response farming" to climate change. Leisa Magazine, 24.

Franzel, S., Degrande, A., Kiptot, E., Kirui, J., Kugonza, J., Preissing, J., and Simpson, B. (2015). Farmer to-farmer extension. Note 7. GFRAS Good Practice Notes for Extension and Advisory Services. Lindau, Switzerland, Global Forum for Rural Advisory Services. Retrieved from http: //www.g-fras. org/en/download. html.

Chapter 15

Role of Local Level Weather Forecasting and Agrometeorological Advisory Services for Climate Smart Agriculture

Trisha Manna[1], *Sachin Govindrao Mundhe*[2] *and Wajid Hasan*[3]

[1]*Department of Agricultural Meteorology and Physics, Bidhan Chandra Krishi Viswavidyalaya, Mohanpur – 741252, Nadia, West Bengal*
[2]*KVK-Durgapur, Amravati, Maharashtra*
[3]*Krishi Vigyan Kendra, Jehanabad, BAU, Sabour, India*

Climate change has already imperilled the livelihood of the communities who are closely related to natural resources by increasing the frequency of extreme weather events like drought, floods, storms, cyclones which are most challenging issues in our age. Climate smart agriculture aims to improve food security and helps the farming communities to adapt climate change and mitigation by appropriate practices and policies. Most of the small holder farmers who basically rely on traditional methods to predict upcoming weather condition are facing a huge challenge because the usual trends of season or weather variabilities are now suddenly changing and becoming more unpredictable. Localised weather report are very much important to take farm based decisions such as when to plant, when to irrigate, when to spray chemicals, when to harvest etc. So the weather forecasts must be not only timely, but also very accurate and should ideally be issued for small areas to maintain accuracy. The

National Centre for Medium Range Weather Forecasting (NCMRWF) under the Ministry of Earth Sciences (MoES), Government of India in collaboration with India Meteorological Department (IMD), Indian Council of Agricultural Research and State Agricultural Universities has been providing Agrometeorological Advisory Services (AAS) at block level or district level farming community based on location-specific medium-range weather forecast (MRWF). The major achievement of the study was to increase awareness among farmers about the adoption of weather based advisories and minimize the economic losses due to climate hazards.

Keywords: *Climate smart agriculture, AAS, Medium range weather forecast, Crop wise economic impact assessment.*

Introduction

Agriculture is considered as the backbone of the economic growth for many countries, especially the developing countries that generate three quarters of the world's agricultural value (FAO 2013). This sector provides livelihoods to almost 2.5 billion people, often among most vulnerable populations living in poverty. The Intergovernmental Panel on Climate Change (IPCC) has highlighted that future agricultural growth will be directed by climate change. Frequency and intensity of extreme events such as drought, heavy rainfall, flooding, high night temperatures, and water scarcity in dry regions are likely to increase significantly by the end of the century. Besides these, greenhouse gas emissions from agriculture, due to burning of crop residues, are a principal contributor to climate change. Hence, there is a strong need to initiate a parallel shift in agricultural development approaches and practices to mitigate the effects of climate change. To alleviate some of the complex challenges posed by climate change, agriculture has to be "climate smart", that is sustainably increase agricultural productivity and incomes, adapt and build resilience to climate change, and reducing greenhouse gas emissions, whatever possible. Climate smart agriculture involves agricultural production based on proper conservation of soil health and moisture analysis on a particular piece of land well before crop cultivation to know what crop and which variety to be grown. CSA also needs local (district or block level) medium range or short range weather forecast to take precautionary measures by the farmers to prevent crop loss and this has to be done with direct communication with the farmers to make agriculture more productive, less costly and more value addition with the produce before going to the market channel.

What is Climate Smart Agriculture (CSA)?

The Food and Agricultural Organisation (FAO) defines Climate Smart Agriculture (CSA) as an approach that is needed to transform and reorient agricultural systems to effectively support development and ensure food security in a changing climate scenario. It takes into account the diversity of social, economic and environmental contexts, including agro-ecological zones. Its implementation requires identification of climate resilient technologies and practices for management of water, energy, land, crops, livestock, *etc.* at the farm level. Three main pillars of climate smart agriculture (CSA) are (FAO, 2013):

1. Sustainable increase in production of agricultural crops, livestock and farmer's income,
2. Reducing exposure of the farmers to short term risks and
3. Reducing the emission of greenhouse gases and building resilience to climate change.

But many technologies and practices, including those qualifying as climate smart, are not achieving their full potential impact because of low levels of adoption by farmers in developing countries. The key challenge is to maintain promising pilot initiatives so that they can have a substantial impact on poverty.

Approaches of Climate Smart Agriculture

Climate smart agriculture is not a single technology or a single practice; rather it is an integrated approach to synchronize economic, social and environmental activities to address food security and climate change issues. All the components of CSA are soil, water, crop, livestock, weather, energy and livelihood that must be well balanced to each other for better synchronization in creating climate smart agriculture (Figure 15.1).

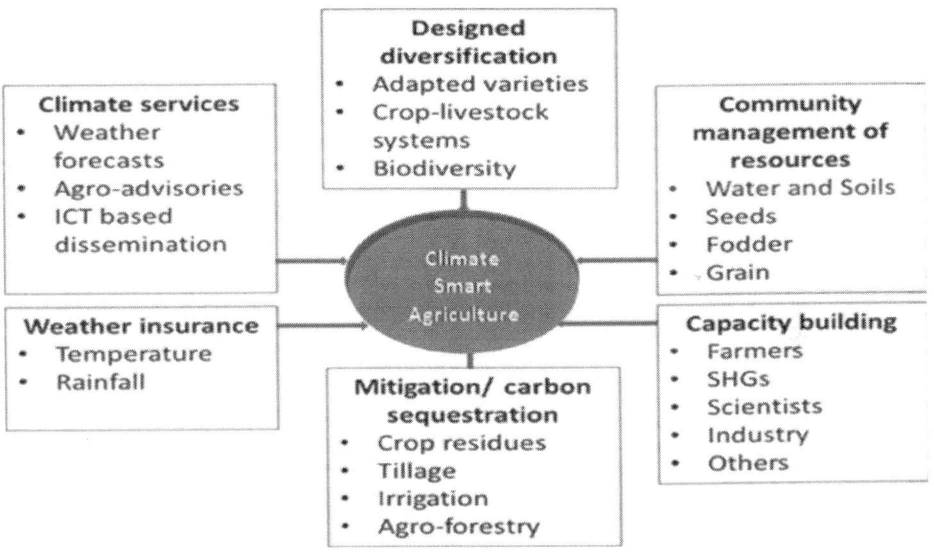

Figure 15.1: Different Components of Climate Smart Agriculture (*Source*: Bhattacharyya *et al.*, 2020).

Crop Smart Agriculture

Crop smart agriculture is implemented to adopt the challenges faced due to climate change. Predicted impact of future climate change on the production of cereal crops showed reduction of 35 per cent, 20 per cent, 50 per cent, 13 per cent and 60 per cent in case of rice, wheat, sorghum, barley and maize respectively (Porter *et al.*, 2014). The way out of these issues is introduction of abiotic stress tolerant varieties, higher input use efficient cultivars, use of short duration cultivars, mixed cropping, intercropping, changing in planting time and planting methods *etc.* The entire goal under crop smart agriculture is to promote such crop management practices that targets positive feedback towards sustainable food production and negative feedback towards changing climatic conditions.

Water Smart Agriculture

Improving water harvesting, water use efficiency and water retention are essential for increasing productivity against increasing untimely rainfall. Agriculture in India is mainly rain-fed and rain fed crops account for 48 per cent of total area under food crops and 68 per cent under non-food crops. In-situ moisture conservation, ground water recharge measures, water harvesting and recycling technologies, bunding and levelling, making of dams and watersheds, micro-irrigation technology (sprinkler and drip) *etc.* are the main constructive ideas for water smart agriculture.

Weather Smart Agriculture

It is a known fact that weather plays a dominant role in year to year variations in crop production; both in rain fed or irrigated farming. Though complete avoidance of farm losses due to weather is not possible, losses can be minimized to a considerable level by making adjustments through timely agricultural operations based on accurate weather forecasts. Hence a weather based agro advisory will better interpret that how the weather parameters of the present and in future will affect crops, livestock and farm operations and suggests actions to be taken. India Meteorological Department has started issuing block-level weather forecast since 2014. This project called Gramin Krishi Mausam Sewa (GKMS) delevers Agromet Advisory Services (AAS) at district or block level used to prepare bi-weekly weather forecasts since 2008, thus making use of the medium range weather forecast and disseminating those agro advisories to the farm level users through bulletins (Figure 15.2a-b), text messages (in pre registered mobile numbers only), voice calls, whatsapp groups and also through radio, television, print media and state agricultural university website.

Nutrient Smart Agriculture

To avoid unnecessary and untimely application of plant nutrients is a key approach towards climate smart agriculture by mitigating emission of greenhouse gas, restoring soil physico-chemical and biological properties, improving soil health. The major approaches includes green manuring, organic fertilizer, composting, vermicomposting, site specific integrated nutrient management, use of leaf colour chart for nitrogen application, issuing soil health card for better nutrient management *etc.* leaf colour chart (LCC) based nutrient management in wheat and maize; SPAD reading based nitrogen management in rice are some of the examples of nutrient smart agriculture.

Climate Smart Village (CSV) Selection: First Step Towards CSA

CSV is a community approach where farmers, researchers, stakeholders

NATIONAL CENTRE FOR MEDIUM RANGE WEATHER FORECASTING
District Agromet Advisory Service
Regional Research & Technology Transfer Station (RRTTS)
Western Undulating Zone , Bhawanipatna
Dist : Kalahandi, Orissa – 766001
(Orissa University of Agriculture & Technology)
Letter No.85(27)/RRTTS, Dt.25.10.2013

Weather Condition during Last Week 18.10.13 to 24.10.13

Rainfall (mm)	61.6
Maximum temperature (^{0}C)	26.0° C to 33.0° C
Minimum temperature (^{0}C)	20.0° C to 21.0° C
RH (%)	Avg. Maximum RH 91.2% and avg. minimum RH 79.6%
Cloud cover (Octa)	The sky remained partly cloudy from 18.10.13 to 20.10.13 and mainly cloudy on 21.10.13 to 24.10.13.

WEEK NO.43

Bulletin No.85/2013 **AGROMET ADVISORY BULLETIN** **Friday/Dt.25.10.2013**

Bhawanipatna

Weather Forecasting Up to: 30.10.2013
Kalahandi District: According to IMD Bhubaneswar 28.0mm cumulative rain of light intensity is expected from 26.10.2013 to 30.10.13. Sky will remain over cloudy on 26.10.13 to 28.10.13 and mainly cloudy on 29.10.13 and partly cloudy on 30.10.13. Wind will blow from N-E and S-E direction with speed of 003- 005 km/hr. Maximum temperature will vary around 28-32^{0}C and minimum temperature will be around 18-20^{0}C. Maximum RH will vary between 96-98% and minimum RH will vary between 66-72%.

Weather parameter	26/10	27/10	28/10	29/10	30/10
Rainfall (mm)	7	3	7	7	4
Max. Temperature (^{0}C)	28	29	30	31	32
Min. Temperature (^{0}C)	20	19	20	18	18
Total cloud cover (octa)	8	8	8	6	3
Max. Relative Humidity (%)	98	96	98	98	97
Min. Relative Humidity (%)	72	72	66	66	68
Wind speed (Kmph)	004	003	004	003	005
Wind direction(deg)	135	93	93	49	34

Agro met - Advisory (For Kalahandi District)

1. Go for transplanting of *rabi* vegetable crops.
2. Keep water in paddy field upto 2.5 to 5.0cm upto milking stage. Maintain saturation at grain filling stage. Drain water at ripening stage.
3. Hharvest maize and cotton and Carry the harvested produce to safe place and stop irrigation to paddy and sugarcane as light rain is expected during this period.

Kharif paddy:

1. The direct seeded paddy crop is at ripening stage. Drain water from field at ripening stage. Harvest the crop when 80% of grain ripens.
2. The transplanted paddy crop is at thirteen to fifteen week stage. The medium and late duration paddy is at flowering to milking stage and early duration paddy is at grain filling to ripening stage. Drain water from field at ripening stage. Harvest when 80% of grain ripens.
3. Stem borer attack has been noticed during this period in paddy. Spray with monocrotophos / chloropyriphos @2ml/lit water after.
4. To control BPH spray buprofezin 25% SC 800ml/ha.
5. Sheath blight infestation occurs during this period. To control sheath blight in paddy spray hexaconazole 1000ml/ha.

Sugarcane:

1. The sugarcane plant is at grand growth stage (thirty three to thirty five week stage).
3. Red rot disease in sugarcane occurs at grand growth stage. Bunding of affected field should be done to avoid movement of rain or floodwater from affected field to non affected field. Destroy trashes from the infected plant by burning and spray carbendazim @2g/lit of water. Severely affected plants or canes should be uprooted and destroyed.

Figure 15.2a: Samples of Agro-advisory Bulletins.

and policymakers collaborate to select the most appropriate technological and institutional interventions on the basis of local conditions to increase productivity and incomes to achieve climate resilience and enable climate mitigation. The CSV approach brings together the farmers, policymakers, researchers and local organisations (KVKs or gram panchayet) to work on different kinds of climate smart package and practices with a view to adapt agriculture to climate change (Aggarwal *et al.*, 2013). The CSV approach was first initiated in Haryana and Bihar

AAS Bulletin for Prakasam district issued by GKMS,
RARS, Lam, Guntur-34,
Acharya N.G. Ranga Agricultural University

BULLETIN No-53 /2015 ISSUED ON 30-06-2015

Weather forecast for next five days for Prakasam district

Weather Parameter	01/07	02/07	03/07	04/07	05/07
Rainfall (mm)	3	0	0	0	0
Max Temperature (deg C)	37	35	34	35	36
Min Temperature (deg C)	28	26	25	26	27
Total cloud cover (octa)	7	6	7	5	7
Max Relative Humidity (%)	72	68	64	66	60
Min Relative Humidity (%)	46	42	38	34	36
Wind speed (kmph)	22	24	26	30	30
Wind direction (deg)	270	270	270	270	270

AGRO METEOROLOGICAL ADVISORY FOR PRAKASAM DISTRICT, A P

Crops	Stage/ Condition	Pest and Disease/problem	Weather based Agro advisories
Paddy	*kharif* season	Seed born and soil born diseases Varietal selection & land leveling, filed preparation etc.,	• Presently received rains suitable for field preparation and nursery preparation, collect suitable and good quality seed materials from reputed firms/ farmer friends. varieties suitable are BPT-5204, cotton dorasannaalu, vijetha, IR-64 and Bavapuri sannaalu • Presently received rains may useful for sowing of nurseries • Strictly follow the seed treatment with carbandism @ 2.5 gramms per kg seed • Under directed seeded rice, immediately after seed sowing apply pendimethaline @ 5 ml per liter of water. • If crop is at 3-4 leafs stage, Apply to 2% urea spry on standing nursery crop to overcome present hot or dry weather condition

Figure 15.2b: Samples of Agro-advisory Bulletins.

in 2011. Later 27 CSVs from Nilokheri, Indri, Gharaunda and Nissing blocks in Karnal district of Haryana were piloted for CSA.

Keeping the above facts Technology Demonstration Component has been undertaken as network project of Indian Council of Agricultural Research (ICAR) under National Innovations in Climate Resilient Agriculture (NICRA) in 121 KVKs in 11 zones of India. The project is implemented by Krishi Vigyan Kendras (KVKs) at district level, regionally coordinated by the Agricultural Technology Application Research Institutes (ATARIs) with overall planning and monitoring and coordinated by Central Research Institute for Dryland Agriculture (CRIDA), Hyderabad. Based on primary scientific analysis about the agro-climatic conditions in-situ problems of the soil and water of that particular village, action plan for each selected village was prepared by the multi-disciplinary team of scientists involved in the process.

Weather Forecasting Tools towards CSA

The weather forecast and early warning systems is very much important to minimize the threats of agricultural losses due to weather aberration. Many small holder farmers in developing countries still face huge challenges for climate change induced weather events. Most of those farmers believe traditional methods to know weather predictions. Localised weather data is important for farm based decisions like when to plant. A complete weather based agro advisory (supported with agrometeorological database, crop conditions, real time weather, research results on crop-weather relationships) can give an interpretation of how the weather parameters of the present and in future will affect crops, livestock and farm operations. For meeting this requirement, 100 automatic weather stations (AWS) was installed in 100 most climate change vulnerable districts of India under the NICRA project districts. The 100 locations were identified representing various climatic vulnerabilities. Subsequently the Agrometeorologists develop the Agromet advisory bulletins with the help of Subject Matter Specialists at KVKs using the field level crop information blended with weather forecasts and communicated to the Field Information Facilitator (FIFs) who pass on the bulletins to farmers. Mass media often suits simpler messages while direct interactions with the farmers through farmer field schools are often simpler for complex knowledge (Figure 15.3).

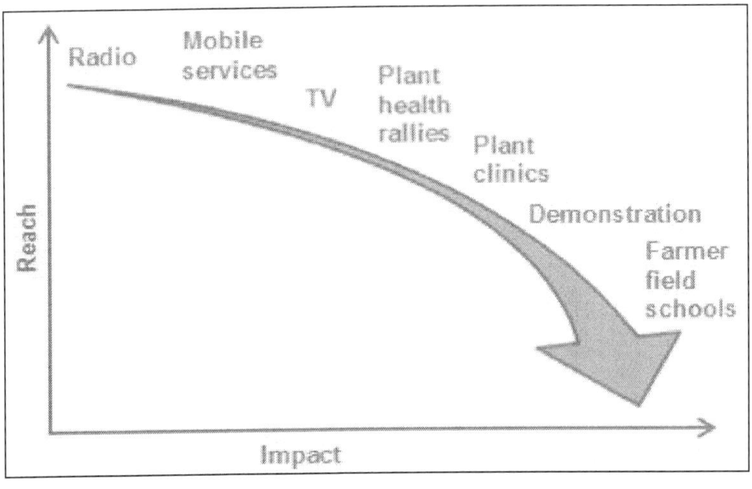

Figure 15.3: Impact of Extension Approach (Saha *et al.*, 2016).

Choice of approach has influence on the ability of extension services to contribute to food security and income, adaptation and resilience, and climate change mitigation. They generally act as the interface between farming community and resource persons of KVKs or NICRA project professionals. The feedback obtained from the farmers' side is also evaluated for improving as well as expanding services for the benefit of farming community. The working procedure has been depicted in Figure 15.4.

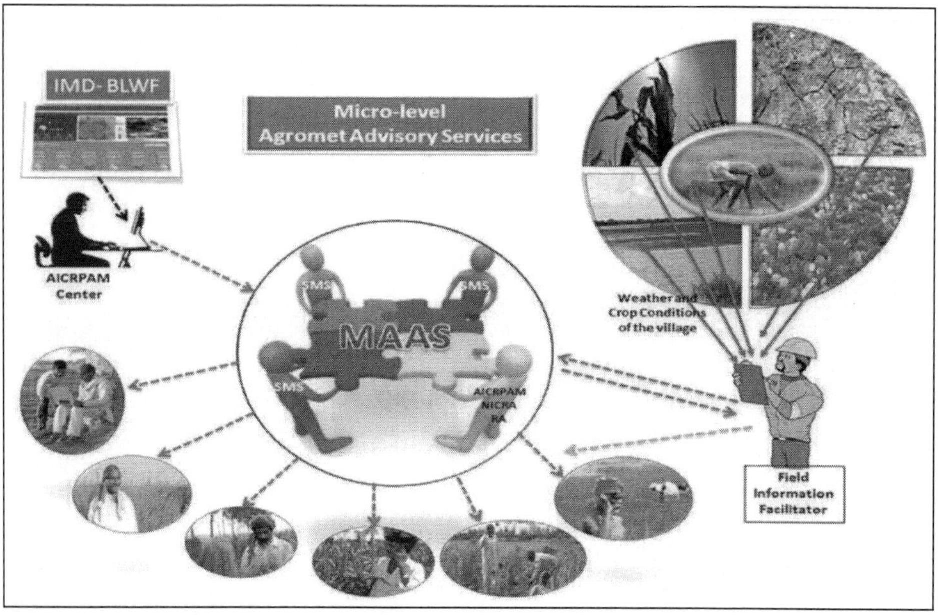

Figure 15.4: Working Procedure of Micro Level Agro Advisory Service.

An extensive analysis was undertaken among the AAS user and non user farmers and the comparative results of the benefit: cost (B: C) has been presented in Table 15.1 (NICRA Research highlights 2016-18).

Table 15.1: B : C Ratio of AAS User and Non User Farmers (*Source*: AICRPAM-NICRA Annual Report, 2016-2018)

Region	*Village*	*Crop*	*B:C Ratio*	
			AAS-farmer	*Non-AAS farmer*
2016-17				
Akola, Maharashtra	Yelgaon	Soybean	2.17	1.76
Bangalore, Karnataka	Patrenahalli and Nayanahalli	Grape	9.21	5.45
Palampur, Himachal Pradesh	Dhamrol	Maize	2.74	0.73
Parbhani, Maharashtra	Shekta	Cotton	2.26	1.88
Udaipur, Rajasthan	Nakli	Maize	1.93	1.28

Region	Village	Crop	B:C Ratio	
			AAS-farmer	Non-AAS farmer
Anantapur, Andhra Pradesh	Yagantipalle	Pigeon pea	3.1	2.48
Kovilpatti, Tamil Nadu	Allikundam	Okra	2.41	1.98
2017-18				
Ramanathapuram, Tamil Nadu	Malangudi	Chilli	1.99	1.46
Kanpur Dehat, Uttar Pradesh	Baghpur	Rice	1.71	1.50
Sirsa, Haryana	Farwain Kalan	Cotton	1.71	1.33

The higher B: C ratio obtained by the AAS users is certainly because of the timely implementation of weather based agro advisories with more accuracy to the forecasted medium range weather forecast that has made different farming operations (irrigation, applying fertilizers, plant protection measures *etc.*) easier and minimizing financial loss.

Impact Analysis of Weather Forecasting and Agro Advisory Towards Climate Smart Agriculture

The main impact analysis indicators mostly used are (a) ability of the weather forecast to influence decision of farmers and adopt appropriate measures to reduce weather-induced losses, and (b) economic and other benefits due to farmers due to change in their farm management decisions due to timely weather forecast and the benefits are in the form of farming cost minimization and with reduction in yield loss. The difference in input use, cost management and yield are analyzed for statistical significance. All the parameters over which these comparison studies are done are given in the Table 15.2.

Table 15.2: List of Farm Level Impact Analysis Parameters

Sl.No.	Impact Area	Parameters
1.	Awareness about agro advisory services	Percentage of AAS user and non user farmers
2.	Usefulness of weather forecast	Percentage of farmers found AAS bulletin useful or not, either forecasted weather matches the original weather condition
3.	Information disseminated	Percentage of farmers using forecasted weather information
4.	Impact on farmer's decision taking about farm input use management	Field preparation, irrigation, spraying of plant protection chemicals, labour charges between AAS user and non user farmers
5.	Cost of cultivation	Difference in total paid out cost, Benefit: cost ratio of user and non user group of farmers
6.	Profit generated	Net profit generation due to improved farm management decisions and due to gaining higher yield

Procedure of Weather Forecasting under District Level Agro-met Advisory Services

India Meteorological Department (IMD) is continuously improving the real time meteorological observations and provides the current and future weather data for the users. Upper air observations comprises data from radiosonde, pilot-balloon, Doppler radar *etc.* In numerical weather prediction model the Doppler radar data in digital form can directly be assimilated which permits the forecaster to analyse different weather situations like thunderstorm, hailstorm, cyclone, heavy rainfall and now we can predict very precisely the track and the actual point of landfall of tropical cyclone. Besides all these general meteorological observations, IMD is also maintaining a network for agro –meteorological observations in farm environments. The automatic weather stations (AWS) have already been installed in 127 Agrometeorological Field Units (AMFU) all over the country. All these data are being received in AWS laboratory, Pune, India on regular basis. IMD has implemented Multi-model Ensemble (MME) based district level forecasts since 2008 as a part of agromet advisory service where five numerical weather prediction (NWP) models has been included namely: (a) IMD GFS T574, (b) ECMWF T799, (c) JMA T899, (d) UKMO and (e) NCEP GFS. The model outputs are interpolated at uniform grid size resolution of 0.25°× 0.25° lat/long. The high resolution gridded rain gauge data are produced regularly at National Centre of IMD, Pune. The ensemble forecasts (day 1 to day 5) are then used for generating district level forecasts by taking average of

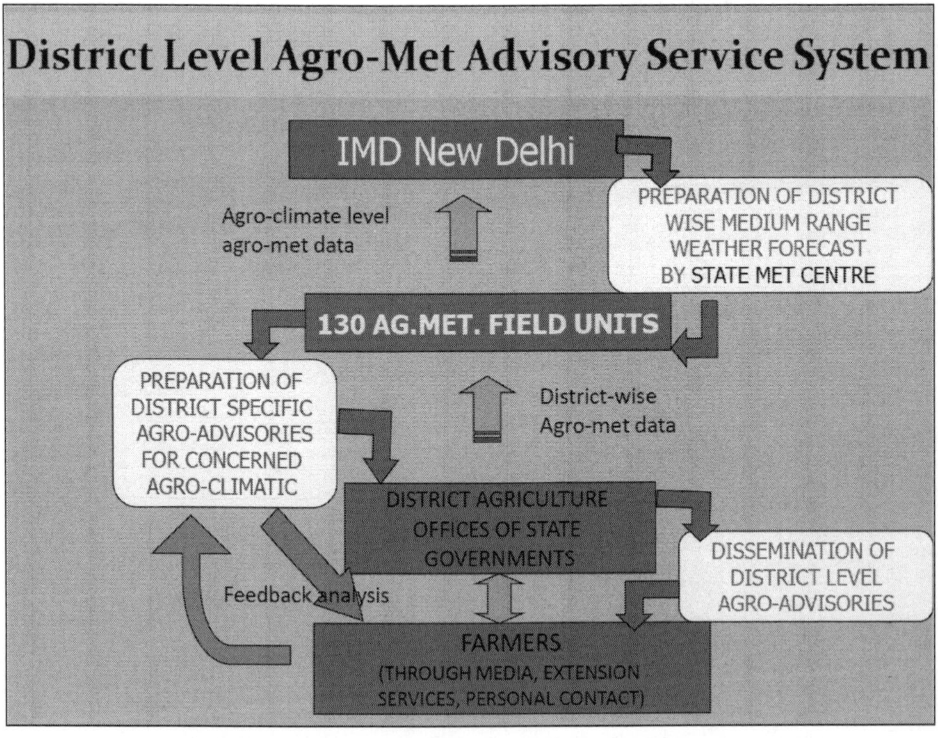

Figure 15.5: Working Process of District Level Agro Advisory Service.

all grid points falling over a particular district. The products comprises of forecast of seven weather parameters *viz.* maximum and minimum temperature, rainfall, wind speed, wind direction, relative humidity and cloudiness. These forecasted products are disseminated to Regional Meteorological Centres of IMD situated at different states and after value addition to those forecasts the medium range weather forecasts are communicated to 130 Agro-met Field Units (AMFU) on Tuesday and Friday every week. The whole procedure has been depicted in Figure 15.5.

Crop-wise different Case Studies of Economic Impact Assessment of Agromet Advisory Services (AAS)

1. Cotton

Bt cotton var Ajeet-155 was grown in Kanshivani, Akola district NICRA village during *kharif* season, 2018 (Vijay *et al.*, 2019). During the whole crop growing period several AAS bulletins were issued which was followed by a group of farmers (AAS-users). Table 15.3 illustrates us the B: C ratio obtained by the AAS user and non-user group of farmers. The higher profit obtained by the AAS user farmers is attributed to timely decision making for farm activities according to crop based agro meteorological advisories, timely and need based application of irrigation, timely harvest of the crop in rain or other weather calamity free time and its safe drying and storage. From the table it is clear that there is a clear difference in seed cotton yield and their final pricing (due to quality of the harvested product) among user and non user farmers.

Table 15.3: Analysis of B : C Ratio in Cotton Cultivation in Kanshivani NICRA Village by AAS User and Non User Farmers (Name of the farmers also mentioned)

Input Details (Rs. ha^{-1})	*AAS Farmers*			*Non AAS Farmers*		
	Punjaji Waghmare	*Pramod Shelke*	*Chandu Waghmare*	*Ramdas Pathak*	*Manik Waghmare*	*Avdhutrao Kale*
Land preparation	4350	3950	3600	4850	3800	4350
Fertilizer	5280	5260	5260	5380	5160	5160
Seed cost	4620	4720	4820	4815	4915	4715
Planting	2700	2800	2800	2800	3240	2800
Gap filling	1420	1420	1420	1420	1200	1420
Weeding	3650	4010	4370	4730	4250	3830
Hoeing(*with furrow opening)	*1900	*1900	*1900	1700	1700	1700
Plant protection	4270	4025	4325	4860	4860	3910
Irrigation	900	900	900	0	0	0
Spraying of 2 per cent Urea and 2 per cent DAP	1000	1000	1000	0	0	0
Miscellaneous	1300	1200	1400	1800	1600	1700
Harvesting cost	6500	6050	6500	5750	5750	5000
Cost of cultivation	37890	37235	38295	38105	36475	34585

Input Details (Rs. ha⁻¹)	*AAS Farmers*			*Non AAS Farmers*		
	Punjaji Waghmare	*Pramod Shelke*	*Chandu Waghmare*	*Ramdas Pathak*	*Manik Waghmare*	*Avdhutrao Kale*
Seed cotton yield	14.8	13.9	14.8	13.1	13.1	11.8
Price	71040	66720	71040	62880	62880	56640
Net Profit	33150	29485	32745	24775	26405	22055
Benefit cost (B:C ratio)	1.87	1.79	1.86	1.65	1.72	1.64

Source: AICRPAM, NICRA Annual report 2018-19.

2. Soybean

Soybean variety JS-335 was grown during *kharif* season in 2018 in the NICRA village Warkhed of Akola district in 5.5 acres of land (Vijay *et al.*, 2019). During the crop duration, a number of AAS bulletins were issued. Later the B: C ratio of the AAS user and non user farmers were analyzed (Tables 15.4a and 15.4b) according to the feedback collected from the farmers. Higher profit by the AAS user group of farmers was obtained due to timely management of farm operations according to the issued weather forecast, timely weeding and hoeing for better crop performances,

Table 15.4a: Analysis of B : C Ratio in Soybean Cultivation in Warkhed NICRA Village by AAS User and Non User Farmers (Name of the farmers also mentioned)

Input Details (Rs. ha⁻¹)	*Narayan Avhade*	*Prakash Avhade*	*Sahadev Tople*	*Mahadev Kondankar*	*Shivdas Rathod*
Land preparation	3810	4260	2960	3810	3430
Fertilizer cost	5255	4565	4565	5255	5255
Seed cost	4839	4935	4305	4906	4705
Planting cost	1350	2700	2500	1350	1350
Weeding	1300	1600	1900	1300	1900
*Hoeing	940	900	900	940	940
Plant protection	1855	2400	2000	1920	1855
Irrigation	600	800	800	600	800
Foliar spray of 2 per cent urea	450	450	450	450	450
Miscellaneous	0	1100	1100	0	0
Harvesting cost	2000	2700	2700	2000	2000
Threshing cost	2272	2600	2585	2183	1591
Cost of cultivation	24671	29010	26765	24714	24276
Seed yield	16.9	19.2	19.1	17.43	14.05
Price of soy bean	47320	53760	53480	48804	39340
Net Profit	22649	24750	26715	24090	15064
Benefit cost ratio	1.92	1.85	2.00	1.97	1.62

Source: AICRPAM, NICRA Annual report 2018-19.

immediate drainage of excess water after rainfall, timely taken pest management practices, harvesting on time, so less loss at the time of harvest or storage. Due to the same reasons, 2 to 3 times more income has been generated from soybean farming (var. JS-9650) by the AAS user farmers in Kheda village under NICRA project in Rajsamand district in Udaipur, during *kharif* season in 2019 (Table 15.4c).

Table 15.4b: Analysis of B : C Ratio in Soybean Cultivation in Warkhed NICRA Village by AAS Non User Farmers (Name of the farmers also mentioned)

Input Details (Rs. ha^{-1})	*Mahadev Dange*	*Anil Avhade*	*Omkar Waghmare*	*Madhukar Shinde*	*Sunil Avhade*
Land preparation	3660	4810	3810	3810	4760
Fertilizer cost	4665	5255	5405	4805	5330
Seed cost	4415	5676	5040	4839	5375
Planting cost	2320	1350	1350	1350	1350
Weeding	1900	1150	1750	1450	1150
Hoeing	800	940	940	900	940
Plant protection	2100	1855	1855	2395	1775
Miscellaneous	1200	0	0	1100	0
Harvesting cost	3200	1700	2300	2000	2000
Threshing cost	2000	1364	1527	1746	1355
Cost of cultivation	26260	24100	23977	24395	24035
Seed yield	15.2	12.7	11.75	14.2	12.2
Price of soy bean	42560	35560	32900	39760	34160
Net Profit	16300	11460	8923	15365	10125
Benefit cost ratio	**1.62**	**1.48**	**1.37**	**1.63**	**1.42**

Source: AICRPAM, NICRA Annual report 2018-19.

Table 15.4c: Analysis of B : C Ratio in Soybean Cultivation in Kheda by AAS Non User Farmers

Input Details	*AAS Farmers*	*Non-AAS Farmers*
Field preparation and sowing cost (ha^{-1})	Rs. 9375	Rs.9375
Seed cost (ha^{-1}) @50/per kg for AAS and @30 kg for non-AAS	Rs. 3750	Rs.2250
Seed treatment	-	-
Fertilizer cost (ha^{-1}) N:P (40:40 kg/ha)	Rs.2400	Rs.2160
Herbicide + labour cost (ha^{-1})	Rs.1800	Rs.1800
Labour cost (Drainages and application of fertilizer, and irrigation) (ha^{-1})5 men days	Rs. 1500	Rs.1500
Harvesting (Cutting, Threshing and Transport) (ha^{-1}) 36 man days	Rs.10800	Rs.10800
Cost of cultivation (ha^{-1})	Rs.29625	Rs.27885
Soy bean grain yield (q ha^{-1})	20	10
Soybean stover yield (q ha^{-1})	20	24

Input Details	AAS Farmers	Non-AAS Farmers
Total income (ha^{-1})	Rs.64000	Rs.34800
Net profit (ha^{-1})	Rs.34375	Rs.6915
Benefit cost ratio	1.16	0.25
Cost of grain and stover = Rs 3000 and Rs 200		

Source: AICRPAM, NICRA Annual report 2018-19.

3. Wheat

Banpurwa village of Bahraich district, Uttar Pradesh is a popular wheat growing region of India. Farmers were categorized in small, medium and large and provided with agromet advisory services during wheat growing season. Selected farmers received agromet advisory through personal contact, newspaper, SMS *etc.* The average yield harvested by AAS user and non user farmers is given in Table 15.5 (Vijay *et al.*, 2019).

Table 15.5: Comparison of Wheat Yield Harvested by AAS User and AAS Non User Farmers in Banpurwa Village, District Bahraich, UP

Sl.No.	Farmers Category	Average Yield (kg/ha)		Increased Yield (per cent)
		AAS	Non-AAS	
1.	Large Farmers	4079.0	3648.0	11.8
2.	Small Farmers	3811.0	3454.0	10.3
3.	Marginal Farmers	3563.4	3328.0	7.0
	Average	**3833.3**	**3476.6**	**9.7**

Source: AICRPAM, NICRA Annual report 2018-19.

The results clearly revealed that besides better quality seed and timely management of field operations, timely providence of agromet advisory has been proven responsible for an overall increase of wheat yield by 9.7 per cent by the AAS user farmers over non user farmers. The comparison of B: C ratio between AAS user and non user farmers has been depicted in Table 15.6.

Table 15.6: Comparison of B : C Ratio by AAS User and Non User Farmers in Wheat Farming in Village Banpurwa, District Bahraich, UP

Sl.No.	Status of Farmer	Cost of Cultivation (Rs)	Gross Return (Rs)	Net Return (Rs)	B : C Ratio
1	AAS	30929	101000	70071	**1:2.26**
2	Non-AAS	35423	87900	52477	**1:1.48**

Source: AICRPAM, NICRA Annual report 2018-19.

This higher B:C ratio achieved by AAS farmers was made possible by conserving moisture in upper layer of the soil and also due to saving the cost of seed, labour, irrigation, insecticide/pesticides which finally resulted in better performance of the crop and reduction in cost of cultivation, as compared to non AAS farmers.

4. Paddy, Banana and Coconut

Twenty farmers each from the panchayats *viz.* Pananchery and Madakkathara (AAS users) and Puthur and Ollikkara (AAS non user) were selected for the economic impact assessment study on rainfed paddy, irrigated banana and coconut for a period from 2003-04 to 2006-07 by Rao and Manikandan (2008). Grain yield of paddy by AAS user and non users has been produced in Table 15.7. An average increase of 10.27 per cent in case of AAS user farmers over non user farmers with economic gain of Rs. 2772/- per ha irrespective of season has been reported. This was attributed to timely implementation of agro advisories. While leaf folder is an economic pest during *kharif* season in Kerala, AAS user farmers reaped higher yield compared to non user farmers due to timely taking actions against the insect pest following the advisories and thus yield increment was 6.3 to 9 per cent during *kharif*.

Table 15.7: Yield and Economic Gain of AAS User and Non User Farmers in Paddy Farming in Kerala

Season/ Crop	*Yield (q ha⁻¹)*			
	AAS Farmers	*Non-AAS Farmers*	*Per cent Increase in Yield*	*Economic Gain (Rs. ha⁻¹)*
		Kharif-paddy		
2004	28.0	26.0	7.1	2290.0
2005	27.0	25.3	6.3	1776.0
2006	27.8	25.3	9.0	2125.0
Average	27.6	25.5	7.6	2064.0
		Rabi-paddy		
2003-04	36.5	29.5	19.2	2932.0
2004-05	30.0	27.8	7.3	2987.0
2005-06	33.0	28.5	13.6	3581.0
2006-07	31.8	28.8	9.4	3363.0
Average	**32.8**	**28.7**	**12.4**	**3216.0**

Source: Rao and Manikandan (2008).

There was an increase of 10.6 per cent in case of AAS user farmers with economic gain of Rs. 24667/- per ha as compared to AAS non users in banana farming in Kerala. Though the per cent yield increase of banana was relatively lower (6.4 per cent) during 2004-05 *Rabi* season, the economic gain was highest (Rs. 40968/-) (Table 15.8).

Table 15.8: Yield and Economic Gain of AAS User and Non User Farmers in Banana Farming in Kerala

Season/ Crop	*Yield (q ha^{-1})*			
	AAS Farmers	*Non-AAS Farmers*	*Per cent Increase in Yield*	*Economic Gain (Rs. ha^{-1})*
Rainfed-banana				
2004	228.0	204.5	10.3	10479.0
2005	252.5	225.0	10.9	10340.0
2006	245.0	215.0	12.2	15365.0
Average	241.8	214.8	11.1	12061.0
Irrigated banana				
2003-04	317.3	278.8	12.1	37690.0
2004-05	312.5	292.5	6.4	40968.0
2005-06	276.8	244.3	11.7	33060.0
Average	**302.2**	**271.9**	**10.1**	**37239.0**

Source: Rao and Manikandan. (2008).

Like paddy, percentage yield increase was stable irrespective of irrigated farming or rainfed banana farming. In case of AAS user farmers, due to timely acceptance of agro advisory about pseudomonas weevil based on prevailing weather and banana growth stages and occurrences of *sigatoka* leaf spot disease with the onset of southwest monsoon helped in better controlling of these pest and diseases resulted better yield. Banana farmers were advised for propping strengthening for protection against heavy rainfall and high wind. This particular advisory helped a lot for the banana growers to restrain heavy damage from rainfall and high wind. Overall banana farmers appeared to be economically sounder as compared to paddy and coconut farmers due to stable yield and higher market price with taking intensive care of the plantation following agro advisories.

Starting from 2003-04 to 2005-06, coconut farmers who have followed AAS harvested more yield than AAS non users and average yield increment was 11.1 per cent over three seasons (Table 15.9).

Table 15.9: Yield and Economic Gain of AAS User and Non User Farmers in Coconut Farming in Kerala

Season/ Crop	*Coconut Yield (q ha^{-1})*			
	AAS Farmers	*Non-AAS Farmers*	*Per cent Increase in Yield*	*Economic Gain (Rs. ha^{-1})*
Rainfed-banana				
2003-04	13460	11025	18.1	4810.0
2004-05	12343	11443	7.3	2888.0
2005-06	12431	11438	8.0	3644.0
Average	**12745**	**11302**	**11.1**	**3781.0**

Source: Rao and Manikandan. (2008).

Timely agro advisory on proper drought management practices like husk burial, basin mulching, in-situ soil moisture conservation, irrigation management *etc.* with proper weather forecasting helped the AAS user farmers to get more coconut yield. Though husk burial is a labour intensive process and for this AAS farmers have to spend more than AAS non user farmers, but this process could mitigate the adverse effect of prolonged dry spell and can skip almost eight irrigations, and with this the yield increase of coconut was 7-8 per cent for AAS farmers than non AAS farmers.

5. Green Gram, Maize

Recent study with green gram and maize was conducted in Regional Research and Technology Transfer Station, Keonjhar during 2015-16 by Ray *et al.*, 2017. A total number of 60 AAS user farmers and the same number of non user farmers were randomly selected. Total cost of cultivation, crop yield, net returns and variation of net returns by AAS user and non user farmers from green gram were presented in Tables 15.10 and 15.11 respectively. It was revealed that cost of cultivation of green gram for AAS user farmers was more than non user farmers, but grain yield generated was higher (25 per cent) for AAS user farmers than non user farmers.

Table 15.10: Economics of Summer Green Gram by AAS and Non-AAS Farmers during 2014-15 and 2015-16 (Cost: Rs ha^{-1})

Particulars	*AAS Farmers*	*Non- AAS Farmers*
Seed	750	750
FYM	1800	1800
Seed treatment	500	-
Fertilizers and micronutrients	2170	3000
Pesticides	1000	1500
Weedicide	450	450
Human labour	6655	6055
Machine labour	2000	2000
Bullock labour	450	450
Irrigation	1000	500
Spraying of 2 per cent Urea solution at pre and post flowering	400	-
Grain yield (q/ha)	12.5(25)	9.0

Figure in parenthesis indicates per cent increase over control.

Source: Ray *et al.*, 2017.

Table 15.11: Economics of Green Gram Farming (Cost: Rs ha^{-1})

Type	*Cost of Cultivation*	*Gross Return*	*Net Return*	*B : C Ratio*
AAS Farmers	17175	39600	22425	**2.31**
Non AAS Farmers	16505	29700	13195	**1.80**

Source: Ray *et al.*, 2017.

Total cost of cultivation, green gram yield, net return and B: C ratio were 17175 Rs.ha^{-1}, 39600 Rs.ha^{-1}and 22425 Rs.ha^{-1}and 2.31, respectively in case of AAS farmers and 16505 Rs.ha^{-1}, 29700 Rs.ha^{-1}and 13195 Rs.ha^{-1}and 1.8 in case of non-AAS farmers (Table 15.11).

The total cost of cultivation, total yield and net returns from maize was presented in Tables 15.12 and 15.13 respectively. The total cost of cultivation, main product, net returns and B: C ratio were 26998 Rs.ha^{-1}, 60,000 Rs.ha^{-1}, 33002 Rs.ha^{-1}and 2.2 respectively in case of AAS user farmers and 25498 Rs.ha^{-1}, 47000 Rs.ha$^{-1,}$ 21502 Rs.ha^{-1} and 1.84 in case of AAS non user farmers.

Table 15.12: Economics of *Kharif* Maize grown by AAS User and Non User Farmers during 2014-2015 (cost: Rs ha^{-1})

Particulars	*AAS Farmers*	*Non- AAS Farmers*
Seed	600	600
FYM	3000	3000
Fertilizers	3348	3348
Micronutrients	200	-
Pesticides	700	300
Weedicide	600	600
Human labour	14350	13950
Machine labour	3200	3200
Irrigation	500	-
Miscellaneous	500	500
Total cost of cultivation	26998	25498
Grain yield (q/ha)	60.0(21.6)	(47.0)

Figure in parenthesis indicates per cent increase over control.

Source: Ray *et al.*, 2017.

Table 15.13: Economics of *Kharif* Maize grown by AAS User and Non User Farmers during 2014-2015 (Cost: Rs ha^{-1})

Type	*Cost of Cultivation*	*Gross Return*	*Net Return*	*B : C Ratio*
AAS Farmers	26998	60,000	33002	**2.2**
Non AAS Farmers	25498	47,000	21502	**1.84**

Source: Ray *et al.*, 2017.

The cost of cultivation of maize in case AAS user farmers was more as compared to non AAS farmers, but the farmers were getting more grain yield (21.6 per cent) as compared non user farmers. The net return and B: C ratio was more in case of AAS user farmers compared to non user farmers. Net return (Rs.ha^{-1}) in case of AAS user farmers was significantly higher for non user farmers due to adoption of proper integrated nutrient management (INM) and integrated pest management

(IPM) practices. Bi weekly crop based agro advisory helped the farmers to avoid adverse effect of extreme weather events like heavy rain, dry spell, high wind speed *etc.* similar results were reported by Rajegowda *et al.* (2008) where the AAS farmers received an additional benefit of 31.4,24.7,16.2 and 20.6 per cent in finger millet, red gram, field bean and tomato cultivation in Karnataka. Yield increase after adaptation of agro advisory bulletins was 13-15 q ha^{-1} for rice and 10 q ha^{-1} for mango and cashewnut in Konkan region of Maharashtra (Chaudhari *et al.*, 2010).

Conclusion

In this article, strategies to incorporate CSA towards sustainable agricultural development and food security has been discussed with prior importance to weather forecasting and its impact to mitigate the effects of climate change on crop production has also been analyzed. Successful implementation of AAS in different projected NICRA villages has strengthened resilience and adaptive capabilities towards increasing weather related hazards due to climate change. With timely weather based agro advisory, the farmers are now achieving higher crop production, higher income, and higher input use efficiency with higher profit margin. Farmers are now able to save or conserve the natural resources like irrigation water due to proper irrigation scheduling with advance information of rainfall or dry spell *etc.* real time crop health monitoring through satellite based system has also been established in Indian Agricultural Research Institute (IARI) by generating different crop condition indices and moisture stress indices. Beside climate resilience through crop improvement, improved pest and disease management, improvement of soil biological, physical and chemical health, reduction of GHG emission through efficient energy management, timely and accurate weather forecasting and suitable crop based agro advisory is well needed for better adaptation and mitigation from climate change and sustainable development in agricultural sectors.

REFERENCES

Aggarwal, P., Zougmoré, R., and Kinyangi, J. (2013). Climate smart villages: A community approach to sustainable agricultural development. CGIAR Research Program on Climate Change, Agriculture and Food Security (CCAFS), Copenhagen, Denmark.

Chaudhari, J. N., Zagada, M. V., Mahadkar, U. V.and Talathi, M. S. (2010) Assessment of weather based agromet advisories in high rainfall zone of Konkan in Maharastra. In: Proc. the Nation.Seminar on "Agrometeorological Services for Farmers" held at Anand Agricultural University, Gujrat during 10-13 November, 2008. pp. 172-177.

FAO. (2013). FAO Statistical Yearbook 2013. Details available at http: //www.fao.org/3/i3107e/i3107e.PDF

Porter JR, Xie L, Challinor AJ, Cochrane K, Howden SM, Iqbal MM, Lobell DB, Travasso MI (2014). *Food security and food production systems.* pp 485–533

Rajegowda, M.B., Janardhanagowda, N.A., Jagadeesha, N. andRavindrababu, B.T. (2008). Influence of agromet advisory services on economic impact of crops, *Journal of Agrometeorology*, **10**: 215-218.

Rao, P.G.S.L.H.V and N. Manikandan. (2008). Economic impact of agrometeorological advisory services over the central zone of Kerala. *Journal of Agrometeorology* (Special Issue- Part I), 230-234.

Saha, S, Rossi F, David S. (2016). Supporting agricultural extension towards Climate Smart Agriculture: An overview of existing tools. Global Alliance for Climate Smart Agriculture (GASCA)/FAO, Italy, pp. 28-30.

Ray, M., Patro, H., Biswas, S., Dash, R. S., Dash, C. A. (2017). Economic Assessment of Weather Based Agromet Advisories in Keonjhar District, Odisha. *Vayu Mandal*, 43(1): 38-48.

National Innovations in Climate Resilient Agriculture (NICRA), Research Highlights (2016-18). ICAR-Central Research Institute for Dryland Agriculture, Hyderabad. 128p.

Vijaya Kumar P., Bal S.K., Subba Rao A.V.M. (2019). National Innovations in Climate

Resilient Agriculture - AICRPAM Component, Annual Report 2018-19. ICAR-CRIDA, Santoshnagar, Hyderabad-500 059, Telangana, India. p64.

Bhattacharyya, P., Pathak, H., and Pal, S. (2020). Climate Smart Agriculture: Concepts, Challenges, and Opportunities. *Green Energy and technology*, Springer Nature Singapore Pte Ltd.

Chapter 16

Climate Smart Agriculture Needs and Pathways for Food Security

Ishani Mohan and Vanya Bawa

Chandigarh University, Gharuan, Punjab

An integrated approach to landscape management, climate smart agriculture (CSA) tackles the interrelated problems of food security and climate change, such as crops, animals, forests and fisheries. An increasing global population and evolving diets are pushing up food demand. Ocean health declines as crop yields drop off in many parts of the world, and natural resources such as soils, water and biodiversity are dangerously low, and development is struggling to keep up. The negative impacts of climate change are already being felt in the form of rising temperatures, weather fluctuations, agro-ecosystem borders, invasive crops and pests, and more frequent extreme weather events. Increased productivity: rising and improving food production to improve nutrition security and raise incomes, particularly for 75 per cent of the world's poor who live in rural areas and depend primarily on agriculture for their livelihoods. The CSA aims to achieve three outcomes at the same time. Improved resilience: and susceptibility to droughts, pests, diseases and other climate related threats and shocks; and improving the capacity to adapt and develop in the face of longer-term pressures, such as shortened seasons and unpredictable weather patterns. Emissions reduction: achieving lower emissions for each calorie or kilo of food produced, avoiding deforestation in agriculture, and finding ways to remove carbon from the atmosphere. To encourage countries to consider maximum benefit options and those where trade-offs need management, the CSA proposes more systematic approaches to the closely related challenges of food security, development and climate change adaptation/mitigation. The CSA recognizes that country-specific situations and capabilities will affect the implementation of options, as well as providing access to better expertise, aligned strategies, coordinated institutional frameworks and flexible incentives and funding mechanisms.

Keywords: *Food security, Adaptation, Agro-ecosystems, CSA, Mitigation.*

Introduction

Climate smart agriculture (CSA) is an approach to the transformation and reorientation of agricultural systems in order to encourage food security in the light of current climate change realities. Climate smart agriculture (CSA) is an approach to the transformation and reorientation of agricultural systems with a view to promoting food security, taking into account the realities of current climate change. (Rural poverty report *et al.,* 2011). The CSA promotes concerted action on climate resilient pathways by farmers, academics, the private sector, civil society and policymakers across four key areas of action: evidence-building, enhancing local institutional effectiveness, promoting coherence between climate and agricultural policies, and linking climate and agricultural financing(Lipper *et al.,* 2014). Ineffective targeting stymies the contribution of CSA to poverty reduction by excluding many of the poor and/or even those for whom agriculture is not a way out of poverty. As agricultural researchers seeking to ensure the impact of growth, we acknowledge the challenge of recognizing that our CA-relevant research is unlikely to contribute to poverty reduction in some cases. (Hellin-Fisher *et al.,* 2018). Agriculture does not reflect a way out of poverty for many farmers, and they are increasingly dependent on sources of non-agricultural income. Ineffective CSA targeting therefore ignores various livelihood transition capacities for small-scale farming households, which are related to the opportunities and constraints offered by different livelihood pathways, agricultural and non-agricultural. (Hellin-Fisher *et al.,* 2019)

Climate Smart Agriculture (CSA)

Climate smart agriculture in the new realities of climate change can be described as an approach in transforming and reorienting agricultural production (Porter, J. R. *et al.,* 2014). Climate smart farming includes farming practices that improve farm productivity and profitability, which help farmers adapt to the negative impacts of climate change, and mitigate the impacts of climate change.Climate smart techniques, such as locally practiced conservation agriculture, are aimed at the conservation of soil moisture, the preservation of crop residues through soil fertility, soil disruption as minimally as possible, and diversification by rotation or inter-cropping; these practices may increase production and increase food security

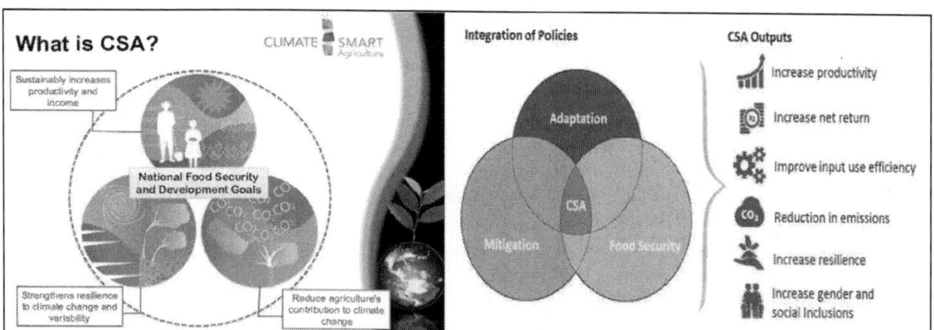

Figure 16.1: Left: CSA and its Goals (Papuso-Faraby *et al.*, 2013) and Right: Outputs through CSA (Arun-Poudel-Bikash-Shirsath *et al.*, 2017).

for farmers. The secret when it comes to CSA is adaptation. "We need to transit to a more sustainable food system and mitigate climate change while at the same time adapt to climate change. Communities that are highly food-insecure or particularly vulnerable to climate change will necessarily priorities adaptation, but many of the changes they might make to enhance resilience will also increase productivity and efficiency of inputs (fertilizer and water use), and even have co-benefits for mitigation" Meryl Richaards, Agroecologist (Hellin-Fisher *et al.*, 2018).

Needs of Climate Smart Agriculture

If we change our approach to agricultural growth and development, planning and investment we risk misallocating human and financial capital, creating agricultural systems that are incapable of fostering food security and leading to climate change. By incorporating climate change into the preparation and execution of sustainable agricultural policies, climate smart agriculture (CSA) will prevent this 'lose-lose' result (Lipper *et al.*, 2014). A wider and more radical agenda would also be needed for climate smart initiatives to encourage the capacity of farm households to create non-agricultural livelihoods (Hellin-Fisher *et al.*, 2019). The concept of climate smart agriculture (CSA) was presented and defined at the first Global Conference on Agriculture Food Security and Climate Change in the Hague as agriculture that sustainably increases productivity, improves resilience, reduces/removes greenhouse gas emissions, and enhances the achievement of national food security and development goals.This represented an attempt to set a global agenda for investments in agricultural research and innovation, joining the agriculture, development and climate change communities under a common brand. Following the second Global Conference on Agriculture Food Security and Climate Change in Hanoi in 2012, the recently published Climate Smart Agriculture Sourcebook further advanced the concept with the intention of benefiting primarily smallholder farmers and vulnerable people in developing countries (Neufeldt-Jahn *et al.*, 2013).

Three objectives are defined for achieving this aim:

1. Increasing agricultural productivity in a sustainable way to encourage equal income rises, food security and growth.
2. Adapting and building resilience to climate change from the farm to national levels.
3. Developing opportunities to reduce GHG emissions from agriculture compared with past trends (Lipper *et al.*, 2014)

Climate Smart Agriculture and Poverty Reduction

It is possible to differentiate farm households on the basis of their asset endowment, *e.g.*, their land quantity, access to key agricultural inputs, *etc.*, combined with features that define the livelihood strategies available to them (GEF assembly report *et al.*, 2014). An awareness of these livelihood pathways informs choices as to where to target CSA and where to build enabling strategies that promote improvements in livelihoods. CSA and agricultural interventions will certainly lead to poverty reduction and increased prosperity (WFP (World Food Programme) *et al.*, 2014). For some farmers coping with little resources and the additional risk of climate change, the agricultural future is bleak. For all farm households, agriculture is not a solution out of poverty. Thus, poverty reduction for some household groups would come from farmers moving out of agriculture and into the non-farm economy. When it comes to small-scale farmers with very small landholdings, CSA is often very troublesome (Hellin, Cox, R.; López-Ridaura *et al.*, 2017). Firstly, land is taken out of cultivation in the case of cross-slope soil protection technologies, such as live plant barriers and stone walls. Second, even if the CSA were to contribute to substantial increases in agricultural productivity, it would be unlikely that the increase (while contributing to food security) would allow farmers to escape poverty (Hellin-Fisher *et al.*, 2019).

A broad range of literature on agriculture, food systems, food and nutritional protection, dietary health, climate change adaptation and mitigation of agricultural GHG (greenhouse gas) emissions was synthesized by Climate Change into a set of suggested policy actions. The commission extended the concept of "safe operating space" beyond the original framing, which focused on biophysical attributes of the planet, to include social ecological systems related to human welfare, agriculture and food security, a safe operating space for agricultural and food systems represents a set of conditions that demonstrably better meets human needs in the short and long term within foreseeable local and planetary limits and holds ourselves accountable for outcomes across temporal and spatial scales (Neufeldt-Jahn *et al.*, 2013). Farming system typologies have been used for nearly two decades to capture the diversity of farming systems (Landais *et al.*, 1998), and are increasingly used to provide guidelines for the development of agricultural innovations and to understand their implications for climate change (Berre *et al.*, 2016; Chopin *et al.*, 2015; Douxchamps *et al.*, 2015; Kuivanen *et al.*, 2016; Pacini *et al.*, 2014; Tittonell, *et al.*, 2014).

In recent years, after a prolonged decline over the last few decades, world hunger and food insecurity have increased. In sub-Saharan Africa, undernourishment is particularly growing, affecting an estimated 23 per cent of the population. The total number of undernourished citizens has risen by more than 22 per cent within six years, partially due to the rapid population growth in sub-Saharan Africa. At the same time, climate change is predicted to be complicated, with models projecting temperature increases across sub-Saharan Africa, changes in rainfall patterns and increases in extreme events, such as heatwaves, droughts, and extreme rainfall events, although with varying confidence.

Framework to Access CSA

Framework to assess the current CSA assessment tools is based on all of the three CSA pillars: Food security, Adaptation, and Mitigation.

Food Security

It is too simplistic to concentrate on production and equate production with food protection. In the literature, there are many examples showing that increased production of crops contributes to reduced food security and even malnutrition (for example introducing a more productive crop could lead to a reduction in the diversity of crops being produced). (Wijk-Merbold *et al.*, 2020). Literally, hundreds of food security concepts exist. The term generally implies that people have equal and sustained physical and economic access to a sufficient amount of safe and nutritious food to meet daily caloric requirements and to maintain an active and balanced lifestyle, regardless of the scale considered, from households to countries to regions. (Lopez-Ridaura *et al.*, 2018).

Adaptation

CSA aims to reduce the exposure of farmers to short-term risks, while also strengthening their resilience by building their capacity to adapt and prosper in the face of shocks and longer-term stresses. Significant factors affecting the coping and adaptation actions of households are considered to be the basic socio-economic characteristics of the sample population (age mean, ethnic context, economic status of households, *e.g.*, farm size or number of livestock, and number of household members) (Rosenstock *et al.*, 2019).

Mitigation

Mitigation is a key pillar of CSA, and this pillar aims at reducing the environmental (climate) footprint of food production as:

1. Agriculture is the most important driver of environmental degradation worldwide
2. Agriculture is a major source of anthropogenic GHGs (Greenhouse gases) especially in LMICs (Low- and Middle-Income Countries).
3. Agriculture can be beneficial for a number of ecosystem services (*e.g.*, biodiversity, carbon sequestration).

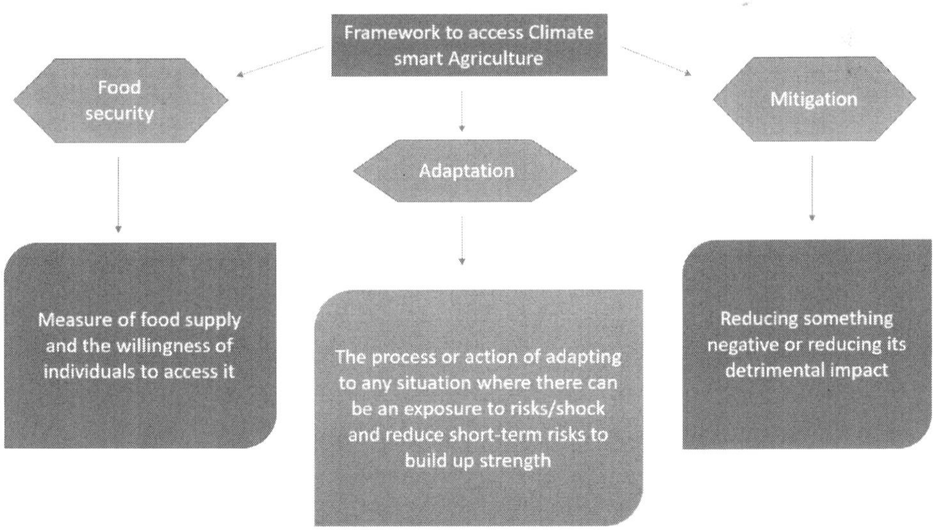

Unless environmental sustainability is explicitly dealt with-in the food security pillar, it should be an important aspect of the mitigation analysies.

Recent Innovations so Far

1. Energy-smart technologies- Technologies that help decrease energy consumption without affecting yield levels during land preparation. These also help reduce the requirements for water for crops.
2. Nutrient-smart technologies- Technologies that save/supplement/prevent the use of chemical fertilizer for crops and enrich the soil with carbon. Green manure-Legume cultivation in a cropping system. Nitrogen economy and soil health/quality are improved by this practice.
3. Weather-smart instruments- Interventions that provide services related to financial protection and weather advisories to farmers.
4. Water - smart technologies- Rainwater may need to be managed by using various tools, such as harvesting it either from rain or from runoff. Another problem that needs more focus is water conservation.
5. Laser land levelling- The land is levelled with a laser in this technology, which is accomplished by a laser leveller. In rice fields, where it is a rice intensification method, this type of technology has been found suitable, 7 to 10 day old seedlings are transplanted with 1-2 seedlings per hill at 20 cm spacing in this technology.

Figure 16.2: Left: Drones Used as a Part of Technology for Crop Spraying; Right: Weather Tracker (Prajapati *et al.*, 2019).

Conclusion

The CSA strategy needs to shift beyond the emphasis on the resilience of food systems even more explicitly. Farmers who have diversified their revenue base in the form of drought, alternate revenue sources are comparatively less vulnerable to environment shocks. The triple victory of increased efficiency, adaptation and mitigation will lead to CSA. We argue that more systematic attention be directed at climate risk management that moves beyond the more conventional adaptation and mitigation discourse, towards an approach that includes livelihood transformation from a broader perspective, *i.e.*, one that does not just focus on rural–agricultural transformation but also identifies (and embraces) where agriculture is not a pathway out of poverty and where support for positive non-agricultural livelihood trajectories are needed for small-scale farmers. To achieve the scale and pace of change required, dressing up food and climate protection at all levels requires concerted action from global to local levels, from research to policy and investment and across the private, public and civil society sectors. The agriculture sector will step onto CSA pathways with the right activities, policies and investments resulting in a short-term decrease in food insecurity and hunger, thus leading to the reduction of climate change as a threat to food security over the longer term. CSA is to effectively promote growth and ensure food security in a changing world, transform and reorient agricultural systems.

REFERENCES

Arun and Poudel, Bikash and Shirsath, Paresh. (2017). Assessment of Climate Smart Agriculture (CSA) Options in Nepal.

Berre, D., Baudron, F., Kassie, M., Craufurd, P., Lopez-Ridaura, S., 2016. Different ways to cut a cake: comparing expert-based and statistical typologies to target sustainable intensification technologies, a case-study in Southern Ethiopia. Exp. Agric. 1–17.

Chopin, P., Blazy, J.M., Dore, T., 2015. A new method to assess farming system evolution at the landscape scale. Agron. Sustain. Dev. 35, 325–337.

Douxchamps, S., Van Wijk, M.T., Silvestri, S., Moussa, A.S., Quiros, C., Ndour, N.Y.B., Buah, S., Somé, L., Herrero, M., Kristjanson, P., Ouedraogo, M., Thornton, P.K., Van Asten, P., Zougmoré, R., Rufifino, M.C., 2015. Linking agricultural adaptation strategies, food security and vulnerability: evidence from West Africa. Reg. Environ.

Hellin, J., and Fisher, E. (2018). Building pathways out of poverty through climate smart agriculture and effective targeting. *Development in Practice*, *28*(7), 974-979.

Hellin, J., and Fisher, E. (2019). Climate smart agriculture and non-agricultural livelihood transformation. *Climate*, *7*(4), 48.

Hellin, J.; Cox, R.; López-Ridaura, S. Maize Diversity, Market Access, and Poverty Reduction in the Western Highlands of Guatemala. *Mt. Res. Dev.* **2017**, *37*, 188–197.

Kuivanen, K.S., Michalscheck, M., Descheemaeker, K., Adjei-Nsiah, S., Mellon-Bedi, S., Groot, J.C.J., Alvarez, S., 2016. A comparison of statistical and participatory clustering of smallholder farming systems – a case study in Northern Ghana. J. Rural. Stud. 45, 184–198.

Landais, E., 1998. Modelling farm diversity – new approaches to typology building in France. Agric. Syst. 58, 505–527.

Lipper, L., Thornton, P., Campbell, B. M., Baedeker, T., Braimoh, A., Bwalya, M. and Torquebiau, E. F. (2014). Climate smart agriculture for food security. Nature climate change, 4(12), 1068-1072.

Lopez-Ridaura, S., Frelat, R., van Wijk, M. T., Valbuena, D., Krupnik, T. J., and Jat, M. L. (2018). Climate smart agriculture, farm household typologies and food security: an ex-ante assessment from Eastern India. *Agricultural systems, 159,* 57-68.

Neufeldt, H., Jahn, M., Campbell, B. M., Beddington, J. R., DeClerck, F., De Pinto, A. and Zougmoré, R. (2013). Beyond climate smart agriculture: toward safe operating spaces for global food systems. *Agriculture and Food Security,* 2(1), 1-6.

Pacini, G.C., Colucci, D., Baudron, F., Righi, E., Corbeels, M., Tittonell, P., Stefanini, F.M., 2014. Combining multi-dimensional scaling and cluster analysis to describe the diversity of rural households. Exp. Agric. 50, 376–397.

Papuso and Jimly Faraby, Seminar on Climate Change and Risk Management, May 6, 2013.

Porter, J. R. *et al.*, in Climate Change 2014: Impacts, Adaptation, and Vulnerability. Part A: Global and Sectoral Aspects (eds Field, C. B. *et al.*) 485–533 (IPCC, Cambridge Univ. Press, 2014).

Rahul Prajapat, Agriculture machine and agriculture technology, Oct 8, 2019

Report on the Sixth Replenishment of the GEF Trust Fund (Global Environment Facility Secretariat (GEF) Secretariat and World Bank, for Fifth GEF Assembly, 2014).

Rosenstock, T. S., Lamanna, C., Namoi, N., Arslan, A., and Richards, M. (2019). "What is the evidence base for climate smart agriculture in East and Southern Africa? A systematic map," in *The Climate Smart Agriculture.*

Rural Poverty Report 2011 (International Fund for Agricultural Development, 2011).

Tittonell, P., 2014. Livelihood strategies, resilience and transformability in African agroecosystems. Agric. Syst. 126, 3–14.

van Wijk, M. T., Merbold, L., Hammond, J., and Butterbach-Bahl, K. (2020). Improving Assessments of the three pillars of Climate Smart Agriculture: current achievements and ideas for the future. *Frontiers in Sustainable Food Systems, 4,* 148.

WFP (World Food Programme). 2014. "Guatemala Overview." Accessed 28 July 2014.

Chapter 17

Restoring Degraded Soils and Using Wasteland and Wetland for Climate Change Resilience

*Satyabrata Sahoo and Vanya Bawa**

UIAS, Chandigarh University, Gharuan, Punjab

Soil degradation, portrayed by decrease in quality and reduction in ecological services, is a significant limitation to accomplishing the necessary in expansion of agricultural goods production. The major soil degradation process involves accelerated soil erosion, reduction in organic carbon percentage in soil (SOC), loss in biodiversity, increasing infertility, nutrition and element imbalance, acidification and salinization. The basic goals is to reduce soil erosion, produce more positive SOC pool, and increase the activity of and diversity of different species in soil biota and to enhance the quality and ability of soil. The land degradation and climate change are quite interlinked which affects global agriculture and life sustainability. Eradicating issues related to land degradation and climate change will help us for building sustainable agro ecosystem on which world's population is dependent. Coastal wetlands are considered as most productive biological system and can offer important environmental types of assistance. However coastal wetlands also considered as one of most vulnerable ecosystems suffering from unnatural activities. The degradation of wetlands draws attention of all developing and developed countries and it requires restoration projects to tackle this. No doubt large number of restoration projects have been conducted worldwide depending upon different restoration goals and different methods. And it's important to evaluate the success rate of restoration, although it's challenging because of current disagreement on definitions and concepts of restoration evaluation. However it's possible by establishing a success indicator system for guiding wetland restoration progress. In this paper we are going to discuss about degraded lands and wetland, their status in global scenario, some techniques of restoring soil quality include conservation agriculture,cover cropping and continuous vegetative cover such as mulching. The steps followed to reduce the losses and increase in soil, water and nutrient efficiency. And will discuss how land degradation is likely interact with climate change

Introduction

Land degradation and climate change affecting global agriculture and human food security for many years and tackling these issues is crucial for growing population and for sustainable agriculture. According the data, in 2014 out of 5.5 billion populations, one billion small landholders cultivate in their own field and a major portion of the population relies directly on agriculture. The rapid increase in land degradation affects world's 33 per cent of whole land which is nearly equal to 500 million hectare (Mha).According to National Remote Sensing Agency (NRSA) in the year 2009 there was 46.70 million ha wasteland in India.Which includes scrubland (17.99 Mha), degraded forests (9.91 Mha), barren rocky area (5.94 Mha) and sandy area (3.02 Mha. Land degradation adversely affects a country's economic development where a majority population depends directly on agriculture. Global estimation shows total range of degraded lands ranges from 1 Gha to over 6 Gha. Soil degradation or soil erosion directly related to soil health which declines soil quality and ecosystem functioning and its services. Soil degradation is up 4 types;-(i) Physical (ii) chemical (iii) biological (iv) ecological. And due to this 37.25 million km^2 (>25 per cent) global land area has been affected and the soil quality, biological and economical productivity has been reduced (ELD Initiative 2015). Land degradation affecting whole world's ecosystem including agricultural lands, arid and semi arid rangelands, pasturelands and agro forestry systems (Miettinen *et al.*, 2014 and croplands (Karamesouti *et al.*, 2015). Currently 40 per cent lands of developing countries has been already degraded and it's expected that this may be increase up to 78 per cent of total global dry land expansion by the year 2100 (Huang *et al.*, 2015). The increase in risks of climate change also boosting the growth of dry land areas.

Links between Land Degradation, Climate Change, and Adaptation Planning

Land degradation and climate change both are interlinked which have both biotic and abiotic factors interference, human influence, impact and responses (Herrick *et al.*, 2013). Land degradation is well defined by reduction or loss of biological and economic productivity which may be due to human activity and land use or from one or more processes. Some of the processes may be responsible are (i) erosion of soil by air or water (ii) degeneration of physical, chemical and biological or economic properties of soil (*e.g.* salinization) (iii) a land without vegetation for long duration (UNCCD 1994). These cycles may likewise be affected by social, financial, and political components that empower or force land-use pressures while neglecting to adjust the stock of biological system administrations with farming production requests (D'Odorico *et al.*, 2013).The connectedness of natural and social frameworks, land debasement typically brings about a decrease in agro-socio-environment versatility, which is characterized as the capacity of a system to keep up the structure needed to support fundamental framework capacities through times of pressure or irritation (Reed and Stringer 2016). Declining strength of agricultural and social frameworks can expand tension on biological cycles, prompting a twisting of debasement as soil assets are exhausted and vegetation networks change. Environmental change can compound and quicken land debasement through different methods, including quickened soil disintegration, expanded evapotranspiration rates, dry season, and changes in biodiversity, bugs, and infections. The tradition of verifiable land debasement may then further impact the extent and heading (positive or negative) of the effects of environmental change on agro ecological systems. The definition of soil quality or soil health includes structure, organic carbon, infiltration rate and nutrition availability, shows the status of soil relatives to its potential (Seybold *et al.,* 1999). Enhancements in soil health are normally connected with diminished affectability to environmental change. Soil health is influenced via land corruption essentially through disintegration, yet in addition by physical, compound, and organic changes. Changes in vegetation because of land use and the board may happen simultaneously with declining soil health and may influence scavenge and crop production reactions to environmental change (Bestmelmeyer *et al.,* 2015).

Land degradation has adverse impacts on financial aspects related to agro ecological systems. And further changes in quality and quantity of ecosystem services will influence the occupations across related industries. These progressions eventually assessment to influence lands the board and land debasement. In light of such reports, land degradation further impacts adaptations alternatives. Tending to land debasement now, as an expectant variation technique, is possibly a profoundly successful way to deal with building gainful and versatile agroecological systems for what's to come. Different reactions are needed across neighborhood, territorial, and public scales to assemble the flexibility and diminish the vulnerability of agroecological systems to land debasement and environmental change.

Coastal Wetlands

Coastal wetlands are situated near the edges of landmass formed by tidal forces, fresh water sources, sediment transport and biota (Barbier *et al.,* 2013). Coastal wetlands are viewed as among the most beneficial biological systems and can offer significant types of assistance, for example, storm buffering, insurance from windstorm and shore disintegration, fishery creation, water cleaning and biodiversity conservation(Newton *et al.,* 2012). At present, the interest in biological reclamation is fortified as rebuilding can alleviate environmental change and biodiversity depletion (Nilsson and Aradottir, 2013). Beginning during the 1990s, wetland rebuilding and re-creation turned into a "hotspot" in the environmental research fields (Thormann and Bayley, 1997; Visser *et al.,* 1997; Zedler and Kercher, 2005).

Methodologies of Coastal Wetland Restoration

1. Active Restoration

Active restoration expects people to control and intercede at regular basis in order to reestablish, reproduce and improve the local area structure and environmental measures in wetlands (Wagner *et al.,* 2008). Improvements in the geography of coastal wetlands, rechanneling the water flows through different water control facilities like dams or divisions, soil transplantation and artificial vegetation is mostly preferred (Zhang *et al.,* 2009). Most common and effective

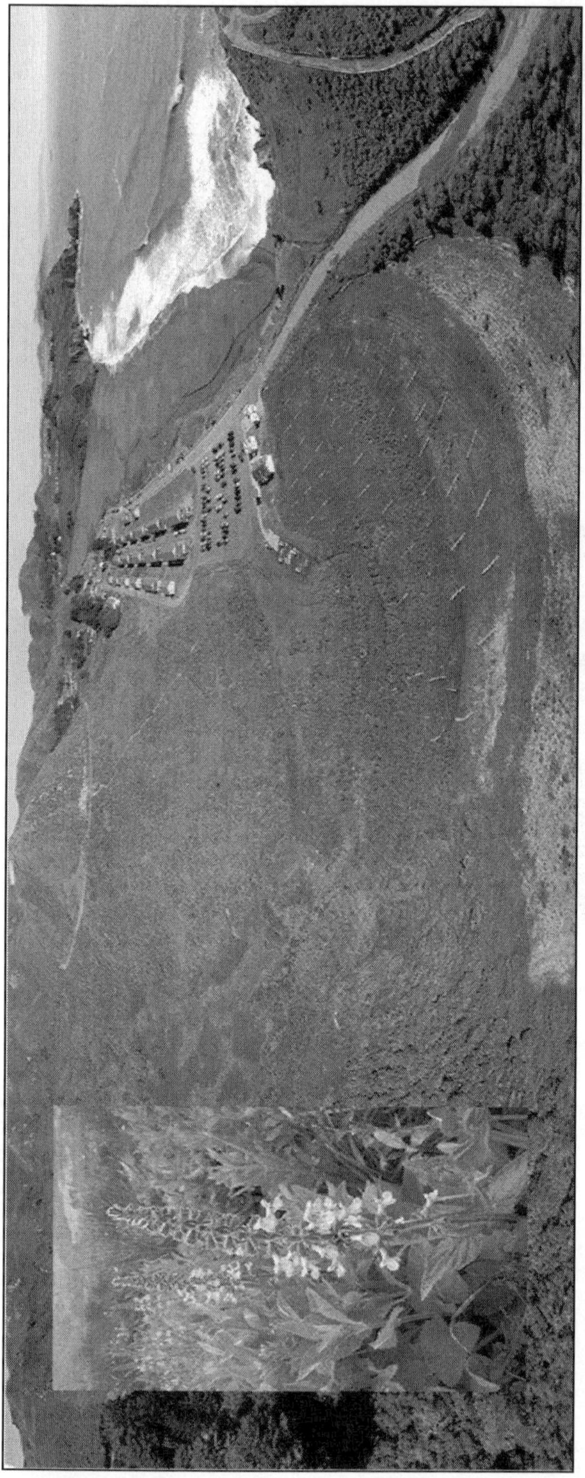

Figure 17.1: Parking Lot Fill was Removed and Eroding Ditches were Backfilled in 2013 to Restore a Wet Meadow at Rodeo Beach, Golden Gate National Recreation Area, California. Wetland hydrology has been restored, and native wetland plants like coastal hedgenettle (inset) are now thriving in the former parking lot area Rodeo Beach photo by Cris Benton, Coastal hedgenttle photo by Evan Wolf.

(*Source*: https://www.nps.gov/subjects/wetlands/restoring-wetlands.htm).

method followed by mangrove reclamation projects are replanting mangroves to debased or damaged mangrove wetland (Ren *et al.*, 2011; Day *et al.*, 2013). In a salt swamp reclamation program in the New England district of the United States, hydroperiod change through flowing stream rebuilding demonstrated successful (Burdick *et al.*, 1996). Not long ago, an enormous achievement including coastal interdune wetland reclamation in Central Veracruz, Mexico, was accounted for. The fundamental methods included were maintaining the flooding level and changing environmental example by hindering Echinochloa (an intrusive species) treated as native species recovery (Lopez Rosas *et al.*, 2013). It is additionally worth referencing that a huge rebuilding project has been led in a debased coastal Phragmites australis wetland in the Nature Reserve of the Yellow River Delta (in China) since 2001. The reclamation estimates included structure dams and banks, freshwater assortment and water redirection from the Yellow River.

2. Passive Restoration

Passive restoration includes complete eradication of influencing factors which are responsible for coastal wetland debasement or destruction and reestablishing the degraded wetlands to a stable healthy natural condition (Wagner *et al.*, 2008). Passive reclamation centers on the upgrade of ecohydrological cycles to remake the hydro geomorphology for wetland self-establishing (Mitsch and Wang, 2000, Hunter *et al.*, 2008; Jarzemsky *et al.*, 2013). The reclamation of degraded marshes can be mitigated by conversion of croplands into marshes and grasses. In the estuarine wetlands of Loushan (China), the contamination and pollution of wetlands was controlled through the configuration of the sewage channel organization, which fortified the hydrological network among wetlands and the adjoining ocean (Zhang and Sun; 2009). Moreover, the improvement of wetland hydrological availability and the recovery of underground hydrology can likewise mitigate abundant nitrogen, phosphorus and microbes, accumulate residue maintenance and lower soil disintegration, all of which add to wetland rebuilding somewhat detailed that wetland protection could be reinforced through differentiating the biological organizations, particularly by setting up higher hydrological connectivity (Richardson *et al.*, 2011; Cui *et al.*, 2012).

3. Creation

The creation involves the process of turning upland or sub tidal living space on destination where no wetlands existed before into a wetland (Simenstad *et al.*, 2006; Mitch and Gosselink, 2007; Broome and Craft, 2009). The wetland establishment requires satisfactory local conditions, especially suitable hydrology (NRC, 2001). To moderate the degradation of flowing swamp, new marshes are regularly made on locales where no wetlands existed already. The fundamental cycle included reasonable plan, building up hydrology, soil and vegetation. A coastal freshwater swamp was accounted for to be made through freshwater surges (Fitzsimmons *et al.*, 2012).

Figure 17.2: Top left: Kortright Farm in 2008. The wetland was created in a previously fallow agricultural field. Top middle: In 2009, the wetland was regarded, tile drains removed, and wetland buffer planted. Top right: By 2011, emergent vegetation was established from the existing seedbank; agricultural plots were created in the area surrounding the wetland (background). Bottom: Aerial photographs of the Kortright Farm before (left) and after (right) restoration, in 2005 and 2009 respectively. (*Source*: https://trca.ca/conservation/restoration/wetlands/)

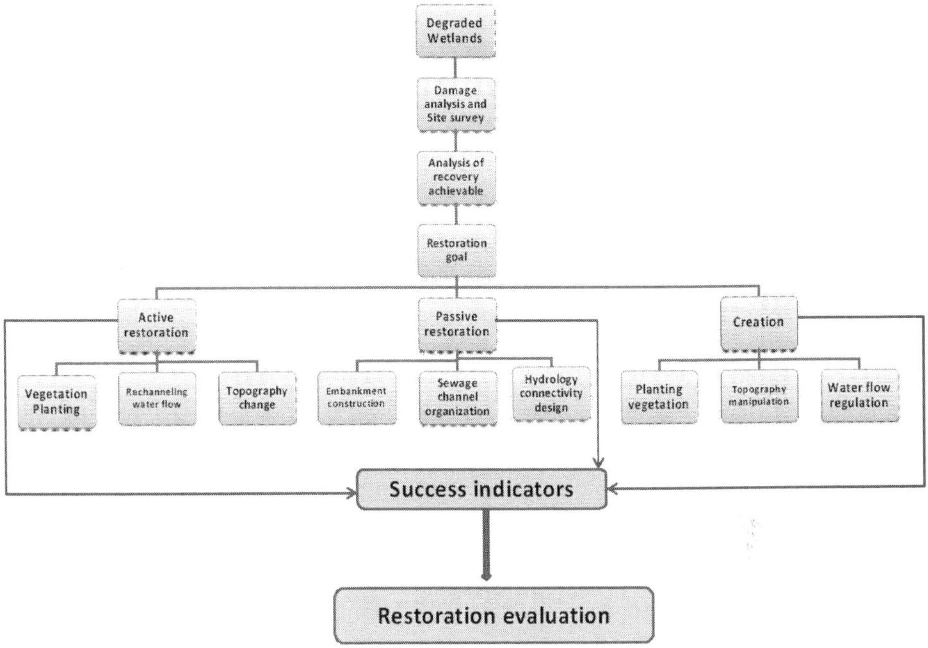

Remote sensing (RS), geographic information system (GIS) and global position system(GPS) techniques offer new specialized help for wetland restoration evaluation (Kevin *et al.*, 2009). Rozas *et al.* (2007) researched the dynamic changes of wetland region and species population size dependent on GIS, elevated pictures and population models and found that species population multiplication was not emphatically influenced by wetland rebuilding as per their prosperity assessment of reestablished wetlands. Kauffman-Axelrod and Steinberg (2010) organized 530 potential wetland rebuilding destinations utilizing GIS, which added to the better reclamation and the executives of wetlands dependent on powerful observing of wetland region and vegetation. Despite the fact that Government have invested a lot of energy into coastal wetland protection practices, most wetland rebuilding projects are zeroed in the restoration and regulation of vegetation and there is an absence of efficient examination on the components of coastal wetland debasement and ecohydrological measures, particularly hydrological and natural availability at an enormous scope.

Conclusions

Soil degradation can be physical, chemical, biological or ecological which includes events like distortion in soil structure, crusting, compaction, soil disintegration, water imbalances, acidification, salinization, element toxicity and deficiencies, nutrition deficiency, reduction in soil bio diversity, decline in microbial activities, interruption in natural cycling, decline in carbon sink capacity and many more. Soil debasement prompts decrease in environment services and capacities

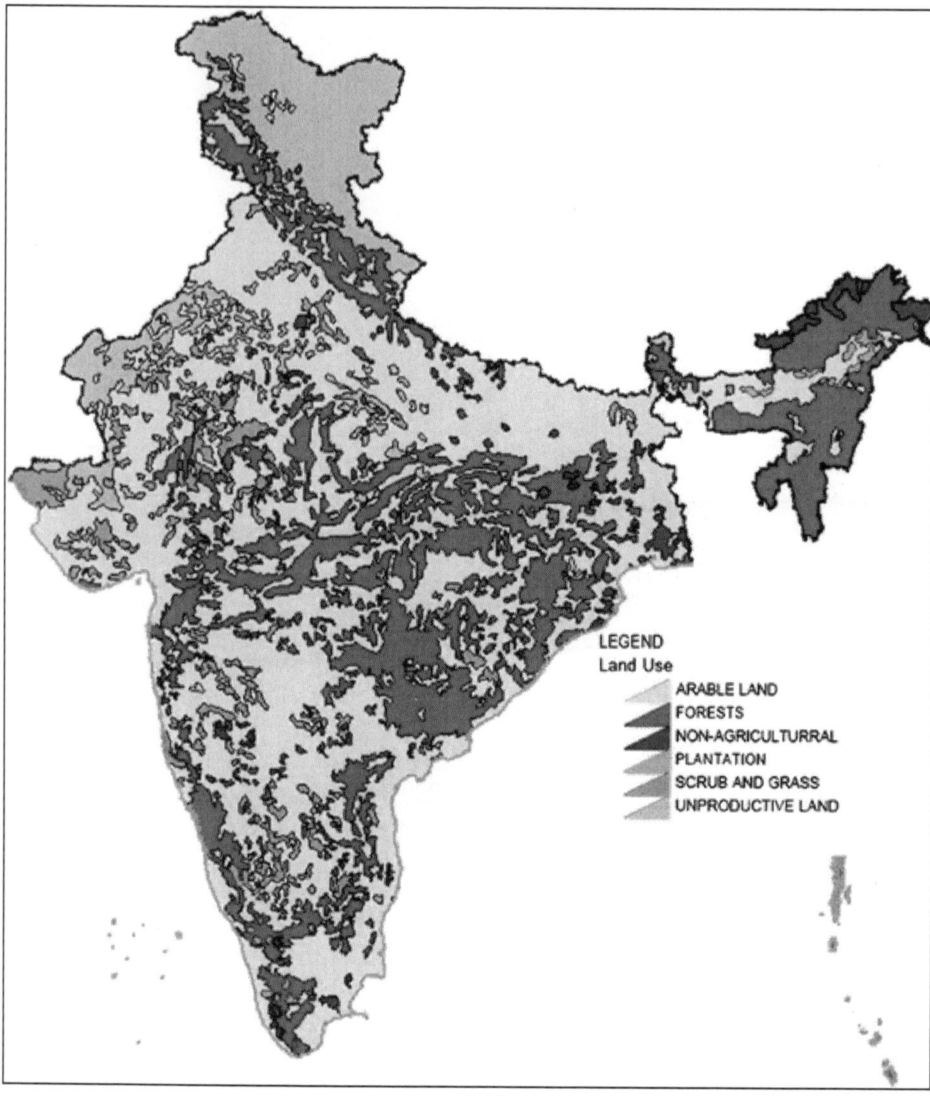

Figure 17.3: Land Use and Land Cover in India. (*Source*: National Institute of Hydrology (2009), Accessed from India WaterPortal.org.)

important to human and preservation of nature. Tending to land debasement is fundamental for building economical agro ecological frameworks that are climate resilient, save biodiversity, and meet worldwide advancement objectives. Future agro ecological frameworks will rely upon the present-day developments goals and policy options. However, it's the responsibility of researchers, administrators and policy makers to reduce the difference of understanding the land degradation-climate change yet should likewise fortify coordinated efforts to address variation related

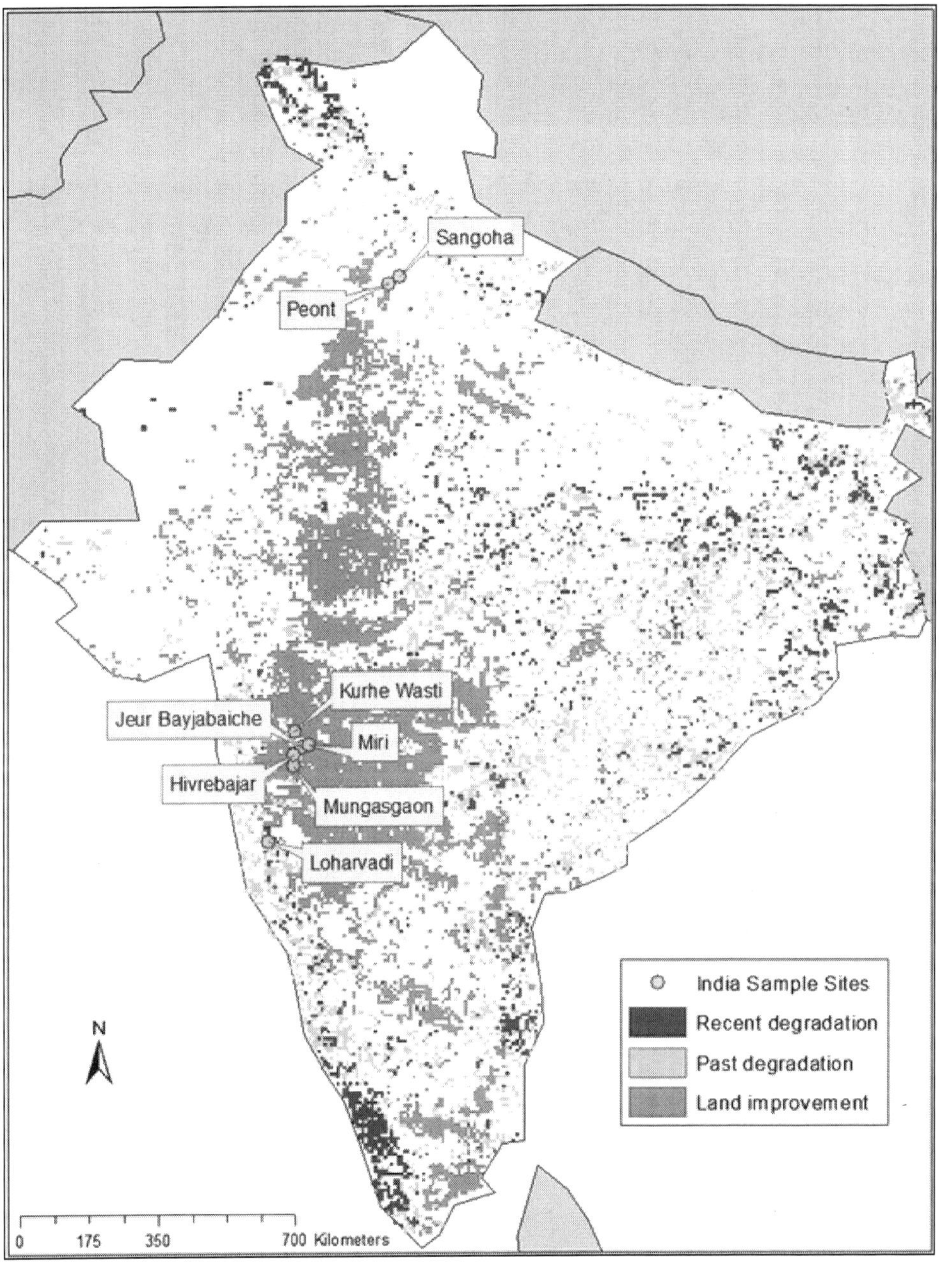

Figure 17.4: Selected Ground Truthing Cites.
(*Source* FGD. Note: Dark red indicates pixels that demonstrate both long - term degradation as well as degradation in recent (2000 – 2006) years, green pixels indicate sites with improved land (Reference: https://link.springer.com/ chapter/10.1007/978-3-319-19168-3-15

difficulties at nearby to worldwide scales. Here, we have introduced four staggered activities that can assist with coordinating area degradation countermeasures into environmental change transformation planning. Reclamation methods focused on various sorts of coastal wetlands have been embraced in rebuilding projects with variable outcomes. Such indicators are there which evaluates the success rate of wetland restoration projects. Nonetheless, it is hard to analyze whether the coastal wetland reclamation is fruitful due to the absence of uniform success measures, long haul observing information and monetary help. Hence, it's highly necessary to develop an appropriate uniform success indicator which links aquatic environment and eco-hydrological indicators to evaluate the success rate of coastal land degradation.

REFERENCES

Barbier, E.B., 2011. Coastal wetland restoration and the deepwater Horizon oil spill. Vanderbilt Law Rev., 64

Broome, S.W., Craft, C.B., 2009. Tidal Marsh Creation. Coastal Wetlands: An Integrated Ecosystems Approach. Elsevier, pp. 715–733.

Burdick, D.M., Dionne, M., Boumans, R.M., Short, F.T., 1996. Ecological responses to tidal restorations of two northern New England salt marshes. Wetl. Ecol. Manag. 4, 129–144.

Cui, B.S., Zhang, Z.M., Lei, X.X., 2012. Implementation of diversified ecological networks to strengthen wetland conservation. Clean Soil Air Water 40, 1015–1026.

D'Odorico P, Bhattachan A, Davis KF, *et al.*, 2013. Global desertification: drivers and feedbacks. Adv Water Resour 51: 326–44.

Day, J.W., Lane, R.R., Robert, R., 2013.Wetland restoration using mangroves in southern Louisiana. In: SPE Americas E and P Health Safety Security and Environmental Conference. Society of Petroleum Engineers, http: //dx.doi.org/10.2118/163789- MS

Edrisi, S. A., Tripathi, V., and Abhilash, P. C. (2018). Towards the sustainable restoration of marginal and degraded lands in India. Tropical Ecology, 59(3), 397-416.

Fitzsimmons, O.N., Ballard, B.M., Merendino, M.T., Baldassarre, G.A., Hartke, K.M., 2012. Implications of coastal wetland management to nonbreeding waterbirds in Texas. Wetlands 32, 1057–1066.

Herrick JE, Beh A, Barrios E, *et al.*, 2016. The Land-Potential Knowledge System (LandPKS): mobile apps and collaboration for optimizing climate change investments. Ecosyst Health Sustain 2: e01209.

Huang J, Yu H, Guan X, *et al.*, 2015. Accelerated dryland expansion under climate change. Nat Clim Change 6: 166–71.

Hunter, R.G., Faulkner, S.P., Gibson, K.A., 2008. The importance of hydrology in restoration of bottomland hardwood wetland functions. Wetlands 28, 605–615

Jarzemsky, R.D., Burchell II, M.R., Evans, R.O., 2013. The impact of manipulating surface topography on the hydrologic restoration of a forested coastal wetland. Ecol. Eng. 58, 35–43

Karamesouti M, Detsis V, Kounalaki A, *et al.*, 2015. Land-use and land degradation processes affecting soil resources: evidence from a traditional Mediterranean cropland (Greece). Catena 132: 45–55.

Kauffman-Axelrod, J.L., Steinberg, S.J., 2010. Development and application of an automated GIS based evaluation to prioritize wetland restoration opportunities. Wetlands 30, 437–448

Kevin, J., Yannick, L., Paul, V., Eric, V., Doug, T., Diego, F.P., 2009. Monitoring and assessment of wetlands using Earth Observation: The Globe Wetland project. J. Environ. Manag. 90, 2154–2169.

Lal, R. (2015). Restoring soil quality to mitigate soil degradation. Sustainability, 7(5), 5875-5895.

López-Rosas, H., Moreno-Casasola, P., López-Barrera, F., Sánchez-Higueredo, L.E., Espejel-González, V.E., Vázquez, J., 2013. Interdune Wetland Restoration in Central Veracruz, Mexico: Plant Diversity Recovery Mediated by the Hydroperiod, Restoration of Coastal Dunes. Springer, New York, USA, pp. 255–269

Miettinen J, Stibig H-J, and Achard F. 2014. Remote sensing of forest degradation in Southeast Asia – aiming for a regional view through 5–30 m satellite data. Global Ecol Conserv 2: 24–36

Mitsch, W.J., Gosselink, J.G., 2007. Wetlands. John Wiley \ and Sons, Inc., Hoboken, N. J, pp. 270–382.

Mitsch, W.J., Wang, N., 2000. Large-scale coastal wetland restoration on the Laurentian Great Lakes: determining the potential for water quality improvement. Ecol. Eng. 15, 267–282

Newton, A., Carruthers, T.J., Icely, J., 2012. The coastal syndromes and hotspots on the coast. Estuar. Coast. Shelf Sci. 96, 39–47.

Nilsson, C., Aradóttir, Á.L., 2013. Ecological and social aspects of ecological restoration: new challenges and opportunities for northernregions. Ecol. Soc. 18, 35

Reed MS, Buenemann M, Atlhopheng J, *et al.*, 2011. Cross-scale monitoring and assessment of land degradation and sustainable land management: a methodological framework for knowledge management. Land Degrad Dev 22: 261–71.

Ren, H., Wu, X.M., Ning, T.Z., Huang, G., Wang, J., Jian, S.G., Lu, H.F., 2011. Wetland changes and mangrove restoration planning in Shenzhen Bay, Southern China. Landsc. Ecol. Eng. 7, 241–250.

Richardson, C.J., Flanagan, N.E., Ho, M., Pahl, J.W., 2011. Integrated stream and wetland restoration: a watershed approach to improved water quality on the landscape. Ecol. Eng. 37, 25–39.

Rozas, L.P., Minello, T.J., Zimmerman, R.J., Caldwell, P., 2007. Nekton populations, long-term wetland loss, and the effect of recent habitat restoration in Galveston Bay, Texas, USA. Mar. Ecol. Prog. 344, 119–130

Seybold CA, Herrick JE, and Brejda JJ. 1999. Soil resilience: a fundamental component of soil quality. Soil Sci 164: 224–34.Bestelmeyer BT, Okin GS, Duniway MC, *et al.*, 2015. Desertification, land use, and the transformation of global drylands. Front Ecol Environ 13: 28–36.

Simenstad, C., Reed, D., Ford, M., 2006. When is restoration not?: Incorporating landscape-scale processes to restore self-sustaining ecosystems in coastal wetland restoration. Ecol. Eng. 26, 27–39.

Stringer LC, Dyer JC, Reed MS, *et al.*, 2009. Adaptations to climate change, drought and desertification: local insights to enhance policy in southern Africa. Environ Sci Policy 12: 748–65.

Thormann, M.N., Bayley, S.E., 1997. Aboveground plant production and nutrient content of the vegetation in six peatlands in Alberta, Canada. Plant Ecol. 131, 1–16.

Visser, J.M., Sasser, C.E., Chabreck, R.H., Linscombe, R.G., 1999. Long-term vegetation change in Louisiana tidal marshes, 1968–1992. Wetlands 19, 168–175.

Wagner, K.I., Gallagher, S.K., Hayes, M., Lawrence, B.A., Zedler, J.B., 2008. Wetland restoration in the new millennium: do research efforts match opportunities? Restor. Ecol. 16, 367–372.

Webb, N. P., Marshall, N. A., Stringer, L. C., Reed, M. S., Chappell, A., and Herrick, J. E. (2017). Land degradation and climate change: building climate resilience in agriculture. Frontiers in Ecology and the Environment, 15(8), 450-459.

Zedler, J.B., 2001. Handbook for Restoring Tidal Wetlands. CRC Press, Washington, DC

Zhang, J., Sun, Y.L., 2009. Eco-environmental quality and ecological restoration: a case study in wetland of the Loushan River Estuary, Jiaozhou Bay, Qingdao. In: International Workshop on Education Technology and Training and 2008 International Workshop on Geoscience and Remote Sensing, vol. 1, pp. 186–189

Zhao, Q., Bai, J., Huang, L., Gu, B., Lu, Q., and Gao, Z. (2016). A review of methodologies and success indicators for coastal wetland restoration. Ecological Indicators, 60, 442-452.

Chapter 18

Climate Smart Agriculture and Hi-Tech Farming

Rishav Paul and Vanya Bawa

University Institutes of Agricultural Sciences, Chandigarh University, Gharuan, Punjab

Climate smart agriculture is a strategy that helps drive efforts to adapt and reorient agricultural systems in order to promote production and food security in a changing climate in an efficient and sustainable way. 'Agriculture' includes the production of crops and cattle, fisheries and forest management.CSA is definitely not another advancement framework; it is method for characterizing which creation frameworks and supporting foundations are in a perfect world prepared to adjust to environmental change issues at specific areas, safeguarding and improving agribusiness' capability to economically encourage food security. As high-tech farming mainly refers to agricultural methods using the latest technology, like hydroponics and aerophonics. That is similar to growing vegetables in water or in the air only without the use of soil and there are also orchid farms and ornamental fish. It is also called agrotechnology. This paper gives a brief introduction about climate smart agriculture and hi tech farming and how it's going to be useful in future including focusing on climate smart agriculture and some possible approaches which can be used in future when the climate will be changing dynamically.

Keywords: *Precision farming, Hi-tech vertical farming, Solar fence, Arduino board.*

Introduction

Climate change is already hindering agricultural growth. Climate change affects crop production in several regions of the world, with negative effects more common than positive, according to the Intergovernmental Panel on Climate Change (IPCC), while developing countries are highly vulnerable to further negative impacts. Increases in the frequency and severity of weather events are already occurring and are predicted to intensify in many countries, such as drought, heavy rainfall, floods and high maximum temperatures. It is projected that average and seasonal maximum temperatures will continue to rise, with higher overall average rainfall. These results would not, however, be uniformly distributed. Water shortages and drought in already dry regions are also expected to increase by the end of the century (Lipper *et al.*, 2014).

In many places around the world, climate change is emerging as a major threat to agriculture, food security and the livelihoods of millions of people (IPCC, 2014). Several studies show that the production of agriculture could be significantly affected by temperature increases,changes in rainfall patterns (Prasanna, 2014; Mall *et al.*, 2006) and variations in the frequency and intensity of extreme climate events such as floods and droughts (Brida and Owiyo, 2013; Singh *et al.*, 2013). The predicted effects on cereal crop yields in different regions of both historical and future climate change suggest that the yield loss can be up to -35 per cent for rice, -20 per cent for wheat, -50 per cent for sorghum, -13 per cent for barley, and -60 per cent for maize depending on the location, future climate scenarios and projected year (Porter *et al.*, 2014). Changes in the suitability of crops and related biodiversity in agriculture, declines in the productivity of input usage and the proliferation of pests and diseases are some of the primary causes of the effect of climate change on agriculture (Zabel *et al.*, 2014; Norton, 2014).In order to ensure the food and livelihood security of farming populations, agricultural development processes need adaptation to those changes. To mitigate mild to extreme climatic threats in agriculture, there are many possible adaptation options. Climate smart agricultural (CSA) technology, activities and services are regarded as adaptation alternatives that sustainably increase efficiency, enhance resistance to climate stress and reduce greenhouse gas emissions.CSA broadly focuses on developing resilient systems of food production that contribute to food and income security in the context ofreformist environmental change and fluctuation (Vermeulen *et al.*, 2012; Lipper *et al.*, 2014).

Hi-tech farming aims at providing productive water, fertilizer and control agents at the right time and in real-time. High-tech farming can reduce the loss of the elements in the case of high-cost fertilizer and reduce the cost of production by using them precisely in high-tech farming. The irrigation system framework in India has given financial advancement a high priority.To empower rural computerization to thrive and convey its maximum capacity, numerous new ideas are being created. We ought not just consider the ramifications of building up another single innovation to exploit these advances, however should take a gander at the more extensive issues for the full improvement of a framework (Vardhan C.H *et al.*, 2019).

Precision Greenhouse Management

To track the atmosphere, there is a temperature/humidity sensor mounted inside the greenhouse. There are three drip irrigation pumps, a cooling pad and a fogging system, and two temperature and humidity monitoring aero exhaust fans, respectively. To constantly track temperature and humidity, a local wireless sensor network (LWSN) using IEEE 802.15.4 ZIGBEE modules is used.The corresponding actuators will switch ON/OFF to regulate the system whenever any irregularity is identified in the sensed data. In this way, the entire process of sensing and actuation is precisely made (Akshay c. *et al.*, 2012).

Farm and Food Framework Issues: Practical Increase, Agroecosystem Board and Food Frameworks

At the farm and food s ystem level, there is considerable research on climate change and agriculture, including topics such as farming practices to mitigateagrarian GHG discharges, choice and variation of harvests and domesticated animals to new environment systems, ranchers' dynamic and life cycle evaluations(de Graaf MA *et al.*, 2006). As shown by the topics of the six sessions in theme 1, the pattern has been to apply disciplinary science that tells fundamental issues and answers for agriculture Initially centred on improved agricultural productivity and food security, economical heightening has now moved to a more extensive scope of focuses with a few financial, moral and ecological perspectives (Godfray *et al.*, 2014). The CSA's integrative test is to more readily comprehend the compromises and decisions that farmers need to make for more prominent multifunctionality and environmental change strength. As environmental change arranging can be exceptionally farming, item and setting explicit, especially because of outrageous occasions, the CSA is focused on better approaches for participating in participatory exploration and maker organizations (FAO 2013).

Crop Physiology and Genetics under Climate Change

Responding to climate change impacts (*e.g.* changes in the availability of nutrients and the acquisition of plant nutrients,higher CO_2 focuses and temperatures, water shortfalls and flooding) that influence the conclusion of the yield hole among potential and genuine creation will require continuation of existing 'best administration practices' alongside enhancements in agronomic administration practice. The degree to which advances in crop physiology and genetics in response to more frequent climate shocks will continue to support higher agricultural production is uncertain.Though effective harvest transformation to new preparing areas can be a solid pointer of expected outcomes, a lot higher CO_2 focuses and temperatures are conditions outside our present scope of encounters(George t.2014). Molecular approaches and genetic engineering would promote better understanding and modulation of physiological processes responsible for crop growth and production, as well as the breeding of stress-adapted genotypes, however there are social controversies concerning the use of some of these technologies. High-throughput stages for phenotyping and thorough harvest models will prompt quicker hereditary asset investigation, empowering both quality revelation and a superior physiological comprehension of how crop improvement can increment ecological pressure resilience (Steenwerth K.L *et al.*, 2014).

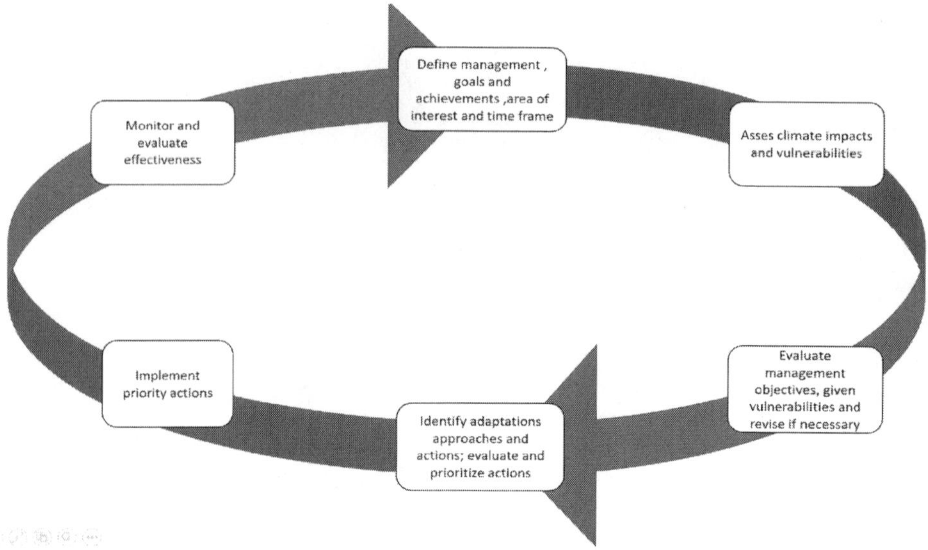

Figure 18.1: General Adaptation Planning Cycle that can be Used to Incorporate Climate Change Considerations into Management Decision-making.

Implementation of Hi-tech Vertical Cultivation and Solar Fence Oriented Security with Voice Input System

Proposed System

A proposed conspire with two viable components. They are the receptors of dampness and the siphon of engine/water. The Arduino Board is then customized with the Arduino IDE applications. The dampness sensor's motivation is to feel the dampness level in the dirt. The information was sent by the sensors to the data set to store the qualities in it. Once more, the qualities are sent by means of the WIFI port to the android telephone. This technique is utilized to keep missing information from being communicated from gadget to gadget when sending the sensor esteems. Also, the motor/water pump provides the plants with water. Using Arduino IDE software, the Arduino Board is programmed.The moisture sensor tests the amount of moisture in the soil and sends the signal to the Arduino if watering is necessary. The motor/water pump provides water to the plants before the optimal moisture level is achieved (Saifudaullah Md *et al.*, 2013).

Working Principle

Based on easy-to-use hardware and software, Arduino is an open-source hardware stage. Arduino sheets will understand inputs, light on a sensor, a finger on a catch or a message from Twitter, and change it into a yield, trigger a generator, turn on a LED, distribute something on the web. By sending a progression of directions to the microcontroller on the board, you will instruct the board. You utilize the Arduino programming language (in light of wiring) and Processing-based Arduino Applications (IDE) to do as such. Arduino has been the brain behind thousands of initiatives over the years, from everyday objects to sophisticated science tools(Saifudaullah Md *et al.*, 2013).

Passive Infrared (PIR)

The PIR sensor's most frequent use is as a 'area' sensor. It's all technically the same sensor and logic,regardless of whether it's to recognize 'somebody moving in the front yard' or 'somebody moving in the restroom' or 'somebody traveling through an entryway' or even 'somebody opening the brew cooler' There is a straightforward electronic gadget that is helpless to 'heat' or the infrared light transmitted by hot or warm items. In its least difficult structure, with a dark plastic 'window' on the top, it would appear that an old metal semiconductor. The 'rationale' of the PIR sensor is that inside the 'field' of its view, it should recognize 'huge change' from the typical degree of warmth (Manakant Intarakamhaeng 2008). It should have the option to figure out what 'ordinary' is by the circuits that control it, and afterward close a switch when the typical field changes, as when a human strolls before it. Moderate changes inside the field should likewise have the option to 'endure' and recall that as the new ordinary(Vardhan C.H 2019).

Precision Farming

PF is an integrated system of agricultural management that includes several technologies. Computers, GIS, GPS, spectral imagery, aircraft and vari-rate controllers *etc.* are often the technological tools. (Das *et al.*, 2017).

Need for Precision Farming

- ✰ Decline in the total productivity.
- ✰ Diminishing and degrading natural resources.
- ✰ Stagnating farm incomes.
- ✰ Lack of eco-regional approach.
- ✰ Declining and fragmented land holdings.
- ✰ Global climatic variation.
- ✰ Limited employment opportunities in non-farm sector.

Objectives of Precision Farming

- ✰ To increase production efficiency
- ✰ To improve product quality
- ✰ More efficient input use
- ✰ To reduce ecological degradation
- ✰ Energy conservation

Elements of Precision Farming

A) Information
B) Technology
C) Decision support (management)

Tools and Equipment

1. Global Positioning System (GPS)
2. Geographic Information System (GIS)
3. Grid Sampling
4. Variable Rate Technology (VRT)
5. Yield Maps
6. Remote Sensors
7. Proximate Sensors
8. Computer Hardware and Software
9. Precision irrigation systems
10. Leaf Chlorophyll Content

Figure 18.2: Left: Robots Used for Vegetable Harvesting; Right: Drones Used for Crop Field Analysis. (*Source*: Simon Birrell *et al.*, 2019).

Latest Technologies in Precision Farming

1. Robots: For quite a while, self-steering tractors have been operating like a plane on an autopilot. The tractor does the vast majority of the work, with the farmer stepping in for emergencies. Technology is advancing towards GPS-customized driverless devices to spread manure or furrow land(Pérez-Ruiz M *et al.*, 2012).
2. Smart phone applications: In precision farming, mobile phone and tablet technologies are increasingly becoming well known. Several helpful applications previously introduced, including the camera, GPS, and accelerometer, are supported by cell phones. There are likewise applications made dedicated to various horticulture and agriculture applications, for example, field mapping, pursuing creatures, acquiring climate and yield data, and the sky is the limit from there. They are efficient, reasonable, and have high computing power(Pérez-Ruiz M *et al.,* 2012).

3. Machine Learning: It is generally used for automatons, robots, and gadgets for the web of things. The contribution of information from each of these sources is taken into consideration. The PC at that point forms this data and sends the fitting activities back to these devices. This takes robots into account to transmit the ideal measure of manure or to give the ideal amount of water legitimately to the soil for IoT gadgets. (Thakur *et al.*, 2020).

A solar-fuelled machine that distinguishes weeds and obviously executes them with a portion of herbicide or lasers is incorporated in various developments. As of now, agricultural robots, otherwise known as AgBots, exist, but advanced harvesting robots are being developed to recognize ripe fruit, conform to its shape and size, and carefully pluck it from branches (Pérez-Ruiz M *et al.,* 2012).

Drones and satellite imagery: In precision farming, drone and satellite technology are utilized. This occurs regularly when the drone takes top notch (high quality) images while the master (bigger) image is captured by satellites. Aerial imagery with information from satellite records will be used by light airplane pilots to foresee potential yields based on the current amount of field biomass (Gemtos TA 2016).

Figure 18.3: Left: Agbots Used in Agriculture; Right: In the above Satellite Image, Green Shows Healthy Crops and Red Shows Unhealthy Crops. (*Source*: Tove Danovich, 2016; Matteo Luccio, 2014).

Conclusion

This study provides insights as to why climate smart agriculture and hi-tech farming are so important in the coming years. As the climate is getting diverse day by day and the level of CO_2 is increasing rapidly, to overcome these situation CSA approaches are applied. Hi technology farming decreases our time, energy consumption and labour cost in the agricultural operations. The use of hi technologies like robots and drones are going to be very beneficial in the coming days. This paper provides approaches towards a better and developed agriculture.

REFERENCES

Akshay, C., Karnwal, N., Abhfeeth, K. A., Khandelwal, R., Govindraju, T., Ezhilarasi, D., and Sujan, Y. (2012, December). Wireless sensing and control for precision Greenhouse management. In *2012 Sixth International Conference on Sensing Technology (ICST)* (pp. 52-56). IEEE.

Brida, A.B., Owiyo, T., 2013. Loss and damage from the double blow of flood and drought in Mozambique. International Journal of Global Warming 5 (4), 514–531.

Das U, Pathak P, Meena, Mallikarjun N. Precision Farming a Promising Technology in Horticulture. International Journal of Pure and Applied Bioscience 2017; 6: 1596-1606.

de Graaf MA, van Groenigen KJ, Six J, Hungate B, van Kessel C: Interactions between plant growth and soil nutrient cycling under elevated CO_2: a meta-analysis. Glob Chang Biol 2006, 12: 2077–2091.

Food and Agriculture Organization of the United Nations (FAO): ClimateSmart Agriculture Sourcebook. Rome: FAO; 2013.

Gemtos TA, Zude-Sasse M, Fountas S, Abu-Khalaf N. Applications of Precision Agriculture in Horticultural Crops.European Journal of Horticultural Science 2016; 81: 78-90

George T: Why crop yields in developing countries have not kept pace with advances in agronomy. Glob Food Sec 2014, 3: 49–58.

Godfray HC, Garnett T: Food security and sustainable intensification. Philos Trans R Soc Lond B Biol Sci 2014, 369: 20120273.

IPCC, 2014. Summary for policymakers. Climate Change 2014: Impacts, Adaptation, and Vulnerability. Part A: Global and Sectoral Aspects. Contribution of Working Group II to the Fifth Assessment Report of the Intergovernmental Panel on Climate Change. Cambridge University Press, Cambridge, United Kingdom and New York, NY, USA, pp. 1–32.

Jagan Gopu, A., and Ravikumar, R. (2016). A study on precision farming technologies and adoption in Tamil Nadu

Khatri-Chhetri, A., Aggarwal, P. K., Joshi, P. K., and Vyas, S. (2017). Farmers' prioritization of climate smart agriculture (CSA) technologies. *Agricultural systems, 151*, 184-191.

Lipper, L., Thornton, P., Campbell, B. M., Baedeker, T., Braimoh, A., Bwalya, M. and Torquebiau, E. F. (2014). Climate smart agriculture for food security. *Nature climate change*, *4*(12), 1068-1072.

Manakant Intarakamhaeng, "The Model Farm Management Automation Technology with RFID", 2008, Office of Science and Technology

McCarthy, N., Lipper, L., and Branca, G. (2011). Climate smart agriculture: smallholder adoption and implications for climate change adaptation and mitigation. *Mitigation of Climate Change in Agriculture Working Paper*, *3*(1), 1-37.

Norton, R., 2014. Combating climate change through improved agronomic practices and input-use efficiency. Journal of Crop Improvement 28 (5), 575–618.

Pérez-Ruiz M, Slaughter DC, Gliever J, Upadhyaya SK. Automatic GPS-based intra-row weed knife control system for transplanted row crops. Computers and Electronics in Agriculture 2012; 80: 41-49.

Porter, J.R., Xie, L., Challinor, A.J., Cochrane, K., Howden, S.M., Iqbal, M.M., Lobell, D.B., Travasso, M.I., 2014. Food security and food production systems. In: Field, *et al.* (Eds.), Climate Change 2014: Impacts, Adaptation, and Vulnerability. Part A: Global and Sectoral Aspects. Contribution of Working Group II to the Fifth Assessment Report of the Intergovernmental Panel on Climate Change. Cambridge University Press, Cambridge, UK and New York, USA, pp. 485–533.

Prasanna, V., 2014. Impact of monsoon rainfall on the total foodgrain yield over India. J. Earth Syst. Sci. 123 (5), 1129–1145

Saifudaullah Md, Bahrudin Bin and Kassim. Rosni Abu, "Development of Fire Alarm System using Raspberry Pi and Arduino Uno.", 2013, Faculty of Electrical Engineering University Teknologi MARA Selangor

Simon Birrell *et al.*, 'A Field Tested Robotic Harvesting System for Iceberg Lettuce.' Journal of Field Robotics (2019). DOI: 10.1002/rob.21888

Steenwerth, K. L., Hodson, A. K., Bloom, A. J., Carter, M. R., Cattaneo, A., Chartres, C. J. and Jackson, L. E. (2014). Climate smart agriculture global research agenda: scientific basis for action. *Agriculture and Food Security*, *3*(1), 1-39.

Thakur, N. Precision farming-Concepts and Practices.

Vardhan, C. H., Krishnan, V. C., Sathish, V., and Kumar, A. S. (2019). Implementation of Hi-tech vertical farming and solar fence security with voice feedback system.

Chapter 19

Climate Smart Agriculture and Hi-Tech Farming

Yengkhom Linthoingambi Devi[1] and Aditya Pratap Singh[2]

[1]College of Agriculture, Indira Gandhi Krishi Vishwavidyalaya, IGKV, Raipur, Chhattisgarh
[2]Bidhan Chandra Krishi Vishwavidyalaya, Mohanpur, West Bengal

According to an estimate of FAO, The population of the world will be increased by one- third of the present population by the year 2050 and the required food for food security would be increased by 60 per cent. The cumulative effect of climate change over the last decades also gave adverse effect on the productivity. Besides this, one important concern is having a high profession risk of being a farmer. Farmer in India considered farming to be very risky and farmers tend to migrate their profession due to unsecured nature of the weather impact on the farming system. This has direct impact on socio-economic process. Thus by using the available knowledge and experience, it is possible to make agriculture a sustainable livelihood means. So, Climate smart agriculture refers to the achievement of sustainable development goals. With the increased in population with less area for cultivation of crops, high-tech farming system provides a boon to the agriculture sector by giving good amout of products by using less area by using advanced techniques. So, Hi- tech farming refers to the process of agriculture by using the latest technologies. It is a capital intensive agriculture because a large amount of sum is required towards the purchase of specialized equipments, maintainance of assets, training of labours etc. Some of the high-tech agricultural activity which help in productivity enhancement are: Soil-less agriculture, Protected cultivation under greenhouse, Vertical farming. Thus, by using farming with high technology the demand for the increased production of crops could be fulfilled to some extend. Also, the production of high value crops has increased the income of farmers and also the availability if such crops even in the off-season time or year-round availability is possible due to high tech farming system. This type of framing must be adopted in the present generation to provide quality product to the growing population. Thus, using high-tech farming helps in obtaining the sustainable development gaol of the country

Keywords: *Climate smart agriculture, Soil-less agriculture, Protected cultivation, Vertical farming*

Introduction

Climate Smart Agriculture

Climate Smart Agriculture (CSA) is an approach to manage agricultural system respond effectively to climate change in order to help the farmers. The CSA approach fulfilled triple objectives which are-1) sustainably increase the productivity, increase the incomes, and lastly adaptation to change in the climate and reducing greenhouse gas emission wherever possible.

As we know that the majority of the world's poor live in rural areas with agriculture being their source of income. So, developing the potential to lift the income an productivity from smallholder crop, livestock, fish and forest production systems will be the factor for elevating the global food security over the next twenty years. Climate changes like change in the precipitation pattern, higher temperature, rising sea levels and frequent extreme weather events is expected to hit developing countries the hardest. All of the above pose a risk for agriculture, food and water supplies. Thus by using the available knowledge and experience, it is possible to make agriculture a sustainable livelihood means. So, Climate smart agriculture refers to the achievement of sustainable development goals. It combines social, economical and environmental development to face challenge of providing sustainable (a) livelihood to farmers (b) food security to hungry millions, and (c) eradication of poverty.

Thus, Climate smart Agriculture as given by the Food and Agricultural Organisation of the United Nations (FAO) is, "agriculture that sustainably increases productivity, enhance resilience or adaptation, reduces/remove greenhouse gas or mitigation where possible and enhance achievement of national food security and development goals".

CSA consists of mainly 3 pillars:

1. **Productivity**: increasing the productivity of agriculture and income sustainably from crops, livestock and fish, without having negative impact on the environment. This, in turn, will raise food and nutritional security.
2. **Adaptation**: CSA helps in strengthening the resilience by strengthening their capacity to adapt and prosper in the face of long term stresses. CSA aims to reduce the short term risk of farmers. Thus it helps in adapting and building resilience to climate change.
3. **Mitigation**: CSA should help to reduce the emission of greenhouse gas wherever and whenever possible. It envisages to achieve this through

increased cropping, by reducing rain fed areas through integrated water and river basin management and expansion of agriculture on wasteland, wetland, degraded fallow areas and introducing urban agriculture.

CSA needs to act together at the local level because-

1. We need to explain or translate the climate change to the village level in such a way that Climate change is bound to happen from year to year but there are available solutions for this.
2. There are many already existing programmes and plans but these need to exploit in the village level because their existance seems nothing if we don't exploit from the village level.
3. Farmers in these generations are all literate and can understand any new technologies and techniques if they were informed. So giving chance to them to develop and prepare their own production plan will help them implementing the CSA by themselves.
4. Technologies keep on adding year after year along with plans prepare by using the available knowledge. But understanding of the technologies is decreased from top to middle to bottom. Only chosen few will be benefited and gap will still prevailed in same village with same land and water resources among the farmers thus forming one farmer very successful, average ones fail and some commits suicide or migrate to urban centres.

Hi-tech Farming/Agriculture in India

Hi- tech farming refers to the process of agriculture by using the latest technologies. It is a capital intensive agriculture because a large amount of sum is required towards the purchase of specialized equipments, maintenance of assets, training of labours *etc.*

Some of the high-tech agricultural activity which help in productivity enhancement are given below:

1. Soil-less agriculture:
2. Protected cultivation under greenhouse:
3. Vertical farming

1. Soil-less Agriculture

Generally plants are growing in the soil because they require the nutrients for their growth and development. But we can grow the plants without the soil if we could provide all the required nutrients by pre-mixing into water reservoir along with high level of oxygen efficiently to the plant root.

Advantages of Soil-less Agriculture

1. Faster growth: the plants can grow faster which means that it could have more cycles per year along with more yield and more food.

2. Extreme decrease in water and nutrient use: Hydroponic culture can typically reduce the use of water by 90 per cent less than the soil based method if we do recycling and closing the loop of the water.
3. Controlled environment agriculture: CEA is a method of agriculture that optimizes the temperature, humidity, airflow, and light within in the growing environment of a farm. Growing in a controlled environment also vastly improves a farm's ability to predict crop timing, grow high quality plants, and maintain high food safety standards. No birds flying over head or animals burrowing in the field, and pesticides can be avoided.
4. Location: The farm can and should be significantly closer to the end consumer, decreasing the carbon footprint of delivery and increasing the freshness of the product. Since there is no dependence on soil fertility, this kind of growing can be done anywhere.
5. Use significantly less space: With stacked, or vertical systems, the yield per square foot greatly increases. This is crucial in order to grow enough food for growing populations without needing to expand farmland into wetlands, forests, or other important natural ecosystems.

Types of Systems

Within the hydroponic methods several different systems are there with several purpose which could be used according to our purpose of work. It provides a system from a small grower to a large massive commercial grower.

So, the different system of hydroponic techniques are-

Deep Water Culture (DWC)

In this system, floating plants are used on recirculating water. Several ways are found to do this but the most often being the use of Styrofoam insulation board (4ft

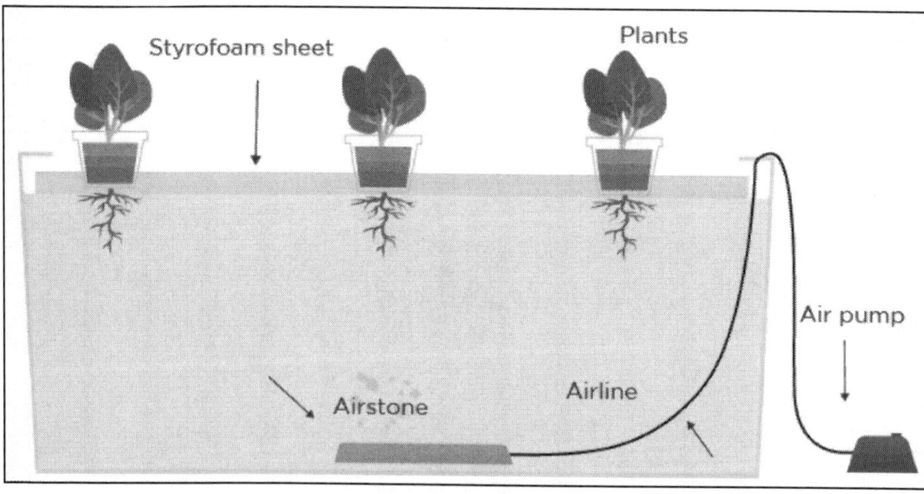

Figure 19.1: Deep Water Cultivation.

× 8ft). Holes need to be drilled in the styrofoam fixed insulation board to allow the roots of the germinated plant from the pot. The roots of the plant will be suspended in the well-oxygenated nutrient solution until the roots of the plant reach about 6-18 inches length until harvest. The growing condition is perfect for those leafy green plants as these plants does not require to have much longer and much root support. The system holds a large volume of water which slows down any swing or changes in the chemistry of the nutrient solution.

Nutrient Film Technique (NFT)

Here in this system, channels are made on slight angle and the plants will grow in the slightly angled channels. Nutrient solution is applied to the plant through pipes by using nutrient pump which is found embedded in the nutrient solution. After getting saturated water by the root of the plant the nutrient water will be drained off from one side of the slanted chamber to the reservoir of nutrient solution. Thus, the nutrient solution and the growing plants are connected by channels or troughs which are set up on slight angle for drainage purpose and provide a shallow stream of water to the roots. As per the growing of short-statured plant it is best as such DWC plants but, these system hold much less water as compare to DWC and they are more easily cleanable, stackable and customizable for your grow space.

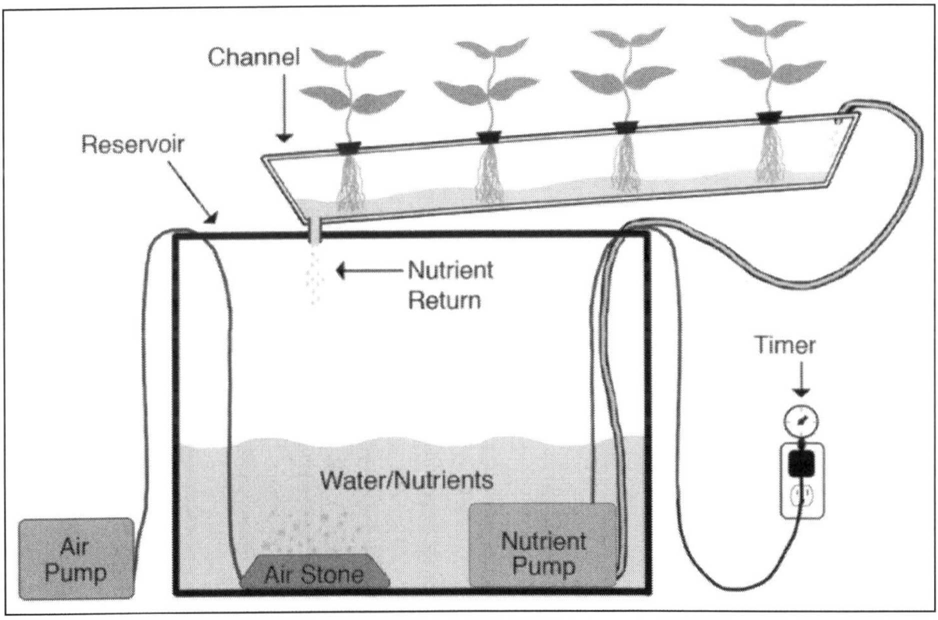

Figure 19.2: Nutrient Film Technique.

Ebb and Flow (Flood and Drain)

Ebb and flow systems can be as basic as a small plastic bucket with some expanded clay pellets or other rock media that you hand water and drain. It can also be as complex as adding a large media filled bed onto an aquaponic system and flooding it with the liquid waste from the system. In every instance, the grow tray is temporarily flooded with solution every few hours, submerging the roots before returning to the reservoir. Because of the root support and oxygen levels they can provide, ebb and flow systems are great for growing pretty much anything, but especially fruiting crops. You just have to be sure that however it's being done, the setup can support the weight of the media and water and your containers drain completely. This article from Upstart University shows how easy it is to get started with Bato Buckets. Ebb and flow systems are low maintenance and produce high yields, but like NFT systems, a pump failure can quickly become catastrophic for your plants.

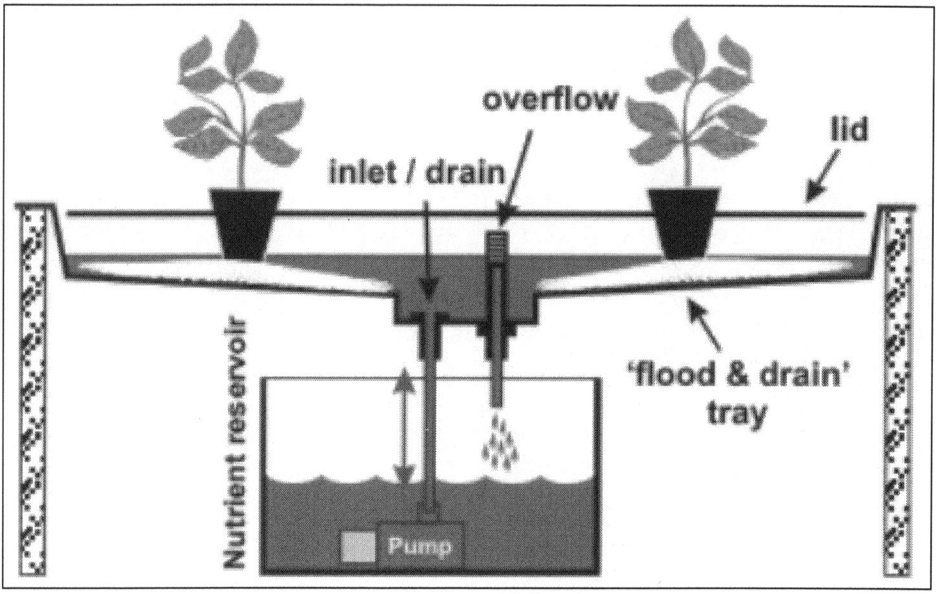

Figure 19.3: Ebb and Flow Method.

Drip System

Drip system use a simple technique, whereby each plant will get the nutrient solution delivered to its base by a water pump system. Here, the excess solution can either be collected again in the reservoir or not collected. The system works well in those growing media like coco coir, peat moss, or rock-wool which have high water retention. It is a low maintenance system along with high output if the drip system works perfectly but sometimes problem occurred due to the clogging of the drip lines. So in order to avoid this problem, instead of using the organic materials we can use the synthetic nutrients as organic clog lines much faster.

Aeroponics

Here in this system the plants are grown suspended in the air and the nutrient solution will be applied to them in the form of mist. One benefits of this system is its ability to provide precise nutrient delivery and water usage. But its major defect

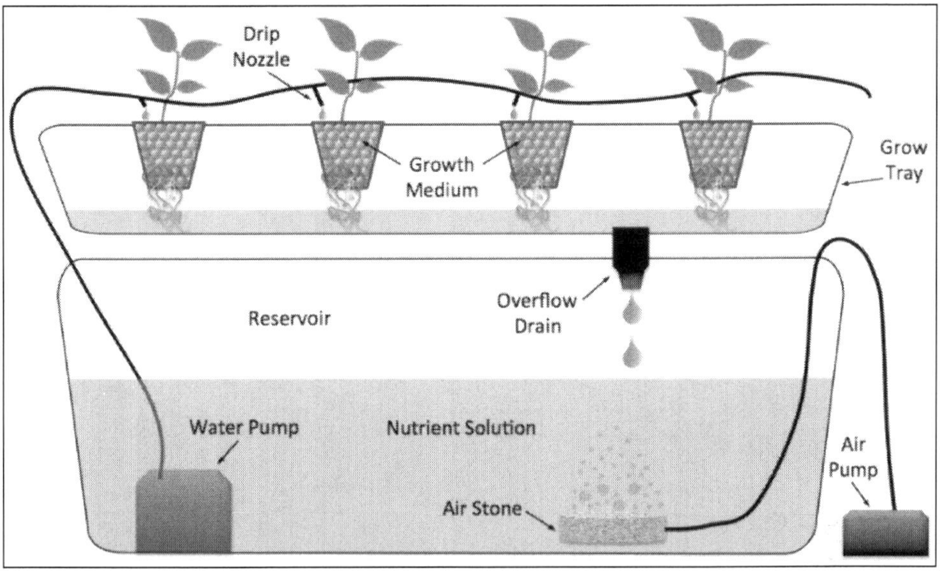

Figure 19.4: Drip System.

is having having high risk of clogging issues as the emitters have very tiny holes. A-frame aeroponic systems are common with PVS frame made tall cones, which gives the root chamber a perfect environment.

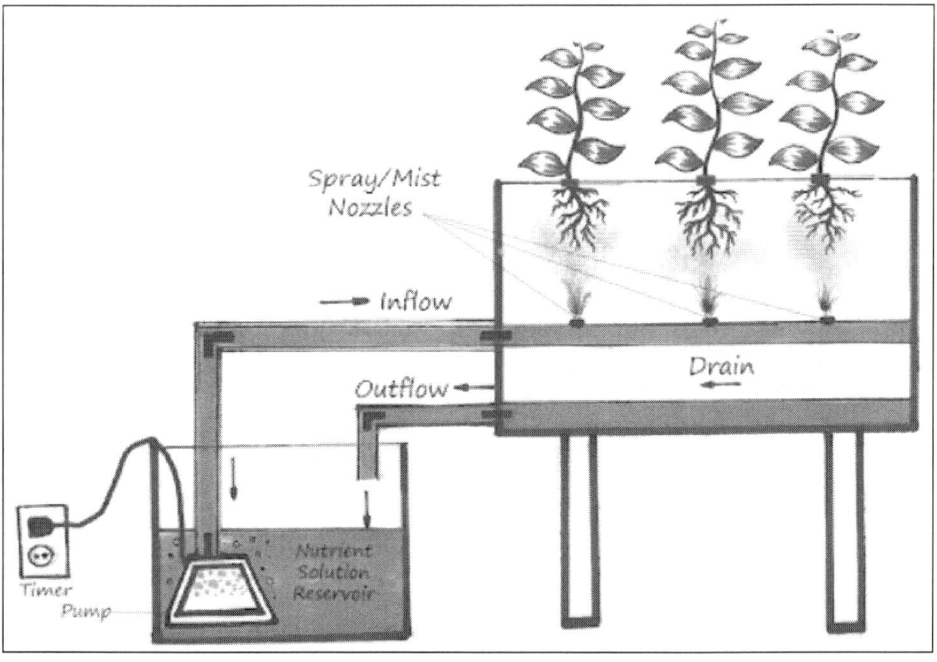

Figure 19.5: Aeroponics.

Wicking System

This is s system in which the wicks are used to deliver the nutrient solution to the tray and from the tray the roots of the plant will be taking the solution through its capillary action. It has no moving parts and works great on a hobby scale, but is not the most efficient when it comes to nutrient use.

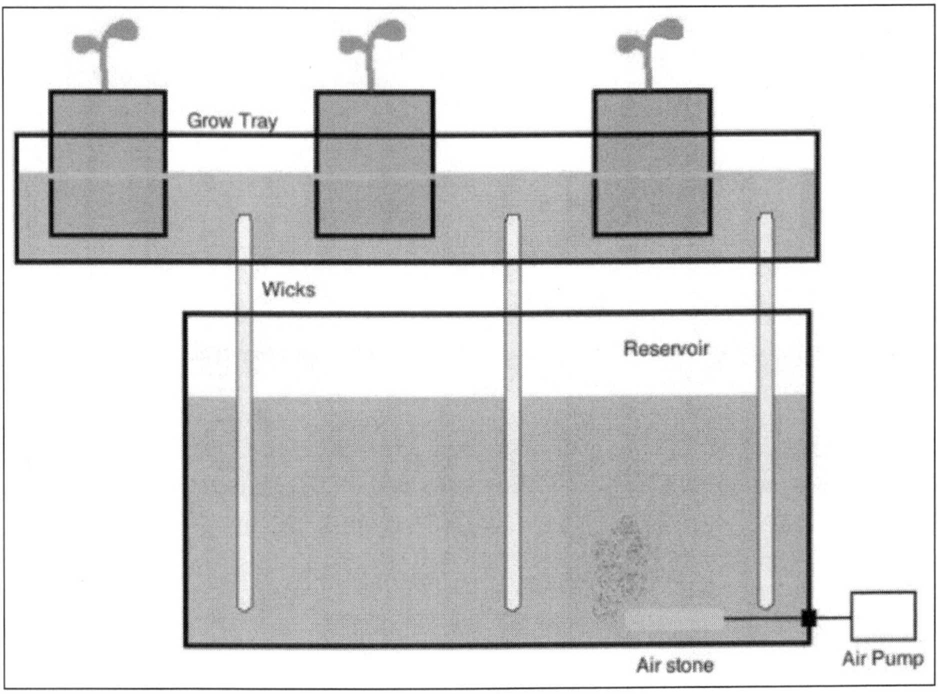

Figure 19.6: Wicking System.

2) Protected Cultivation Under Greenhouse Condition

Components and Features of a Greenhouse Based Hi-tech Agriculture System

Polyhouse or greenhouse made of tight, cheap, flexible and transparent polythene enables the cultivation of vegetables and other crops in any season round the year by controlling the humidity and temperature in the poly house. These structures prevent the escaping of thermal radiation which increase the temperature and energy and thus helps in the process of photosynthesis.

Equipment's required: In case we want to make a permanent poly-house structure, fibre and steel made glass are galvanised hallow pipe-having glass. And another way is to use transparent polythene sheet structure is used for making permanent poly-house structure. Irrigation is provided by sprinkler unit system, while for controlling the air temperature ventilators are required.

Roof of the poly-house can be made from plastic film, nylon, acrylic, vinyl, poly carbonate and polyethylene film. Majority of the rooftop are made from poly-film of size 200 microns to 800 gauge thickness along with ultraviolet protective film. The framework of the poly-house made from G.I pipe.

Watering system: the best form of watering plants in greenhouse is micro irrigation which can be either micro sprinklers or drip irrigation. In micro-sprinkler, water is released under high pressure through the nozzles which are arranged on a supporting stand at about 1 feet height. The equipments which is necessary for drip irrigation system are i) a pump unit to generate 2.8kg/cm^2 pressure ii) PVC tubing with dripper or emitters

Hi-tech Greenhouse

To overcome some of the difficulties in medium-tech greenhouse, a hi-tech greenhouse where the entire device, controlling the environment parameters, are supported to function automatically.

In hi-tech greenhouse system, both solar and photovoltaic systems was used. Sustainable and environment friendly used of pumps, UV filters and lighting through the use of modern solar technology was used. One way of sustainable and environment friendly way of using greenhouse is the use of solar aquaponic greenhouse system which allows the simultaneous production of vegetables, fish and solar power. This system of greenhouse leads to production of lower CO_2 production. It was first used by the chinese in the 6th century AD by rearing of ducks, fish in their rice field. And the use of fish in the system of aquaponic was first used at the New Alchemy Institute in the 1970s where the tilapia fish was raised in large circular transparent containers. The main purpose of growing the plants along with the fish was to clean the water bodies for the fish to survive because the quality of water in the Arks' fish tanks would deteriote quickly and regular cleaning was required. This is the first modern mainstream example of aquaponics.

Aquaponics and Organic Hydroponics

Hydroponic technique is has been found to be used back to thousands of years ago. As we can see its practice at the Floating Gardens in China and the Hanging Gardens of Babylon where the practice of aquaponics was observed in which raising of fish and plants in the same water source. The fish gave the fertilizer or nutrient to the plant and the roots of the plant filter the water for the fish. The fish release ammonia as its excretory material through their gills and waste, these ammonia are converted to nitrites through naturally occurring nitrifying microorganisms and eventually into nitrates, which is the plant food. So, aquaponic methods exploit the use of three important things that is plants, fish and beneficial bacteria.That same nitrifying bacteria in aquaponics is also needed in organic hydroponic applications. Aquaponics are more closely related to growing organically in hydroponics than to synthetic hydroponics. The reason for this is that maintaining a healthy and vibrant colony of bacterial colony helps in sanitising and flushing the majority of the water, thus can improve the overall health of the system. But the ammonia produced by the fish needs to be converted fully into nitrate so fully mature water needs to be obtained before using. Oxygen is the defining element of these systems so it is important to have a good flow rate and biofiltration to remove any oxygen consuming waste to avoid anaerobic activity.

The Sections or Portions of the Greenhouse for Aquaponics

1. The rearing of the fish was done on the rear of the greenhouse so that it won't get the sunlight in order to minimize temperature shifts. These water from the fish tank will be circulated to those plants which are growing at the front of the greenhouse to facilitate fertilization organically.
2. Vents are present in the transparent south wall as well as at the top back of the roof and its main purpose is maintain a thermostable environment and it is allowed to open and release the hot air as needed on summer days.

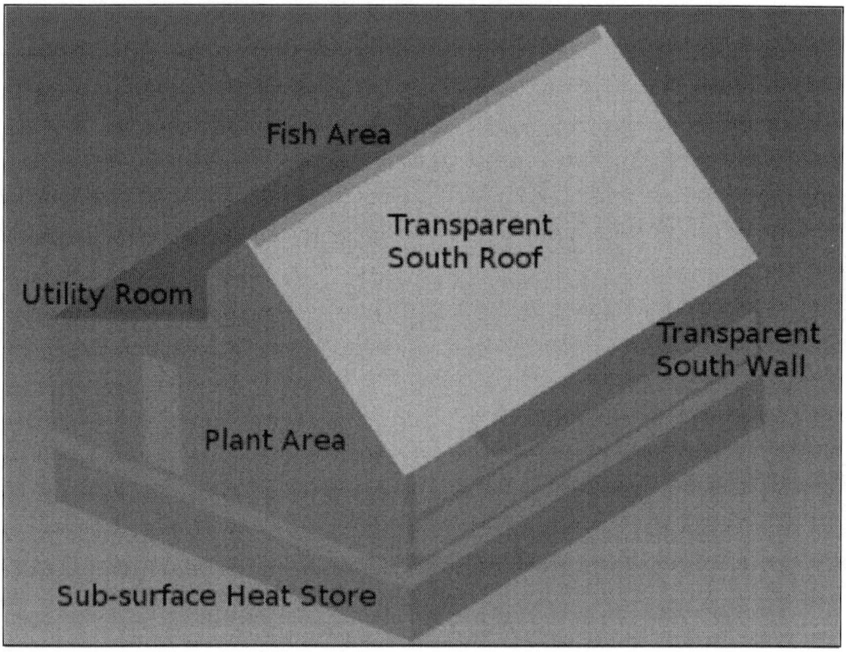

Figure 19.7: Aquaponic Greenhouse.

3. The south surface which faces the transparent roof as well as the vertical south wall are triple glazed for heat retention.
4. The front of the greenhouse is the area where the plants are grown beneath the glazed roof. It is separated from the fish area by a dividing wall and 36 heat-retaining water barrels.
5. The foundation of the greenhouse is made in such a way that the entire greenhouse floor is filled with one metre of rip-rap rock beneath them, which will make the foundation concrete. A fan is blown inside the greenhouse to maintain an stable interior temperature during day and night

6. At the rear side of the greenhouse an utility room is there for the processing of fish and also for the washing as well as processing of the greens.
7. The parts of the roof like the side, rear and back portion are insulated thickly to retain the heat.

3. Vertical Farming

Vertical farming is a proposed indoor, urban farming technology which involves the large scale production of agricultural products in multi-storey buildings. So, the cultivation practice can exploit the use of soil, hydroponic or aeroponic technique or a combination of them.it helps in production of food in challenging time where the arable land for cultivation of rare or not available. Thus, Vertical farming can be defined as farming of fruits, vegetables, grains *etc.* in the middle of a building where different floors have different purpose which means that one floor is meant for a particular type of crop, another floor for another crop and so on or we can used all the floor for the same crop. The concept of vertical farming is a large scale extension of urban agriculture.

Figure 19.8: Vertical Farming.

The main goals of vertical farming is:

1. To supply the sustainable food source for urban centres.
2. To allow the agro land to natural landscape
3. To provide a sustainable organic farming technique
4. To provide black or grey water remediation
5. Appropriate unused and abandoned urban spaces

6. End food contamination
7. To get year round food production
8. End reliance on pesticides, herbicides and petro- based fertilizers.
9. Create sustainable urban space

Systems Used in Vertical Farming

1. Hydroponics: cultivation of plant through continuous supply of oxygenated nutrient water.
2. Nutrient- flow technique
3. Float stem rectangular reservoirs filled with water
4. AquaponicsL a combination of hydroponic and aquaculture. In this system fish waste is used as nutrients for plants.
5. Aeroponics: a method in which nutrient rich mist are provided to the exposed root into the air chamber

Advantages of Vertical Farming

1. Reliable harvest: Since the vertical farming approach is practised in fully controlled condition, there is no risk of having threat due to change in the weather condition. Thus, we can have a reliable harvest from the crop which we were growing.
2. Low-energy usage: The method use the high energy efficiency LED lighting system, which gives minimum power usage for maximum plant growth.
3. Low-labour cost: Nowadays labour charges becomes high and requiring more labour may such up the amount of money required for maintenance. Vertical farming is the form of farming where the requirement of labour is less as it required only during sowing, harvesting and packaging of the material besides this, any other defects or errors related to the cultivation will be sent via message so not so many labour requirement is there.
4. Low-water usage: Vertical farming use less water as compare to other methods of farming and it is said that it require water about 10! less than that used in the open field condition and 20! less than that used in the hydroponic condition as it is performed under the closed system and it provides less transpiration from the plant and the transpire water is allowed to reuse again.
5. Reduced washing and processing: Since vertical farming is practised in closed condition, the requirement for attack of pest and disease is less which make the plant less required for washing and processing practices,
6. Increased growing area and space saving: Vertical farming provides a multilevel design of farming which allow 8 times more growing area than the conventional hydroponic and other greenhouse method. It is estimated that every acre used for vertical farming is equivalent to 4 acres of horizontal farming.
7. Toxic water runoff doesnot take place

Disadvantages of Vertical Farming

1. The cost of production as well as the maintenance of vertical farming is quite high as compare to other method of cultivation.

High Value Crops (HVC)

High value crops are those crops which on cultivation gives a significantly higher productivity value as well as good net income per unit of resource used for production, compared to other competing activities. Diversification in high-value crops has become an important way of increasing the income of the farmer since the demand for high-value crops has increased nowadays as compare to the staple crops. Some activities related to HCV cultivation are:

- Cultivation of the fruits and vegetables in the urban peripherals.
- Off-season and exotic vegetables are grown under greenhouse condition for export purpose (*e.g.* Asparagus,celery, bell pepper, sweet corn, green and lima beans).
- Cultivation of flowers in the open as well as greenhouse condition for local and export purpose.
- Based on the needs and demands of industry or under contract farming arrangement, exclusive production of the crops can be performed. Eg for medicinal and aromatic purpose
- Cultivation of mushroom
- High-tech nursery unit for fruit crops, flowers and vegetables, *etc.*

High Value Agriculture for Small Farmers

Even though the high value crop is mainly associated with big farmers, there is still a scope for small and marginal farmers by adopting and utilizing those benefits given by government. Some of the government interventions are given below:

Govt. Interventions

The important promotional interventions of Government departments towards high value agriculture sector include:

- Incentives or subsidy support are provided under National Horticulture Mission and other programmes of National Horticulture Board for adopting high value agriculture production technologies.
- Facilitation of export and promotional interventions of Agricultural and Processed Food Products Export Development Authority (APEDA)/ Marine Products Export Development Authority (MPEDA)
- Promotional interventions of Commodity Boards
- Programmes of the state government like precision farming with focus on high value crops.

Initiatives of NABARD

- Guidance are given to banks for high-tech agriculture projects including techno-financial appraisal, investment assessments, risk management and monitoring systems.
- Projects are being developed for high-tech and high value agriculture sector activities.
- Financial support are also landed to the state government under RIDF for infrastructure development.
- Landing hands as consultancy services to prospective entrepreneurs.

Strategies for Promotion of Hi-tech Agriculture in India

- The participation of the farmers especially small and marginal farmers should be encouraged for better improvement of farm income.
- Studying and learning the successful models for replication
- Indentification of crop specific location and establishment of potential mapping.
- Development with Group approach such as Farmers' Societies, Producer Companies, JLGs/SHGs, *etc.*
- Providing and assessing good infrastructural needs for post-harvest handling and transport logistics, warehousing, marketing *etc.*
- Preparation of Sector/activity specific credit linked Development Plans (Area Development Plans/Banking Plans)
- ICT enabled services such as market information, technology inputs and credit access

Conclusions

Hi-tech farming facilitate the uses of recent technologies and equipments to get a high quality of agricultural produce. Use of herbicide, insecticide and pesticide is less in this method so quality product is obtained from high-tech farming. It has high output and high market value because the product obtained is good and demanding for better health conscious people. Physical input to the system are kept minimum to the system, but it is capital intensive since a large sum of money is required to obtain the specialized equipment, maintainance, training of labour *etc.* thus, the construction process is an expensive method but the output is good. One demerit of high-tech farming is not all persons can initiate it because of its high initial cost. In india, not so many farmers are rich enough to establish such a high-tech farming system even though such a farming helps them with better return in their income. Several initiative and schemes are being initiated by the government so, farmers should take the advantage of those scheme to start such system. Thus, using high-tech farming helps in obtaining the sustainable development gaol of the country.

REFERENCES

FAO. 2013a Climate –Smart Agriculture: Sourcebook. Rome, Italy: Food and Agriculture Organisation of the United Nations.

High-tech farming in India, www.nabard.org/auth/writereaddata/CareerNotices

Jose, K.J. Breathing highrises: Vertical farming as an emerging option for the urban resources management in future cities. M. Arch. Lovely Professional University, Jalandhar, Delhi, 14-16.

Neelands, P. (2019). The year-round solar aquaponics greenhouse project. MSc Thesis, Yorklands Green Hub, 5-16.

Vermeulen, S.J., Campbell, B.M., Ingram S.J.I. (2012). Climate change and food system. Annual Review of Environment and Resources, 37: 195-222.

Shelat, K.N (2014) Climate Smart Agriculture- the way forward the Indian Perspective, NCCSD,CRIDA, Hyderabad, pp.4-7

Tenkil, R. (2015). Model Bankable Project on Hi-Tech A griculture, NABARD, Thiruvananthapuram, pp.5-7

Chapter 20

Soil and Climate Change in Adaptation and Mitigation Option Towards Better Crop Production

Mallepu. S. Likhitha Reddy[1], *Shivani Kumari*[2] *and Sentiyanger Jamir*[3]

[1]*Department of Agronomy, UBKV, West Bengal*
[2]*Department of Agronomy, SASRD, Nagaland University, Medziphema, Nagaland*
[3]*Department of Agronomy, SHUATS, Prayagraj, Uttar Pradesh*

Environmental change is, in principle, the ideal theme for worldwide conversation. It is a developing worldwide issue and concern, calling for deliberate endeavors by the created just as agricultural nations. The main factor that is liable for the environmental change is the "nursery impact". The environmental change will influence different yields during the whole developing time frame attributable to outrageous climate occasions. The versatility of yields to environment fluctuation would secure huge measurement as successive event of extraordinary climate conditions may direct concerning how to adapt up to the circumstance. The environmental change will incorporate limit heat just as chilly waves, expanded changeability regarding rabi and kharif crop creation, expanded limit whether occasions when all is said in done, whimsical beginning just as advances of storm, and so forth Since ranch creation is probably going to be antagonistically influenced, the environmental change will bring about expanded water the executives works on, improving area, soil and sustenance the board rehearses, decrease in shortcoming in water use, conservation and improvement of plant and animal hereditary assets, acclimation to the example of food utilization, advancement of eco-accommodating use transformation, variation to environment based methodologies of danger the board, move in trimming design, transformation to coordinated cultivating framework and incorporated agro ranger service frameworks, and so forth There are zones where transformation program have effectively been created and these incorporate harvest improvement, hazard financing, dry season sealing, fiasco the board, and so on Adjusting farming frameworks to environmental change is critical on the grounds that its effect is now obvious and the patterns will proceed regardless of whether emanations of GHG outflows are settled at current levels.

Keywords: *Climate change, Environment, Global warming, Greenhouse effect etc.*

Introduction

Climate is one of the primary determinants of agrarian creation. All through the world there is critical worry about the impacts of environmental change and its fluctuation on farming creation. Analysts and executives are worried about the possible harms and advantages that may emerge in future from environmental change impacts on farming, since these will influence homegrown and worldwide approaches, exchanging design, asset use and food security. The Climate change is any adjustment in environment over the long haul that is credited straightforwardly or by implication to human action that modifies the arrangement of the worldwide air notwithstanding regular environment inconstancy saw throughout practically identical time-frames (IPCC, 2007). Since climatic variables fill in as immediate contributions to farming, any change in climatic components will undoubtedly essentially affect crop yields and creation. Studies have shown a critical impact of progress in climatic factors on the normal harvest yield (Dinar *et al.*, 1998; Mall *et al.*, 2006; Cline, 2007; Seo and Mendelsohn, 2008). In agricultural nations, environmental change will cause yield decreases for the main harvests and South Asia will be especially hard hit (IFPRI, 2009). Numerous examinations in the past have shown that India is probably going to observe one of the greatest horticultural profitability misfortunes on the planet as per the environmental change design noticed and situations anticipated. Environmental switch projections made up to 2100 for India show a general expansion in temperature by 2-4 °C with no generous change in precipitation amount (Kavikumar, 2010). In course of time where the mechanical upheaval happened in western nations and utilization of the non-renewable energy sources expanded quickly, on the opposite side the normal buffering framework for environmental change woodlands, were obliterated aimlessly for need of fuel, grain and lumbers in the agricultural nations. These components were increased by the human exercises in the previous 250 years, which enormously affected the environment framework. As per the IPCC the greenhouse gas emanation could make the mean worldwide temperature ascend by another 1.4 °C to 5.8 °C. Effectively the manifestations of environmental change were seen at a quicker rate in the icy and under cold areas through softening of the frozen ice which have threat of submergence of the beach front zones. On account of the inland water source, there is colossal change in both surface just as ground water because of inconsistent precipitation and event of successive dry spells. Numerous investigations (Parry *et al.*, 1999; Olesen and Bindi, 2002; Adams *et al.*, 2003; Tsvetsinskaya *et al.*, 2003; Darwin, 2004).

Most past evaluations of the effects of environmental change on horticulture (and in fact on different areas) have zeroed in on time skylines towards the finish of the twenty-first century, showing the effects of anthropogenic environmental change that could be kept away from by lessening ozone harming substance emanations. In any case, there is likewise a need to evaluate the effects of environmental change over the course of the following not many years, which may now be to a great extent unavoidable inferable from dormancy in the actual environment framework and the time scales over which huge scope change in human social, monetary and political impacts on ozone harming substance outflows could be achieved. Regardless of whether ozone harming substance outflows started to be decreased promptly, there would in any case be some degree of continuous warming for quite a long time and some ocean level ascent proceeding for quite a long time, as the environment framework is delayed to react completely to forced changes. There is moderately little data in the writing accessible on environmental change impacts throughout these time skylines, so we present MOHC environment projections for roughly 2020 and 2050 to place the current writing into setting on these time scales.

Impacts of Climate Change on Agriculture

Worldwide climatic changes can influence farming through their immediate and circuitous consequences for the harvests, soils, domesticated animals and bugs. An increment in environmental carbon dioxide level will have a treatment impact on crops with C3 photosynthetic pathway and along these lines will advance their development and profitability. The expansion in temperature, contingent on the current encompassing temperature, can diminish crop length, increment crop breath rates, change photosynthetic parceling to monetary items, influence the endurance and dispersion of irritation populaces, rush supplement mineralization in soils, decline compost use efficiencies, and increment evapo-happening rate. In a roundabout way, there might be impressive consequences for land use because of snow dissolve, accessibility of water system water, recurrence and power of between and intra-occasional dry spells and floods, soil natural matter changes, soil disintegration, changes in bug profiles, decrease in arable territories because of submergence of waterfront grounds, and accessibility of energy. Similarly significant determinants of food supply are financial climate, including government strategies, capital accessibility, costs and returns, foundation, land changes, and entomb and intra-public exchange that may be influenced by the climatic change.

Reduction in Crop Yield

Ascend in the mean temperature over a limit level will cause a decrease in rural yields. An adjustment in the base temperature is more pivotal than an adjustment in the greatest temperature. Grain yield of rice, for instance, declined by 10 per cent for every 1 °C increment in the developing season least temperature over 32°C (Pathak *et al.*, 2003). The environmental change sway on the efficiency of rice in Punjab (India) has shown that with any remaining climatic factors staying consistent, temperature increments of 1°C, 2°C and 3°C, would lessen the grain yield of rice by 5.4 per cent, 7.4 per cent and 25.1 per cent, separately (Aggarwal *et al.*, 2009).

Shortage of Water

The expanded temperature would bring about more water deficiencies and the interest for water system water would rise. Expansion in air temperature will prompt more potential evaporation happening in the zones south of 40° N. Similarly; water deficiency because of environmental change would result in about 20 per cent net decrease in the rice yields in India.

Irregularities in Onset of Monsoon, Drought, Flood and Cyclone

Indian horticulture is exceptionally reliant on the beginning, retreat and extent of storm precipitation, especially in the rainfed zones of east, north-east and south India. Environment modelers and IPCC records have extended prospects of expanding changeability in Asian Monsoon flow in a hotter world. In spite of development of zone under water system, dry seasons, brought about by insufficient and lopsided dispersion of precipitation, keep on being the main climatic distortions, which impact the agrarian creation in India. The seriousness of a dry season will be strengthened in a hotter world. Extraordinary and continuous flooding because of environmental change would be a significant issue in the Indian subcontinent.

Rise in Sea Level

In South, South East and East Asia about 10 per cent of the provincial rice creation, which is sufficient to take care of 200 million individuals, is from the regions that are vulnerable to 1 m ascent in the ocean level. Direct loss of land joined with less positive water driven conditions may decrease rice yields by 4 per cent if no transformation measures are taken, imperiling the food security of at any rate of 75 million individuals. Saltwater interruption and soil salinization are different worries for horticultural profitability.

Decline in Soil Fertility

Soil temperature influences the paces of natural matter disintegration and arrival of supplements. At high temperatures, however supplement accessibility will increment for the time being, over the long haul natural matter substance will lessen, bringing about a decrease in soil richness.

Loss of Biodiversity

Types of creatures and plants are assessed to vanish at a rate which would be around 100-times quicker than the chronicled record, generally because of human exercises. A point by point appraisal of the 394 types of primates from South America to Indonesia has demonstrated that 29 per cent are at risk for vanishing because of chasing, natural surroundings misfortune and environmental change.

Pests, Weeds and Diseases

As temperature expands, the creepy crawly irritations will turn out to be more plentiful through various between related cycles, including range expansions and phonological changes, just as expanded paces of populace improvement, development, relocation and over-wintering. The environmental change is probably going to adjust the harmony between creepy crawly bothers, their common adversaries and their hosts. The ascent in temperature will support bug advancement and winter endurance. Rising environmental carbon dioxide fixations may prompt a decrease in food quality for plant-taking care of bugs, because of diminished foliar nitrogen levels. The study of disease transmission of plant sicknesses will be modified. The forecast of illness episodes will be more troublesome in times of quickly changing environment and insecure climate. Natural shakiness and expanded occurrence of extraordinary climate may lessen the adequacy of pesticides on focused nuisances or result in more injury to non-target creatures.

Impact of Climate Change on Soils

Environmental change is worldwide wonders and happening ceaselessly since the earth appeared. Environmental change has become a significant logical and policy driven issue during the most recent decade. There are very much stamped cold and sweltering cycles throughout the entire existence of earth's environment, notwithstanding, these progressions have been noticed moderately fast in the last 150-200 years around the planet. Soil is by all accounts more significant for present day human social orders than any other time in recent memory to satisfy the worldwide needs for food and fiber for expanding populace from restricted soil assets. Environmental change is compromising food security universally. Nations like India are more helpless considering the heat and humidity and helpless adapting limit of the little and peripheral ranchers. Environmental change is projected to essentially affect farming through immediate and circuitous consequences for crops, soils, animals and vermin. However, environmental change is a sluggish cycle including moderately little changes in temperature and precipitation throughout extensive stretch of time, by the by these lethargic changes in environment impact the different soil measures especially those identified with soil ripeness. The impacts of environmental change on soils are normal essentially through modification in soil dampness conditions and expansion in soil temperature and CO_2 levels because of environmental change.

Soil Formation

Soil formation is constrained by various components including climatic factors like temperature and precipitation. These boundaries of environment impact the dirt development straightforwardly by giving biomass and conditions to enduring. Primary boundaries of environment that straightforwardly effect on soil arrangement are amount of dynamic temperatures and precipitation-vanishing proportion. They decide estimations of energy utilization for soil development and water adjusts in soil, component of natural mineral collaborations, change of natural and mineral substances and streams of soil arrangements. Stable reformist environment warming lead to irreversible changes in mineral lattice of soils. Changes in outside elements of soil arrangement (temperatures and precipitation) will prompt change of interior elements (energy, hydrological, natural). The environmental change will expand energy of obliteration of soil minerals bringing about improvement of mineral grid because of collection of minerals open minded to enduring. It will lead loss of soil work for ripeness support and more prominent reliance of on mineral manures.

Soil Development

Soil improvement is comprehensively constrained by three fundamental variables for example environment, parent material and vegetation type. The impacts of environmental change on soil advancement are normal mostly through adjustment in soil dampness conditions and expansion in soil temperature and CO_2 levels. Environmental change will impact soil dampness levels by direct climatic impacts (precipitation, temperature consequences for dissipation), environment initiated changes in vegetation, plant development rates, paces of soil water extraction by plants and the impact of improved CO_2 levels on plant happening. Changes in soil water motions may likewise criticism to the actual environment and even may add to dry season conditions by diminishing accessible dampness, adjusting flow designs and expanding air temperatures. Among different elements controlling the cycle of soil improvement, environment assumes a significant part in enduring of rocks and minerals. The factors of environmental change especially temperature and precipitation directs different phases of enduring of rocks and minerals (parent material) bringing about compound and mineralogical changes in soil framing rocks. Water is fundamental for substance enduring to occur and consequently, an increment in precipitation quickens enduring. Similar kinds of essential minerals offer ascent to various optional minerals when the states of enduring contrast. Consequently comparable stone sorts going through enduring in various climatic conditions could offer ascent to particular soil profiles.

Soil Fertility and Productivity

The drivers of environmental change like dampness, temperature and CO_2 are required to effectively affect different soil cycles and properties having significance in soil ripeness and efficiency. Nonetheless, these impacts of the environmental change factors can't been independently; being one factor impact the other and resultant impact would be intricate. Further, every one of these impacts will be profoundly locale explicit, contingent upon the size of the environmental change, soil properties and climatic conditions. India is offered with 9 of the 12 soil orders that exist in world and 15 agro-climatic zones, with different seasons, yields and cultivating frameworks. Since environmental change is a reality, it will in a roundabout way affect soil improvement cycles and properties identified with crop creation impacting the jobs of millions of people groups in the country.

Plant Nutrient Availability and Acquisition

Plant accessibility of supplements in the dirt is an element of soil substance properties just as area of the particle comparative with the root surface and the length of the pathway the supplement should go in the dirt to arrive at the root surface. Expansions in air temperature and changes in precipitation fundamentally affect root zone temperature and dampness systems. It is notable that dirt dampness and temperature are essential determinants of supplement accessibility and root development and improvement and that carbon portion to roots administers supplement securing, it is sensible to expect that cycle results will be intelligent of the changed environment. The nature and degree of the adjustment in these two boundaries will be site-and soils explicit. It affects supplement use proficiency is be principally influenced through direct effects on root surface zone and inundation rate.

Nutrient Transformation in Soil

Plants gather supplements from the dirt arrangement pool, and supplements should be in answer for be versatile in the dirt. Natural change among natural and inorganic pools is emphatically impacted by dampness and temperature, and consequently, worldwide environmental change may unequivocally impact arrangement centralizations of N just as S (Pendall *et al.*, 2004) recommends that expanded CO_2 may not apply a huge direct impact on N mineralization fundamentally however related warming can cause expanded N mineralization, prompting expanded arrangement stage N. Paces of adsorption/desorption responses will quicken with expanded temperature, and changes in soil dampness may additionally adjust responses by modifying the ionic strength of the dirt arrangement. **3.6. Soil carbon dynamics**

It is by and large acknowledged that increments in CO_2 fixation quantitatively and subjectively change the arrival of root inferred compounds. Plants under raised CO_2 decline their distribution of N-rich metabolites and increment the assignment of C rich metabolites to root exudates (Tarnawski *et al.*, 2006). It brings about an increment in microbial action and therefore the CO_2 creation, which has is a likely negative impact on the gathering of natural C in soils and along these lines on possible sequestration of soils. It has been seen that the preparing impact because of the upgraded microbial movement in soil at raised air CO_2 focus has a huge negative criticism on worldwide change measures and will diminish the sequestration capability of soils. A few investigations utilizing C isotope tracers have shown that the creation of CO_2 in the rhizosphere by roots and microorganisms is fundamentally invigorated by raised CO_2 plant development conditions. The incitement of CO_2 breath in the rhizosphere might be a lot higher than the improvement of root biomass. Cheng *et al.* (1998) exhibited that in spite of the fact that plants delivered just 15–26 per cent more biomass under raised CO_2, rhizosphere breathed C expanded by 56–74 per cent when contrasted with encompassing CO_2 medicines.

Response to Mycorrhizal Association

The impacts of raised climatic CO_2 fixation on soil microbial local area structure are frequently described by an expanded mycorrhizal colonization because of the expanded plant interest for supplements, combined with expanded C digestion rates CO_2 improvement should increment mycorrhizal biomass since plant requests for N and P will increment simultaneously with C osmosis rates, and plants will allot more hotosynthates subterranean to the roots and mycorrhizal parasites to help fulfill this expanded supplement interest. More prominent fine root mass and mycorrhizal disease advance upgraded P take-up in mycorrhizal plants developed under raised CO_2 fixations. It appears to be sensible to expect that at raised CO_2 levels, mycorrhizal biomass will increment as C turns out to be generally less restricting and soil supplements become more restricting to plant development. Nonetheless, data accessible in writing isn't generally predictable on this point (Drigo *et al.*, 2008).

Soil Biological Activities

The reaction of soil microorganisms to changes in plant creation under raised CO_2 is exceptionally factor because of altogether different examples of plant C designation in various plant–soil frameworks. Microbial biomass, net N mineralization, microbial immobilization, and net N mineralization under raised CO_2 show a serious level of changeability. In any case, paces of soil and microbial breath are by and large more fast under raised CO_2, demonstrating that upgraded plant development under raised CO_2 builds the measure of C entering the dirt, subsequently animating soil microbial movement. Soil microorganisms are regularly C-restricted and in this way, expanded C accessibility invigorates microbial development and movement. It is for the most part expected that the CO_2 prompted increments in soil C accessibility will increment contagious biomass more than bacterial biomass. It is because of increased concentrations of dissolved organic C in the rhizosphere and to increases in soil water dissolved organic N. Given the important roles played by fungi in organic matter degradation, nutrient cycling, plant nutrition, and soil aggregate formation, shifts in fungal communities might have a strong impact on soil functioning.

Adaptation and Mitigation Strategies

Agriculture can adjust to environmental change by receiving ranch the executives rehearses that limit the antagonistic impacts of expanding or diminishing precipitation and temperatures or other limit climate conditions. Numerous administration level variation choices are accessible to constrict the impacts of environmental change on crop creation, including zero culturing, holding crop buildups, broadening fallows, expanding the variety of creation, adjusting sums and timing of outer sources of info (manures, water), just as more extensive agronomic administration methodologies (for example modifying planting thickness, column dispersing, planting time; acquainting new germplasm safe with warmth or dry season pressure). Horticulture can add to environmental change moderation through ranch the board rehearses that decrease ozone depleting substance outflows (carbon dioxide, nitrous oxide, methane) and upgrade soil carbon sequestration. Emanations of carbon dioxide can be diminished through decreased biomass consuming and more productive energy use. Outflows of methane can be decreased through improved homestead the executives rehearses that incorporate improved administration of domesticated animals waste and water in rice paddies. Nitrous oxide emanations can be diminished through improved administration of N composts including suitable sort, rate and strategy for application and soil the board (shirking of soil compaction).

Different ranch the executives practices can improve soil carbon stocks and support soil utilitarian dependability. Preservation agribusiness innovations (least soil unsettling influence, cover yields and harvest revolutions including vegetables), soil protection measures (for example shape cultivating) and supplement renewal methodologies can reestablish soil natural matter by giving a defensive soil cover and a climate helpful for energetic plant development. At times, notwithstanding, an adjustment in the agrarian creation framework might be required. for example consistent grain trimming being supplanted by ley cultivating or by the presentation of agroforestry frameworks. The worldwide soil carbon pool surpasses biomass pools by a factor of four or five, without considering that new soil corruption has prompted misfortunes of between 30 per cent and 75 per cent of their predecessor soil natural carbon. Around the world, in this manner, a dirt carbon increment offers incredible moderation potential. Carbon sequestration alludes to the capacity of carbon in a steady strong structure. It happens through immediate and backhanded obsession of air CO_2. Direct soil carbon sequestration happens by inorganic substance responses that convert CO_2 into soil inorganic carbon mixtures, for example, calcium and magnesium carbonates. Direct plant carbon sequestration happens as plants photosynthesize environmental CO_2 into plant biomass. Along these lines, a portion of this plant biomass is in a roundabout way sequestered as soil natural carbon (SOC) during decay measures. The measure of carbon sequestered at a site mirrors the drawn out balance between carbon take-up and discharge components. Numerous agronomic, ranger service, and protection works on, including best administration rehearses, prompts a helpful net addition in carbon obsession in soil.

Agro-biological systems can assume a significant part in alleviating CO_2 outflows through biotic C sequestration in soils and vegetation. Due to notable misfortunes of C from soils, assessed to be 41 to 55 Gt, the dirts presently offer a chance for carbon stockpiling. The carbon sequestration capability of a dirt relies upon environment, the sort of vegetation it underpins, the idea of parent material, the profundity of solum, soil waste, the edaphic climate, soil natural matter (SOM) substance and its decomposability and land the executives rehearses. Improved administration of agro-environments can altogether upgrade C sequestration in soils. The executives practices or advancements that increment carbon contribution to the dirt and diminish C shortfall or both lead to net carbon sequestration in soils. Expanded C contribution to agro-biological systems can be accomplished in various manners, for example, determination of high biomass delivering crops, buildup reusing or buildup maintenance by diminished culturing force, utilization of natural materials (for example creature fertilizer, fertilizer, muck, green excrement and so forth), appropriation of agro ranger service frameworks, strengthening of farming through improved supplement and water the executives works on, lessening summer or winter neglected, changing from monoculture to revolution trimming, and changing from yearly harvests to enduring vegetation. Soil carbon misfortune could be diminished by receiving preservation agribusiness and limiting soil unsettling influence, checking disintegration through decreased culturing force, and utilizing bad quality natural sources of info. Mechanical choices that have been discovered to be productive for soil C sequestration in Indian agro-environments incorporate coordinated supplement the executives and developing, crop buildup joining, mulch cultivating and/protection agribusiness, agro-ranger service frameworks, touching administration, decision of editing framework and heightening of horticulture. Incorporated supplement the board including expansion of natural excrements/composts alongside inorganic manures brings about improved soil accumulation (Benbi *et al.*, 2009).

Mitigation Strategies to Climate Change for Better Crop Production

The systems for moderating methane discharge from rice development could be modification in water the board, especially advancing mid-season air circulation by momentary waste; improving natural matter administration by advancing oxygen consuming debasement through treating the soil or fusing it into soil during slow time of year depleted period; utilization of rice cultivars with few inefficient turners, high root oxidative movement and high collect file; and use of aged composts like biogas slurry instead of unfermented farmstead fertilizer (Pathak and Wassmann, 2007). Methane emanation from ruminants can be diminished by modifying the feed piece, either to decrease the rate which is changed over into methane or to improve the milk and meat yield.

The most productive administration practice to lessen nitrous oxide emanation is site-explicit, proficient supplement the board (Pathak, 2010). The outflow could likewise be diminished by nitrification inhibitors, for example, nitrapyrin and dicyandiamide (DCD). There are some plant-inferred organics, for example, neem oil, neem cake and karanja seed extricate which can likewise go about as nitrification inhibitors.

Relief of CO_2 discharge from farming can be accomplished by expanding carbon sequestration in soil through control of soil dampness and temperature, putting to the side excess rural land, and rebuilding of soil carbon on debased grounds. Soil the executives practices like decreased culturing, manuring, buildup fuse, improving soil biodiversity, miniature conglomeration, and mulching can assume significant parts in sequestering carbon in soil. A few innovations, for example, discontinuous drying, site-explicit N the executives, and so forth can be effectively received by the ranchers without extra venture, while different advancements need monetary impetuses and strategy uphold (Wassmann and Pathak, 2007).

Adaptation Strategies to Climate Change for Better Crop Production

To manage the effect of environmental change, the potential variation systems are: creating cultivars lenient to warmth and saltiness stress and impervious to flood and dry spell, altering crop the board works on, improving water the executives, embracing new homestead procedures, for example, asset rationing advancements (RCTs), crop expansion, improving irritation the board, better climate estimating and crop protection and outfitting the native specialized information on ranchers. A portion of these procedures are talked about beneath.

Developing Climate Ready Crops

Advancement of new harvest assortments with better return potential and protection from various anxieties (dry season, flood, saltiness) will be the way to keep up yield soundness. Improvement in germplasm of significant harvests for heat-stress resilience ought to be one of the objectives of reproducing program. Also, it is fundamental to create resistance to different abiotic stresses as they happen in nature. The abiotic stress resistance systems are quantitative characteristics in plants. Germplasm with more prominent oxidative pressure resilience might be abused as oxidative pressure resistance is one model where plant's guard component focuses on a few abiotic stresses. Like the exploration endeavors on transformation of rice from C3 to C4 crop, steps ought to be taken for development in radiation-use proficiency of different yields too.

Improvement in water-use and nitrogen-use efficiencies is being endeavored since long. These endeavors expect more pertinence in the environmental change situations as water assets for farming are probably going to decrease in future. Nitrogen-use productivity might be diminished under the environmental change situations as a result of high temperatures and hefty precipitation occasions causing volatilization and draining misfortunes. Aside from this, for misusing the valuable impacts of raised CO_2 fixations, crop interest for nitrogen is probably going to increment. Accordingly, it is imperative to improve the root proficiency for mining the water and ingestion of supplements. Abuse of hereditary designing for 'quality pyramiding' has gotten fundamental for pool every one of the attractive characteristics in a single plant to get the 'ideal plant type' which may likewise be 'unfavorable environment lenient' genotype.

Ranchers should be furnished with cultivars with an expansive hereditary base. Their variation cycle could be fortified with accessibility of new assortments having resistance to dry spell, warmth and saltiness and accordingly, limit the dangers of climatic deviations. Likewise, improvement of assortments is needed to counterbalance the arising issues of shortening of developing season and different fancies of creation climate. Ranchers could more readily settle their creation framework with crate of innovative choices.

Crop Diversification

Broadening of yield and animals assortments, including substitution of plant types, cultivars, mixtures, and creature breeds with new assortments expected for higher dry season or warmth resistance, is being supported as having the capacity to build efficiency despite temperature and dampness stresses. Variety in the seed hereditary construction and arrangement has been perceived as a successful protection against sickness and irritation flare-up and climatic risks. Additionally, interest for high-esteem food products, like natural products, vegetables, dairy, meat, eggs and fish is expanding a direct result of developing pay and urbanization. This is decreasing the interest for conventional rice and wheat. Enhancement from rice-wheat towards high-esteem items will build pay and result in decreased water and compost use. Nonetheless, there is a need to measure the effects of harvest expansion on pay, business, soil wellbeing, water use and ozone depleting substance emanations. A huge constraint of expansion is that it is exorbitant regarding the pay openings that ranchers forego, *i.e.*, exchanging of harvest can be costly, making crop enhancement regularly less beneficial than specialization. Besides, customs can frequently be hard to survive and will direct neighborhood rehearses.

Changes in Land-use Management Practices

Evolving land-use practices like the area of yield and animals creation, pivoting or moving creation among harvests and animals, moving creation away from minor regions, adjusting the force of manure and pesticide application just as capital and work data sources can help decrease chances from environmental change in ranch creation. Changing the editing succession, including changing the circumstance of planting, planting, splashing, and collecting, to exploit the changing term of developing seasons and related warmth and dampness levels is another choice. Adjusting the time at which fields are planted or planted can likewise assist ranchers with managing the length of the developing season to more readily suit the changed climate. Rancher transformation can likewise include changing the circumstance of water system or utilization of different information sources like composts.

Adjusting Cropping Season

Change of planting dates to limit the impact of temperature increment instigated spikelet sterility can be utilized to lessen yield shakiness, by abstaining from having the blooming time frame to match with the most blazing time frame. Variation measures to diminish the negative impacts of expanded climatic fluctuation as regularly experienced in bone-dry and semi-parched jungles may incorporate changing of the trimming schedule to exploit the wet time frame and to stay away from outrageous climate occasions (*e.g.*, hurricanes and tempests) during the developing season. Trimming frameworks may must be changed to incorporate developing of reasonable cultivars (to check pressure of harvest advancement), expanding crop powers (*i.e.*, the quantity of progressive yield delivered per unit territory each year) or planting various sorts of yields. Ranchers should adjust to changing hydrological systems by evolving crops.

Efficient Use of Resources

The asset monitoring innovations (RCTs) incorporate practices that upgrade asset or info use effectiveness and give quick, recognizable and verifiable monetary advantages, for example, decrease underway expenses; investment funds in water, fuel and work necessities; and convenient foundation of harvests, bringing about improved yields. Yields of wheat in warmth and water-focused on conditions can be raised essentially by embracing RCTs, which limit negative ecological effects, particularly in little and medium-scale ranches. Asset monitoring rehearses like zero-culturing (ZT) can permit ranchers to plant wheat sooner after rice reap, so the harvest heads and fills the grain before the beginning of premonsoon sweltering climate. As the normal temperatures in the district rise, early planting will turn out to be considerably more significant for wheat. Field results have shown that the RCTs are progressively being received by ranchers in the rice-wheat belt of the Indo-Gangetic Plains on account of a few benefits of work saving, water saving, and early planting of wheat. The RCTs in rice-wheat framework additionally effect sly affect moderation of ozone harming substance emanation and transformation to environmental change (Pathak *et al.*, 2009). These methodologies of yield the executives ought to be combined with the proportions of harvest improvement for more extensive variation to environmental change. Soil and water the executives is exceptionally basic for variation to environmental change. With higher temperatures and changing precipitation designs, water will additionally turn into a scant asset. Genuine endeavors towards water protection, water gathering improvement in water system availability, and water-use productivity will get fundamental for crop creation and job the board. Ranchers must be prepared and persuaded for receiving on-ranch water preservation procedures, miniature water system frameworks for better water-use effectiveness, determination of proper yields, and so on Standards of expanding water invasion with progress in soil accumulation, diminishing spillover with utilization of forms, edges, vegetative fences, and so forth and decreasing soil vanishing with utilization of harvest deposits mulch could be utilized for better administration of soil-water.

Relocation of Crops in Alternative Areas

Environmental change as far as expanded temperature, CO_2 level, dry seasons and floods would influence creation of harvests. In any case, the effect will be distinctive across harvests and areas. There is a need to distinguish the harvests and locales that are more touchy to environment changes/changeability and move them in more reasonable territories. For instance, it is caught that expanded temperature would influence the nature of harvests, especially significant sweet-smelling yields, for example, basmati rice and tea. Elective regions that would get appropriate for such harvests according to quality perspective should be distinguished and surveyed for their appropriateness.

Harnessing Indigenous Technical Knowledge of Farmers

Ranchers in South Asia, regularly poor and negligible, are exploring different avenues regarding the climatic inconstancy for quite a long time. There is an abundance of information on the scope of measures that can help in creating advancements to beat environment weaknesses. There is a need to saddle that information and tweak them to suit the cutting edge needs. Conventional environmental information on individuals created and conveyed which have stood the trial of time could give experiences and reasonable choices to versatile measures. Anthropological and sociological investigations have featured the significance of local area based asset the executives and social figuring out how to upgrade their ability to adjust to the effects of future environmental change. Ancestral and slope information frameworks are pregnant with potential native practices utilized for assimilation and preservation of water, supplement and weed the board, crop creation and plant assurance. Their conviction frameworks have adequately helped in climate anticipating and hazard change in crop development. Over the span of their home, the native individuals of Himalayan territory locale through experience, experimentation and gathered information, have concocted methods of diminishing their weakness to regular risks. Studies have shown that their agreement was reasonably advanced in the issue of tremor, avalanche and dry season and they have contrived productive methods of moderating the impact of characteristic or climatic changes.

Improved Pest Management

Changes in temperature and fluctuation in precipitation would influence rate of vermin and illness and harmfulness of significant harvests. It is on the grounds that environmental change will conceivably influence the vermin/weed have relationship by influencing the bug/weed populace, the host populace and the bug/weed-have collaborations.

A portion of the potential transformation systems could be: (I) creating cultivars protection from vermin and sicknesses; (ii) selection of coordinated nuisance the executives with more accentuation on natural control and changes in social practices, (iii) bug anticipating utilizing ongoing apparatuses, for example, reenactment displaying, and (iv) creating elective creation strategies and harvests, just as areas, that are impervious to pervasions and different dangers. The executives of vermin and sicknesses with utilization of safe assortments and breeds; elective normal pesticides; bacterial and viral pesticides; pheromones for disturbing irritation multiplication, and so forth could be received for supportability of rural creation measure. Bioagents have a significant part in bug the board, subsequently practices to advance common adversaries like arrival of hunters and parasites; improving the territory for regular foes; encouraging bug banks and blooming strips; crop revolution and numerous editing ought to be coordinated in bother the executives rehearses. Decrease being used of pesticides will likewise help in lessening fossil fuel byproducts.

Better Weather Forecasting and Crop Insurance Schemes

Climate guaging and early admonition frameworks will be helpful in limiting dangers of climatic enemies. Data and correspondence innovations (ICT) could enormously help the analysts and overseers in creating emergency courses of action. Successful yield protection plans ought to be developed to help the ranchers in diminishing the danger of harvest disappointment because of these occasions. Both formal and casual, just as private and public, protection programs should be set up to help lessen pay misfortunes because of environment related effects. Notwithstanding, data is expected to build out strategies that support successful protection openings. Miniature account has been a triumph among provincial poor, including ladies. Ease admittance to monetary administrations could be a shelter for weak ranchers. Developing organization of versatile communication could additionally accelerate SMS-based financial administrations and assist the ranchers with bettering reconciliation with monetary establishments. In any case, contrasted with miniature money, miniature protection developments and accessibility is restricted. There is a need to create reasonable protection framework, while the provincial poor are to be taught about profiting such freedoms.

Conclusion

Environmental change represents a developing danger to maintainable turn of events. The normal impacts of environmental change could genuinely bargain the capacity of the farming areas to take care of the world, and seriously subvert progress toward destroying appetite, unhealthiness and destitution. Activity is critically expected to set up the rural areas for the possibility of quickly changing ecological conditions. As the horticulture areas are mostly liable for the gathering of ozone harming substance in the environment that are dependable environmental change, it is likewise essential to lessen farming outflows. Indeed, even without environmental change, world agribusiness and food security are face overwhelming difficulties. Populace development and rising livelihoods in a large part of the creating scene have pushed the interest for food and other agrarian items to remarkable levels. Without elevated endeavors to lessen destitution and improve agrarian profitability, some low-pay nations, particularly those in sub-Saharan Africa and South Asia, will think that its hard to guarantee admittance to sufficient amounts of nourishment for all.

REFERENCES

Adams, R. M. McCarl, B. A. and Mearns, L. O. (2003). "The effect of spatial scale of climate scenarios on economic assessments: An example from U.S. Agriculture", Climate Change, 60: 131-148.

Aggarwal, P. K., Singh, A. K., Samra, J. S., Singh, G., Gogoi, A. K, Rao, G. and Ramakrishna, Y. S. (2009). Introduction In Global Climate Change and Indian Agriculture, Ed: PK Aggarwal, ICAR, New Delhi, pp. 1-5.

Benbi, D. K., Senapati, N. (2009). Soil aggregation and carbon and nitrogen stabilization in relation to residue and manure application in rice-wheat systems in northwest India, Nutrient Cycling in Agroecosystems, 87: 233–247.

Brouder, S. M, Volenec, J. J. (2008). Impact of climate change on crop nutrient and water use Efficiencies, Physiologia Plantarum, 133(4): 705–724.

Cheng, W. X., Johnson, D. W. (1998). Elevated CO_2, rhizosphere processes, and soil organic matter decomposition, Plant and Soil, 202(2): 167–174.

Cline, W. R. (2007). "Global warming and agriculture: Impact estimates by country", Peterson Institute of International Economics, NW, Washington, D.C., U.S.A.

Darwin, R. (2004). "Effect of greenhouse gas emission on world agriculture, Food Consumption and Economic Welfare", Climate Change, 66: 191-238.

Dinar, A. R., Mendelsohn, R., Evenson, J., Parikh, A., Sanghi, K., Kumar, J., McKinsey and Lonergen, S. (1998). "Measuring the Impact of Climate Change on Indian Agriculture", Technical Report, The World Bank, Washington, D.C., U.S.A.

Drigo, B., Kowalchuk, G. A., Johannes, A. V. (2008). Climate change goes underground: effects of elevated atmospheric CO_2 on microbial community structure and activities in the rhizosphere, Biology and Fertility Soils, 44(5): 667- 679.

Fauchereau, N., Trzaska, M., Rouault, M. *et al.* (2003). Rainfall variability and changes in Southern Africa during the 20th century in the global warming context, Natural Hazards, 29(2): 139–154.

IPCC. (2007). "Summary for Policymakers", in Solomon, S., D. Qin, M. Manning, Z. Chen, M. Marquis, K.B. Averyt, M. Tignor and H.L. Miller (Eds.), Climate Change: The Physical Science Basis. Contribution of Working Group I to the Fourth Assessment Report of the Intergovernmental Panel on Climate Change, Cambridge University Press, Cambridge, United Kingdom and New York, NY, U.S.A.

Kavikumar, K.S. 2010. "Climate sensitivity of Indian Agriculture: Role of Technological Development and Information Diffusion, in Lead Papers-National Symposium on Climate Change and Rainfed Agriculture, Indian Society of Dryland Agriculture", Ceentral Research Institute for Dryland Agriculture, Hyderabad, February 18-20, 2010, pp.1-18.

Mall *et al.* (2006). IFPRI, "Climate Change – Impact on Agriculture and Costs of Adaptation", Food Policy Report International Food Policy Research Institute, Washington, D.C., U.S.A., pp.30., 2009.

Olesen, J. E. and Bindi, M. (2002). "Consequences of Climate Change for European Agriculture Productiivty, Land Use and Policy", European Journal of Agronomy, 16: 239-262.

Parry, M., Rosenzweig, C., Inglesias, A., Fischer, G. and Livermore, M. (1999). "Climate Change and World Food Security: A New Assessment", Global Environ. Change, 9(l): S51-S67.

Pathak H, Ladha JK, Aggarwal PK, Peng S, Das S, Yadvinder Singh, Bijay-Singh, Kamra, S. K., Mishra, B., Sastri, A., Aggarwal, H. P., Das, D. K and Gupta, R. K. (2003). Climatic potential and on-farm yield trends of rice and wheat in the Indo-Gangetic Plains, Field Crops Res., 80(3): 223-234.

Pendall, E., Bridgham, S., Hanson, P. J, *et al.* (2004). Below-ground process responses to elevated CO_2 and temperature: a discussion of observations, measurement methods and models, New Phytol, 162(2): 311–322.

Pathak, H. (2010). Mitigating greenhouse gas and nitrogen loss with improved fertilizer management in rice: Quantification and economic assessment, Nutr. Cycling Agroecosys, 87: 443-454.

Pathak, H. and Wassmann, R. (2007). Introducing greenhouse gas mitigation as a development objective in rice-based agriculture: I. Generation of technical coefficients. Agril. Syst., 94: 807-825.

Seo, N. and R. Mendelsohn. (2008). "A Ricardian Analysis of the Impact of Climate Change on South American Farms", Chilean Journal of Agricultural Research, vol. 68(1): 69-79.

Tsvetsinskaya, E. A., Mearns, L. O., Mavromatis, T., Gao, W., McDaniel, L. and Downton, M.W. (2003). "The effect of spatial scale of climate change scenario on simulated mazie, winter, wheat and rice production in the Southeastern United States", Climate Change, 60: 37-71

Tarnawski, S. and Aragno, M. (2006). The influence of elevated CO_2 on diversity, activity and biogeochemical function of rhizosphere and soil bacterial communities. In: Nosberger J, Long SP, *et al.*, editors. Managed ecosystems and CO_2 case studies, processes and perspectives, Ecological studies Series, Berlin: Springer; 187: 393–409.

Wassmann, R. and Pathak, H. (2007) Introducing greenhouse gas mitigation as a development objective in rice-based agriculture: II. Cost-benefit assessment for different technologies, regions and scales, Agril. Syst., 94: 826-840.

Chapter 21

Mitigation and Adaptation Strategies for Doubling the Farmers Income under Climate Smart Agriculture

S.G. Mundhe[1], *Wajid Hasan*[2] *and Trisha Manna*[3]

[1]*Krishi Vigyan Kendra, Durgapur, Amravati – 444701, Maharashtra*
[2]*Krishi Vigyan Kendra, Jehanabad, BAU, Sabour, Bihar*
[3]*Bidhan Chandra Krishi Viswavidyalaya, Mohanpur – 741252, Nadia, West Bengal*

This paper reviews various articles and documents on relationship between climate change and agriculture. The two-way relationship of climate change and agriculture is of great significance particular to developing countries due to their large dependence on agricultural practice for livelihoods and their lack of infrastructure for adaptation when compared to developed countries. Agricultural activities are affected by climate change affects due to their direct dependence on climatic factors. In high latitude areas with low temperature, increased temperature due to climate change could allow for longer growing season. Agriculture affects climate through emissions of greenhouse gases (GHGs) such as carbon dioxide, methane and nitrous oxide. These emissions come in large proportion directly from use of fossil fuels, tillage practices, fertilized agricultural soils and livestock manure etc. Globally, the agricultural sector is the major source of employment in most of the countries. Changing climate has adverse impacts on agriculture, hydropower, forest management and biodiversity, yield and productivity of several major crops in India by directly or indirectly influencing water-related parameters viz. evaporation, water runoff, soil moisture, soil productivity, incidence of pest and diseases etc. Several adaptation strategies, government policies and mitigation techniques can acts as to overcome the global threat of increasing temperature it is implications related to climate change.

Keywords: *Mitigation, Adaptation, Income, Climate change and Agriculture.*

Introduction

Globally, climate change (CC) is the most serious environmental threat that adversely affects agricultural productivity. Climate change refers to any change in climate over time, due to natural variability or as a result of human activity (IPCC, 2007). The main cause of CC are greenhouse gases (GHGs) accumulation in the atmosphere, it results into increased greenhouse effects. The agricultural commodities include field crops, horticultural crops, livestock, poultry and fisheries. Agriculture production is sensitive to temperature, increasing carbon dioxide concentration as well as change in precipitation. Impacts of all these forces together imply that agriculture production will respond non-linearly to future climate change. The impacts are complex to understand. However, there is high level of agreement across studies that the impact in all probability is going to be negative for most crop categories. The volume of loss of agricultural produce due to extreme weather events will be more as compared to global warming effects (IPCC, 2012).Based on some projections, changes in temperature, rainfall and weather uncertainties are expected to reduce crop yield in many regions of the developing world, particularly sub-Saharan Africa and parts of Asia. The impact and consequences of climate change for agriculture tend to be more severe for countries with higher initial temperatures, areas with marginal or already degraded lands and lower levels of development with little adaptation capacity.

Asia's population now lives in areas that are projected to become moderate to severe hotspots under the carbon- intensive scenario and almost half of South Asia's population, including India, now lives in the vulnerable areas and will suffer from declining living standards that could be attributed to falling agricultural yields, lower labour productivity or related health impacts. Some of these areas are already less developed, suffer from poor connectivity and are water stressed. India's average annual temperatures are expected to rise by 1.00°C to 2°C by 2050 even if preventive measures are taken along the lines of those recommended by the Paris climate change agreement of 2015. If no measures are taken average temperatures in India are predicted to increase by 1.5°C to 3°C. In India today, approximately 600 million people live in locations that could either become moderate or severe hotspots by 2050 under a business-as-usual scenario, the report says. States in the central, northern and north-western parts of India emerge as most vulnerable to changes in average temperature and precipitation (Muthukumara *et al.*, 2018).

Climate change scenarios include higher temperatures, changes in precipitation, and higher atmospheric CO_2 concentrations. The GHGs effect on agriculture is three ways, increased atmospheric CO_2 concentrations can have a direct effect on the growth rate of crop plants and weeds; CO_2-induced changes of climate may alter levels of temperature, rainfall and sunshine that can influence plant and animal productivity (Table 21.1) and rises in sea level may lead to loss of farmland by inundation and increasing salinity of groundwater in coastal (Muller, 2009). GHGs produce the relatively warm and hospitable environment near the earth's surface where humans and other life-forms have been able to develop and prosper. However, the increased level of greenhouse gases (GHGs) (carbon dioxide (CO_2), water vapor (H_2O), methane (CH_4), nitrous oxide (N_2O), hydro fluorocarbons (HFCs),

Per fluorocarbons (PFCs), and sulfur hexafluoride (SF_6) *etc.*) due to anthropogenic activities has contributed to an overall increase of the earth's temperature, leading to a global warming.

Table 21.1: Changing Climatic Elements, their Projected Impacts on Impact on Agriculture under Climate Change Scenarios

Climatic Element	*Expected Changes by 2050's*	*Confidence in Prediction*	*Effects on Agriculture*
CO_2	Increase from 360 ppm to 450 - 600 ppm (2005 levels now at 379 ppm)	Very high	Good for crops: increased photosynthesis; reduced water use
Sea level rise	Rise by 10 -15 cm Increased in south and offset in north by natural subsistence/rebound	Very high	Loss of land, coastal erosion, flooding, Stalinization of groundwater
Temperature	Rise by 1-2 °C. Winters warming more than summers. Increased frequency of heat waves	High	Faster, shorter, earlier growing seasons, range moving north and to higher altitudes, heat stress risk, increased evapotranspiration
Precipitation	Seasonal changes by ± 10 per cent	Low	Impacts on drought risk' soil workability, water logging irrigation supply, transpiration
Storminess	Increased wind speeds, especially in north. More intense rainfall events.	Very low	Lodging, soil erosion, reduced infiltration of rainfall
Variability	Increases across most climatic variables. Predictions uncertain	Very low	Changing risk of damaging events (heat waves, frost, droughts floods) which effect crops and timing of farm operations

Source: Climate change and Agriculture, MAFF (2000).

The global scenario of climate change (Figure 21.1) showed that the average global surface temperature have increased by 0.74°C since the late 19th Century and is expected to increase by 1.4°C - 5.8°C by 2100 AD with significant regional variations (IPCC, 2007). The atmospheric CO_2 concentration has increased from 280 ppm to 395 ppm, CH_4 concentration increased from 715 ppb to 1882 ppb and N_2O concentration from 227 ppb to 323 ppb from the year 1750 and 2012. The Global Warming Potential (GWP) of these gases *i.e.*, CO_2, CH_4 and N_2O are 1, 25 and 310 respectively.

Climate Change Impact on Agriculture

Whole Agricultural system manifested b the climate and agricultural production completely depends on climatic change. Worldwide numbers of scientists worked on to determine impact of changing climate on food security. Mohandass *et al.* (1995) predicted increase in rice production mainly because in the main season, crops enhanced CO_2 levels more than off-set the negative effects of increased temperatures. Though large decreases were predicted for second-season crops at many of the locations due to high temperatures, the relatively low proportion of total rice produced in this season meant that its overall effect on rice production was small.

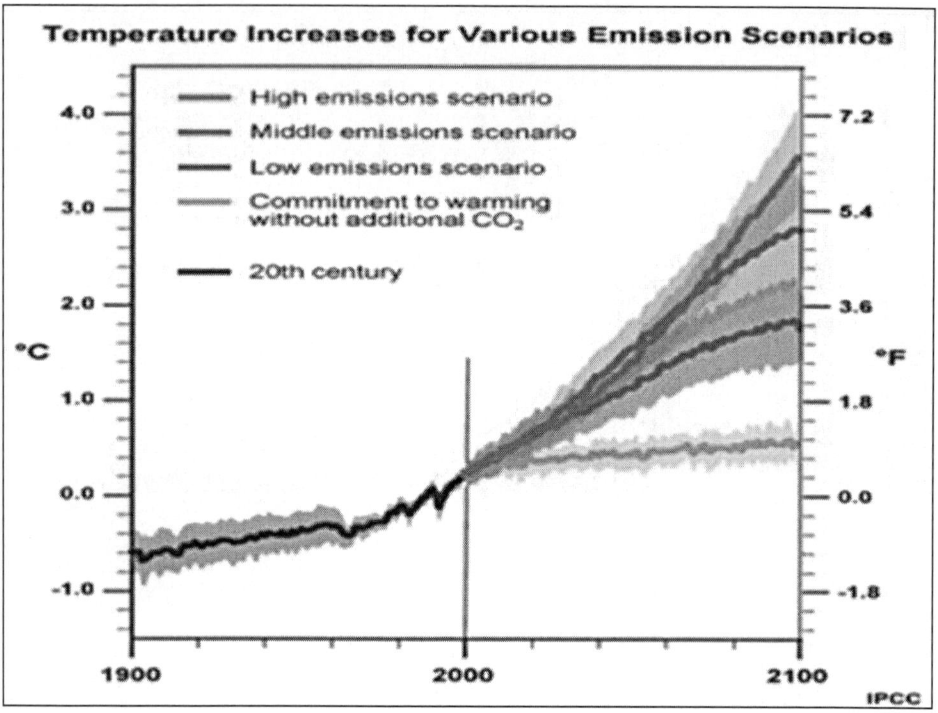

Figure 21.1: Global Scenario of Climate Change (IPCC, 2007).

Impact of Elevated CO_2 and Temperature on Agriculture

Increased concentrations of CO_2 may boost crop productivity, only where moisture is not a constraint. Higher levels of CO_2 can stimulate photosynthesis in certain plants (30-100 per cent). Experimental observations confirm that when plants absorb more carbon grow bigger and more quickly. This is particularly true for C_3 plants *e.g.* Rice, Wheat, soybean *etc.* (so called because the product of their first biochemical reactions during photosynthesis has three carbon atoms). Increased CO_2 tends to suppress photo-respiration in these plants, making them more water-efficient. The response of C_4 plants like maize, sorghum and sugarcane would not be as dramatic. The impact on yields of low-latitude crops Howden *et al.* (2008)is difficult to predict while the mid-latitude yields may be reduced by 10-30 per cent due to increased summer dryness. The effects of an increase in carbon dioxide would be higher on C_3 crops (such as wheat) than on C_4 crops (such as maize), because the former is more susceptible to carbon dioxide shortage. Moreover, the protein content of the grain decreases under combined increases of temperature and CO_2. For rice, the amylase content of the grain-a major determinant of cooking quality-is increased under elevated CO_2 Yao *et al.* (2010) reported that Within wheat, elevated CO_2 reduces the protein content of grain and flour by 9-13 per cent. Concentrations of Fe and Zn which are important for human nutrition would be lower. According to Ignaciuk and Mason-D'Croz (2014), climate change currently decreases the yield of maize, rice, wheat, potatoes and vegetables and continue to reduce seriously by 2050 globally (Table 21.2). As Lobell *et al.* (2007).study in 12 food-insecure regions of the world reported that climate change could significantly impact agricultural production and food security up to 2030 particularly for sub-Saharan Africa and South Asia due to both changes in mean temperatures and rainfall as well as increased variability associated with both.

Table 21.2: Temperature Effects on Crop Yield, Selected Crops, India

Crop	T_{opt} (°C)	T_{max} (°C) Yield at	Yield at T_{opt} (t/ha)	Yield at 28°C (t/ha)	Yield at 32°C (t/ha)	Per cent Decrease (28°C to 32°C)
Rice	25	36	7.55	6.31	2.93	54
Soyabean	28	39	3.41	3.41	3.06	10
Drybean	22	32	2.87	1.39	0	100
Peanut	25	40	3.38	3.22	2.58	20
Grain: sorghum	26	35	12.24	11.75	6.95	41

Note: T_{opt}: Optimal temperature; T_{max}: Maximum sustainable temperature; t/ha: tonnes per hectare.

Source: Rao (2008).

Impact on Climate Change on C_3 Crops

Climate changes may directly affect rice plant growth through changes in air temperature, precipitation, evapotranspiration, and water temperature (Table 2.0). Shifts in the hydrological regimes of catchments within which paddy cultivation occurs might also be expected, including changes in flood patterns and water availability for paddy irrigation, and drainage conditions.

Increased atmospheric CO_2 concentrations in the future, one of the major drivers of global warming, may also enhance plant growth through the fertilization effect since CO_2 is an essential component of photosynthesis (Watanabe and Kume, 2009). Sinha and Swaminathan (1991) predicted that a 2°C increase in temperature could decrease rice yield by about 0.75 tonne/hectare in high-yield areas and by about 0.06 tonne/hectare (Figure 21.2) in low-yield coastal regions.

Lal *et al.* (1998) used CERED model in northwest India under a doubling of CO_2 levels and predicted that rice yield would increase by 15 per cent. However, a 2°C rise cancelled out the positive effect of elevated CO_2. The combined effect of enhanced CO_2 and temperature increase leads to a 4 per cent increase in rice yield for the irrigation schedule presently practiced. The adverse impact of water shortage would be a net decline in rice yields.

The study by Sinha and Swaminathan (1991) noted also that a 0.5°C rise in winter temperature would lower yields by 0.45 tonne/hectare. The major findings of subsequent studies are presented in tabular form below. One of the major findings

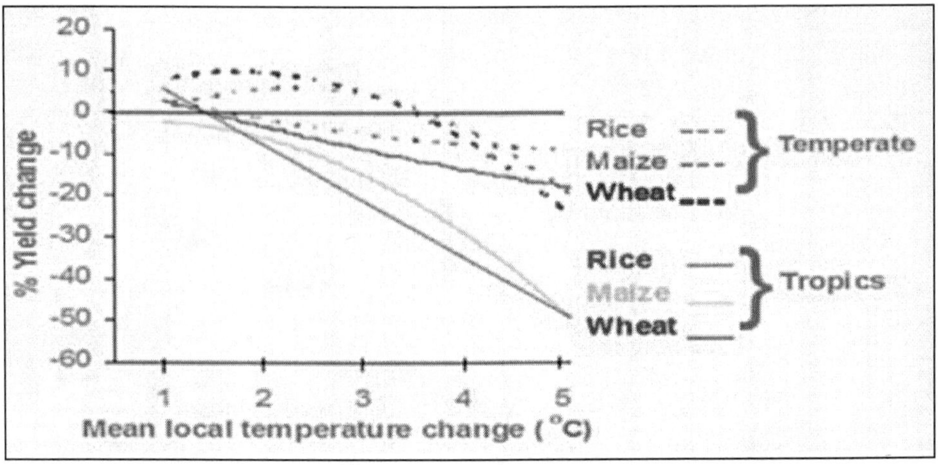

Figure 21.2: Comparison of Temperature Rise on Crop Yields in Temperate and Tropical Regions (*Source*: IPCC, 2007; Nicholls *et al.*, 2008).

of the study of the Indian Council of Agricultural Research (ICAR) on the impact of climate change is that it is likely to lower the potential yields of wheat.55 Thus, the gap between potential yields and actual yields will narrow, with potential yields likely to fall faster than the rise in actual yields. Attri and Rathore (2003) used CERES wheat dynamic simulation model and climate change scenarios to find an increase in wheat yield between 29–37 per cent and 16–28 per cent under rainfed and irrigated conditions, especially in different genotypes under a modified climate. A 3°C increase in temperature or more cancelled out the positive effects of enhanced CO_2.

Impacts of Climate Change on C_4 Crops

Rao*et al.* (1995) used the CERES sorghum model and climate change scenarios generated by the three GCMs: GISS, GFDL and Met Office, UK. Results indicated a decrease in the yield and biomass of rainy-season sorghum at Hyderabad and Akola under all climate change scenarios. Post-rainy-season sorghum grown at Solapur on stored soil water showed a marginal increase in yield. The positive effects of increased CO_2, if any, were masked by the adverse effects of predicted increase in temperature resulting in shortened crop-growing seasons. The study also showed that the effects of climate change on the same crop would depend upon the season in which it is grown.

Chatterjee (1998) studied that for a 1 and 2°C rise in, sorghum yields decreased by 7–12 per cent on average. A further increase in temperature drastically reduced the potential yields by 18–24 per cent on average. An increase in 50 ppm CO_2 increased yields by only 0.5 per cent. The beneficial effect of 700 ppm CO_2 was cancelled out by a temperature rise of 0.9°C.

Gadgil (1995) and Gadgil *et al.* (1999a, 1999b) reported that the incidence of locally triggered pests/diseases (leaf miner and late leaf spot) is low when sowing is postponed to after mid-July.

Hundal and Kaur (1996) predicted that in the case of Punjab (using the PNUTGRO model), with other climate variables held constant, a temperature rise of 1, 2 and 3°C from present-day level would reduce groundnut yields by 8.7, 23.2 and 36.2 per cent respectively.

Impacts of Climate Change on Crops

Aggarwal (2003) studied the impact assessment of temperature over wheat cultivars and showed that an increase of temperature of the order of 3°C or more cancels out the beneficial effects of elevated CO_2 in all the cultivars under study. He stated that the declines of the simulated irrigated wheat yields under current climate and under climate change scenarios. There was no effect of climate change in northern India but yields were reduced in Central India by 10–15 per cent. This reduction in productivity under changed climate unless accompanied with suitable research and policy interventions may reduce wheat production options in central India (Kalra and Aggarwal, 1994; Aggarwal, 2000, Sud, 2008) reported that as per the findings of the study conducted by the Indian Agriculture Research Institute (IARI) New0 Delhi, with every 1p C increase in the temperature throughout the growing period of the crop, the overall wheat production may be lost by 4 to 5 million tones. Hundal and Kaur (1996) reported that the temperature rise of 1, 2 and 3°C from present-day level would reduce maize yield by 10.4, 14.6 and 21.4 per cent respectively. Sahoo (1999) showed that temperature rise decreased maize yield under both irrigated and rainfed conditions and maize yield decreased continuously till 4°C uder CO_2 concentration of 350 ppm. While, under 700 ppm CO_2, grain yield increased by about 9 per cent. The effect of elevated CO_2 concentration on the growth and yield of maize was established, but found to be less pronounced than the effect on wheat, chickpea and mustard crops. The beneficial effect of 700 ppm CO_2 was cancelled by an increase of only 0.6°C. The annual variation of rainfall (*i.e.* quantity) and rainy days (quality) and trend was affected plantations, productivity and production of crops. Since last one and half decade onset and distribution of monsoon are showing irregular behavior and it was very difficult to coincide with appropriate sowing time, (Jaybhaye *et al.*, 2016).

Impact of Climate Change on Population Dynamics of Insect Pests

Increased soil temperature may also lead to an increase in autotrophic CO_2 losses from the soil caused by root respiration, root exudates, and fine-root turnover. Climate change impacts will also impact on rice production through rising sea level rise. Most studies project decreased yields in non-irrigated wheat and in rice, and a loss in farm-level net revenue between 9 per cent and 25 per cent for a temperature increase of 2–3.5°C. Aggarwal and Mall (2002) observed that a 2 °C increase resulted in a 15–17 per cent decrease in grain yield of rice and wheat. Fungal and bacterial pathogens are also likely to increase in severity in areas where precipitation increases. Under warmer and more humid conditions cereals would be more prone to outbreaks of pest and diseases thereby reducing yield. In conformity to this results Jaybhaye *et al.* (*2015*), stated that the leaf miner incidence mostly associated with minimum temperature, evaporation and wind velocity weather parameters and not variety of soybean. Climate change will have a major effect on geographic distribution of insect pests, and low temperatures are often more important than high temperatures in determining geographical distribution of insect pests (Hill 1987). Increasing temperatures may result in a greater ability to overwinter in insect species limited by low temperatures at higher latitudes, extending their geographical range (EPA 1989; Hill and Dymock 1989; Elphinstone and Toth 2008), and sudden outbreaks of insect pests can wipe out certain crop species, and also encourage the invasion by exotic species (Kannan and James 2009). Spatial shifts in distribution of crops under changing climatic conditions will also influence the distribution of insect pests in a geographical region (Parry and Carter 1989). Some plant species may be unable to follow the climate change, resulting in extinction of species that are specific to particular hosts (Thomas *et al.,* 2004). However, whether or not an insect pest would move with a crop into a new habitat will depend on other environmental conditions such as the presence of overwintering sites, soil type, and moisture, *e.g.,* populations of the corn earworm, *Heliothiszea* (Boddie) in the North America might move to higher latitudes/altitudes, leading to greater damage in maize and other crops (EPA, 1989). For all the insect species, higher temperatures, below the species upper threshold limit, will result in faster development and rapid increase in pest populations as the time to reproductive maturity will be reduced considerably. In addition to the direct effects of temperature changes on development rates, improvement in food quality due to abiotic stress may result in dramatic increases in growth of some insect species (White 1984), while the growth of certain insect pests may be adversely affected (Maffei *et al.,* 2007). Pest outbreaks are more likely to occur with stressed plants as a result of weakening of plants' defensive system, and thus, increasing the level of susceptibility to insect pests (Rhoades, 1985). Gadgil (1995) and Gadgil *et al.* (1999a, 1999b) reported that the incidence of locally triggered pests/diseases (leaf miner and late leaf spot) is low when sowing is postponed to after mid-July.

Simulating the Effects of Climate Change also Needs to include Effects such as the Interaction with Other Factors of Pests and Weeds

The impact of increased $CO_2$and temperature and variation in rainfall will be modified by such interactions, while the behaviour of pests and weeds may itself vary with climate change. There is a significant literature on potential competition between C_3 and C_4 crops in the context of enhanced CO_2 levels. Such studies include the competition between C_3 crops and C4 weeds. The IPCC's Third Assessment Report provides a useful summary introduction to these issues. The studies cited there show that the interaction between pests and major food and cash crops could be complex, with elevated levels of CO_2 and temperature and increased or decreased precipitation, setting in motion secondary effects that affect crop–pest interactions. In rice, for instance, model studies show that leaf-blast epidemics are more likely with elevated temperatures in cool, sub-tropical zones than in warm, humid tropics, where such epidemics are inhibited by temperature rise. Another experimental study showed that higher concentrations of CO_2 lowered nitrogen uptake in plant tissues, leading to significantly enhanced damage by pests.

Effect of Climate Change on Effectiveness of Transgenic Crops for Pest Management

Environmental factors such as soil moisture, soil fertility, and temperature have strong influence on the expression of *Bacillus thuringiensis*(*Bt*) toxin proteins deployed in transgenic plants (Sachs *et al.*, 1998). Cotton bollworm, *Heliothisvirescens*(F.) destroyed *Bt*-transgenic cottons due to high temperatures in Texas, USA (Kaiser, 1996). Similarly, *H. armigera* and *H. punctigera* (Wallen.) destroyed the *Bt*-transgenic cotton in the second half of the growing season in Australia because of reduced production of *Bt* toxins (Hilder and Boulter 1999). Cry1Ac levels in transgenic plants decrease with the plant age, resulting in greater susceptibility of the crop to insect pests during the later stages of crop growth (Sachs *et al.*, 1998; Green plate *et al.*, 2000; Adamczyk *et al.*, 2001; Kranthi *et al.*, 2005). Possible causes for the failure of insect control in transgenic crops may be due to inadequate production of the toxin protein, effect of environment on transgene expression, *Bt*-resistant insect populations, and development of resistance due to inadequate management (Sharma and Ortiz 2000). It is therefore important to understand the effects of climate change on the efficacy of transgenic plants for pest management.

Mitigation and Adaptation in Agriculture

To coping the future food demand of increasing population of the world under changing the climate some mitigation and adaptations should require. The mitigation is the global issue and adaptations are the local or ground based issue. Some mitigations and adaptations are revived here, which are useful to doubling the farmer's income by contributing reduction in increasing the global temperature locally.

Mitigation

According to IPCC [2007], mitigation is an intervention to reduce the emissions sources or enhance the GHG sinks, whereas adaptation is the adjustment in natural or human systems in response to actual or expected climatic change or their effects, to reduce harm or exploit beneficial opportunities. Sustainable and organic agricultural systems can help reduce agricultural GHG emissions through energy conservation, lower levels of carbon-based inputs, lower use of synthetic fertilizer and other features that minimize GHG emissions and sequester carbon in the soil. During mitigation and adaptation of climate change through different agricultural activities, there might be many challenges or barriers such as financial, policy and implementation barrier *etc.* In general, Agricultural activity could be a source of GHGs as well as a sink, notably through the storage of carbon in the soil organic matter and in Biomass and influenced by CC.

Adaptation

- ☆ The Agromet Advisory Bulletin (AAB) is a useful weather smart agriculture technological tools that can minimize climatic risks and build a resilient agricultural production. AAB assist farmers in coping with current climatic risks by providing value-added weather services to farmers. Farmers can adapt to climate changes to some degree by shifting planting dates, choosing varieties with different growth duration, or changing crop rotations, (Jaybhaye *et al.*, 2018).

- ☆ Fine tuning sowing window/transplanting date- *e.g.* Early transplanting of Kharif rice in Gangetic West Bengal will help to combat the CC effect, (Anonyms, 2013).Seasonal weather forecasts could be used as a supportive measure to optimize planting and irrigation patterns. Inconformity to the above statement Gadgil, (1995) and Gadgil *et al.* (1999a, 1999b) showed that for rainfed groundnut, the sowing period of 22 June to 17 August is optimum for minimizing the risk of failure. One of the most important agronomic adaptations to overcome abiotic stress is to select an appropriate sowing time for cotton crop for optimizing yield and quality, (Jaybhaye *et al.*, 2016).
- ☆ An Early warning system should be put in place to monitor changes in pest and disease outbreaks. The overall pest control strategy should be based on integrated pest management because it takes care of multiple pests in a given climatic scenario.
- ☆ Participatory and formal plant breeding to develop climate resilient crop varieties that can tolerate higher temperatures, drought and salinity. Developing short-duration crop varieties that can mature before the peak heat phase set in. Similar results are given by Jaybhaye *et al.* (2015), stated that the moderate canopy temperature, moderate canopy temperature deferential (Tc-Ta) showed genotype under rainfed condition in *rabi* sorghum gives moderate grain and fodder yield; it is because of these genotypes possess characters attributing drought tolerance.

Selecting genotype in crops that have a higher per day yield potential to counter yield loss from heat-induced reduction in growing periods. Grow the varieties having greater thermo-tolerance characteristics.

- ☆ Preventive measures for drought that include on-farm reservoirs in medium lands, growing of pulses and oilseeds instead of rice in uplands, ridges and furrow system in cotton crops, growing of intercrops in place of pure crops in uplands, land grading and leveling, stabilization of field bunds by stone and grasses, graded line bunds, contour trenching for runoff collection, conservation furrows, mulching and more application of Farm yard manure (FYM).
- ☆ Efficient water use such as frequent but shallow irrigation, drip and sprinkler and trickle irrigation for high value crops, irrigation at critical stages.
- ☆ Efficient fertilizer use such as optimum fertilizer dose, split application of nitrogenous and potassium fertilizers, deep placement, use of *neem, karanja* products and other such nitrification inhibitors, liming of acid soils, use of micronutrients such as zinc and boron, use of sulphur in oilseed crops, integrated nutrient management.

- ☆ Adopt resource conservation technologies such as no-tillage, laser land leveling, direct seeding of rice and crop diversification which will help in reducing in the global warming potential. Crop diversification can be done by growing non-paddy crops in rain fed uplands to perform better under prolonged soil moisture stress in kharif.
- ☆ Develop a long-term land use plan for ensuring food security and climatic resilience.
- ☆ National grid grain storages at the household/community level to the district level must be established to ensure local food security and stabilize prices.
- ☆ Provide incentives to farmers for resource conservation and efficiency by providing credit to the farmers for transition to adaptation technologies.
- ☆ Provide technical, institutional and financial support for establishment of community banks of food, forage and seed.
- ☆ Intensify the food production system by improving the technology and input delivery system.
- ☆ Provide more funds to strengthen research for enhancing adaptation and mitigation capacity of agriculture.
- ☆ In tea plantations, studies on chemical kinetics of organic manures like vermin-compost (VC), cattle manure (CM) and Neem Cake (NC) used in carbon sequestration experiments confirmed that the rate of mineralization was more in decomposed NC followed by CM and VC.
- ☆ Sustainable policy for import and export agricultural commodities should be made in favours to farmers and growers.
- ☆ Minimum support price should be based on changed seasonal weather condition, natural calamities and actual production cost for wellness of farming community.
- ☆ Provide greater coverage of weather linked agriculture-insurance.

REFERENCES

Abrol, Y. P., Bagga, A. K., Chakravorty, N. V. K. and Wattal, P. N. (1991),Impact of rise in temperature on productivity of wheat in India. In: Impact of globalclimate change on photosynthesis andplant productivity. Abrol, Y. P., Wattal, P.N., Gnanam, A., and Teramura, A. H. (eds) Oxford Publishers, New Delhi,India. Pp. 787-798.

Adamczyk Jr., J.J., Adams, L.C., and Hardee, D.D. (2001), Field efficacy and seasonal expression profiles for terminal leaves of single and double *Bacillus thuringiensis*toxin cotton genotypes.*Journal ofEconomic Entomology* 94: 1589-1593.

Adams, Richard M.,Hurd, Brian H., Lenhart Stephanie, Leary, Neil. (1998), Effects of global climate change on agriculture: an interpretative review CLIMATE RESEARCH Clim Res, Vol. 11: 1,20,21.

Aggarwal and Sinha, 1993 P.K. Aggarwal and S.K. Sinha, Effect of probable increase in carbon dioxide and temperature on productivity of wheat in India, *J. Agric. Meteorol.* 48 (1993), pp. 811–814.

Aggarwal, P. K. (2003), Impact of climate change on Indian agriculture.*JournalofPlant Biology* 30(2): 189-198.

Aggarwal, P. K. (2009), "Executive Summary," in P. K. Aggarwal (ed.), *Global Climate Change and Indian Agriculture: Case Studies from the ICAR Network Project*, IndianAgricultural Research Institute Publication, New Delhi.

Aggarwal, P. K. and Kalra, N. (1994), 'Simulating the effect of climatic factors, genotype and management on productivity of wheat in India', Indian Agricultural Research Institute Publication, New Delhi, India, pp. 156.

Aggarwal, P. K. (2000), 'Application of systems simulation for understanding and increasing yield potential of wheat and rice', Ph.D. thesis, Wageningen University, The Netherlands, pp. 176.

Aggarwal, P. K., and Mall, R. K. (2002), "Climate Change and Rice Yields in Diverse Agro-Environments of India: II. Effect of Uncertainties in Scenarios and Crop Models on Impact Assessment," *Climatic Change*, 52, 3, pp. 331–43.

Aggarwal, P. K., and Sinha, S. K. (1993), "Effect of Probable Increase in CO_2 and Temperature on Productivity of Wheat in India," *Journal of Agricultural Meteorology*, 48, 5, pp. 811–14.

Attri, S. D., and Rathore, L. S. (2003), "Simulation of Impact of Projected Climate Change on Wheat in India," *International Journal of Climatology*, 23, pp. 693–705.

Auffhammer M, Ramanathan V, Vincent J (2012) Climate change, the monsoon, and rice yield in India. ClimChange 111(2): 411–424.

Bouman, A. (2001), Global Estimates of Gaseous Emissions from Agricultural Land. Food and Agricultural Organization, Rome, ISBN: 92-5-104698-1.

Chatterjee, A. (1998), "Simulating the Impact of Increase in Temperature and CO_2 on Growth and Yield of Maize and Sorghum," M.Sc. thesis (unpublished), Indian Agricultural Research Institute, New Delhi;

Chattopadhyay, N. (2008),Climate Change and Food Security in India.International Symposium on Climate Change and Food Security in South Asia, 25-30 August 2008, Dhaka, Bangladesh.

Cheng, Weiguo, Hidemitsu Sakai, Kazuyuki Yagi, Toshihiro Hasegawa. (2009). Interactions of elevated CO_2 and night temperature on rice growth and yield. Agricultural and forest meteorology 149 (2009) 51–58.

Climate Change 2007: Synthesis Report (IPCC 2007).

Committee on Stabilisation Targets for Atmospheric Greenhouse Gas Concentrations (2010),"Climate Stabilisation Targets: Emissions, Targets and Concentrations from Decades to Millennia," National Academies Press, Washington D. C., available at http: //www.nap.edu/ catalog/12877.html, viewed on August 25, 2010.

David B. Lobell, Adam Sibley and J. Ivan Ortiz-Monasterio (2012) "Extreme heat effects on wheat senescence in India", Nature Climate Change, 2(2012): 186, http: //www.nature.com/nclimate/journal/v2/n3/full/nclimate1356.html

Elphinstone, J. and Toth, I.K. (2008),*Erwiniachrysanthemi*(*Dikeya*spp.) - the facts. Oxford, U.K: Potato Council.

Environment Protection Agency (EPA). (1989),*The potential Effects of Global Climate Change on theUnited States*. Vol 2: National Studies. Review of the Report to Congress, US Environmental ProtectionAgency, Washington DC, 261 pp.

FAO (2008) Climate change water and food security. FAO Water Report no.36

Food and Agriculture Organization of the United Nations, Food and AgricultureOrganization of the United Nations Statistical Database (FAOSTAT) (2007).Available at www.faostat.org.Accessed June 5, 2007.

Gadgil, S. (1995), "Climate Change and Agriculture: An Indian Perspective," *Current Science*, 69, 8, pp. 649–59.

Gadgil, S., Abrol, Y. P., and Rao, Seshagiri, P. R. (1999a), "On Growth and Fluctuation of Indian Food Grain Production," *Current Science*, 76, 4, pp. 548–56.

Gadgil, S., Rao, Seshagiri, P. R., and Sridhar, P. R. (1999b), "Modeling Impact of Climate Variability on Rainfed Groundnut," *Current Science*, 76, 4, pp. 557–569.

Government of India, Ministry of Environment and Forests (2008), "National ActionPlanonClimateChange,"availableathttp: //www.moef.nic.in/downloads/home/Pg01-52.pdf, viewed on October 15, 2010.

Greenplate, J.T., Penn, S.R., Shappley, Z., Oppenhuizen, M., Mann, J., Reich, B., and Osborn, J. (2000), Bollgard II efficacy: quantification of total lepidopteran activity in a 2-gene product. In: *Proceedings, Beltwide Cotton Conference* (Dugger, P. and Richter, D., eds.). Memphis, Tennessee, USA: National Cotton Council of America. pp. 1041-1043.

Hafner, Sasha. (2003), "Trends in Maize, Rice, and Wheat Yields for 188 Nations over the Past 40 Years: A Prevalence of Linear Growth," *Agriculture, Ecosystems and Environment*, 97, 1–3.

Hill, D.S. (1987), *Agricultural Insects Pests of Temperate Regions and Their Control.* Cambridge, UK: Cambridge University Press. 659 pp.

Hill, M.G. and Dymock, J.J. (1989), *Impact of Climate Change: Agricultural/Horticultural Systems.*DSIR Entomology Division Submission to the New Zealand Climate Change Program. Auckland, NewZealand: Department of Scientific and Industrial Research. 16 pp.

Howden S. M., S. J. Crimp, and C.J. Stokes (2008).Climate change and Australian livestock systems: impacts, research and policy issues. Australian Journal ofExperimental Agriculture 48, 780 – 788.

http: //www.climate.org,

http: //en.wikipedia.org/wiki/Agriculture_in_India

http: //www.ipcc.ch/pdf/assessmentreport

http: //www.rbi.org.in/

Hundal, S. S., and Kaur, P. (1996), "Climate Change and its Impact on Crop Productivity in the Punjab, India," in Y. P. Abrol, G. Gadgil and G. B. Pant (eds.), *Climate Variability andAgriculture*, Narosa Publishing House, New Delhi, pp. 377–93.

Ignaciuk[i] and Daniel Mason-D'Croz (2014)OECD International Food Policy Research Institute (IFPRI). 22 Dec 2014. 57 pages.No. 70.

IPCC (2007) Climate change: the physical science basis. In S Solomon, D Qin,M Manning, ZChen,MMarquis, KB Averyt,MTignor, HL Miller, eds, Contribution of Working Group I to the Fourth Annual Assessment Report of the Intergovernmental Panel on Climate Change. CambridgeUniversity Press, Cambridge, UK, pp 996

IPCC (2014).Climate change, AR5, Synthesis report.

IPCC, (2012).*Climate Change, Climate Change Impacts, Adaptation and Vulnerability*, Working Group II Contribution to the Intergovernmental Panel on Climate Change Fifth Assessment Report,Summary for Policy Makers.

Jaybhaye, P. R., V. N. Sable, N. S. Harwadikar and P. B. Shinde (2015), Studies on canopy temperature of different Rabi Sorghum genotypes under dry land condition BIOGLOBIA Vol. 2 (1): 2015 : PP19-24.

Jaybhaye, P. R., S. D., Gawali, Mohite, V.K., P. B. Shinde, and B. V. Asewar(2016), Impact of Changing Rainfall Trend on Yield of Mango and Sweet Orange in Marathwada Region. Indian

Jaybhaye, P.R., N.H. Deore and P.B. Shinde (2018),Agromet Advisory Bulletin - A Weather Smart Agriculture Technology option for Adaptation and Mitigation of Changing Climate *Int.J.Curr.Microbiol.App.Sci*(2018) *7*(2): 2644-2653.

Journal of Ecology (2016) 43 (SpecialIssue-1): 250-253.

Kaiser, J. (1996),Pests overwhelm Bt cotton crop. *Nature* 273: 423.

Kannan, R. and James, D.A. (2009), Effects of climate change on global diversity: a review of keyliterature. *Tropical Ecology* 50: 31-39.

Khan, S., Kumar, S., Hussain, M. Z., and Kalra, N. (2009), "Climate Change, Climate Variability and Indian Agriculture: Impacts Vulnerability and Adaptation Strategies," in S. N.Singh (ed.), *Climate Change and Crops, Environmental Science and Engineering*, pp. 19–38.

Kranthi, K.R., Naidu, S., Dhawad, C.S., Tatwawadi, A., Mate, K., Patil, E., Bharose, A.A., Behere,G.T., Wadaskar, R.M., and Kranthi, S. (2005), Temporal and intra-plant variability of Cry1Ac expression in Bt-cotton and its influence on the survival of the cotton bollworm, *Helicoverpaarmigera* (Hubner) (Noctuidae: Lepidoptera). *Current Science* 89: 291-298.

Kripalani R. H., Kulkarni, A., Sabade, S.S. and Khandekar, M. L. (2003),Indian monsoon variability in a global warming scenarios', *Natural Hazards* 29(2): 189-206.

Krishnan, P, D.K. Swain, B. Chandra Bhaskar, S.K. Nayak, R.N. Dash (2007) Impact of elevated CO_2 and temperature on rice yield and methods of adaptation as evaluated by crop simulation studies. Agriculture, Ecosystems and Environment 122 (2007) 233 242.

Kumar, K. K. S. and Parikh, J.: 2001. Indian Agriculture and Climate sensitivity. *Global Environment Change*11: 147-154.

Lal, M. (2001),Climatic Change — Implications for India's Water Resources. Journal of Social and Economic Development, Vol III91)57-87.

Lal, M., Singh, K. K., Rathore, L. S., Srinivasan, G. and Saseendran, S. A. (1998), Vulnerability of rice and wheat yields in NW India to future changes in climate. *Agricultural and ForestMeteorology*89: 101-114.

Lal, M., Singh, K. K., Srinivasan, G., Rathore, L. S., and Saseendran, A. S. (1998),"Vulnerability of Rice and Wheat Yields in NW-India to Future Change in Climate, "*Agricultural and Forest Meteorology*, 89, pp. 101–14.

Lal, M., Singh, K. K., Srinivasan, G., Rathore, L. S., Naidu, D., and Tripathi, C. N. (1999),"Growth and Yield Response of Soybean in Madhya Pradesh, India to Climate Variability and Change," *Agricultural and Forest Meteorology*, 93, pp. 53–70.

Lobell, David B., and Field, Christopher B. (2007), "Global Scale Climate–Crop Yield Relationships and the Impacts of Recent Warming," *Environmental Research Letters*, 2, 1, March, 014002.

Mall, R. K., Gupta, A., and Kumar, K. J. A. (2010), "Climate Change and Water Resources: Future Research Agenda," in B. Venkateswarlu *et al.* (eds.), *Lead Papers: NationalSymposium on Climate Change and Rainfed Agriculture*, Indian Society for Dryland Agriculture, Central Research Institute for Dryland Agriculture, Hyderabad, 18-20F ebruary, available at http: //www.crida. ernet.in/conf-completed/Publications-Climate per cent 20change per cent 20/ leadpapers.pdf, viewed on October 20, 2010.

Mendelssohn, Robert *et al.* (1994), "The impact of climate change on agriculture: A Ricardian Analysis," American Economic Review Vol 84, No4, pp 753-771.

Mohandass, S., Kareem, A. A., Ranganathan, T. B., and Jeyaraman, S. (1995), "Rice Production in India Under Current and Future Climates," in R. B. Matthews, M. J. Kropff, D. Bachelet and H. H. Laar van (eds.), *Modeling the Impact of Climate Change on RiceProduction in Asia*, CA B International, U. K., pp. 165–81

Muller A. (2009). Sustainable agriculture and the production of biomass for energy use.Climatic Change 94, 319 – 331. doi: 10.1007/s10584 - 008 - 9501 - 2, ISSN: 0165-0009, 1573 – 1480.

Muthukumara Mani, Sushenjit Bandyopadhyay, Shun Chonabayashi, Anil Markandya and Thomas Mosier (2018), South Asia's Hotspots The Impact of

Temperature and Precipitation Changes on Living Standards. Internet: www.worldbank.org.

Nicholls, T., Norgrove, L. and Masters,G. 2008. Innovative solutions to newinvaders: managing agricultural pests, diseases and weeds under climate change. In: Agriculture in a Changing Climate. Ed. Brown, A.G. 14 Annual DevelopmentConferences, TheATSECrawford Fund. Pp. 9-13

Parry, M.L. and Carter, T.R. (1989), An assessment of the effects of climatic change on agriculture.*Climatic Change* 15: 95-116.

Pathak, H. *et al.* (2003), "Trends of Climatic Potential and on-Farm Yields of Rice and Wheat in the Indo-Gangetic Plains," *Field Crops Research*, 80, pp. 223–34.

Pathak, H., Ladha, Aggarwal, P. K., Peng, S., Das, S., Singh, Y., Singh, B., Kamra, S. K., Mishra, B., Sastri, A. S. R. A. S., Aggarwal, H. P., Das, D. K., Gupta, R. K. (2003), Trends of climatic potential and on-farm yields of rice and wheat in the Indo-Gangetic Plains. *Field CropsResearch*80: 223-234.

Rai, Mangala (2010), India's preparedness for food security in view of climate change, International Conference on Food Security and Climate Change in Dry Areas,: 6-7

Raneesh KY. (2014) n Impact of Climate Change on Water Resources. J Earth SciClim Change Vol 5: 185.

Rao GD and S.K. Sinha (1994), Impact of climatic change on simulated wheat production in India. In: C.Rosenzweig and I. Iglesias, Editors, *Implications of Climate Change for International Agriculture:* CropModelling Study, EPA, USA (1994), pp. 1–10.

Rao, D. Gangadhar, Katyal, J. C., Sinha, S. K., and Srinivas, K. (1995), "Impacts of Climate Change on Sorghum Productivity in India: Simulation Study," in C. Rosenzweig*et al.* (eds.),*Climate Change and Agriculture: Analysis of Potential International Impacts*, ASA Special Publication no. 59, American Society of Agronomy, Madison, pp. 325–37.

Rao, G. G. S. N. (2008), "Impacts of Long-Term Climate Change on Indian Agriculture,"Presentation at ICRISA T workshop, 7–9 May, available at http: //www.icrisat.org/what-wedo/impi/inception-workshop/16-long-term-changes-crida.pdf, viewed on October 2, 2010.

Rao, M. Srinivasa, Rao, Ranga, G. V., and Venkateswarlu, B. (2010), "Impact of Climate Change on Insect Pests," in B. Venkateswarlu*et al.* (eds.), *Lead Papers: National Symposiumon Climate Change and Rainfed Agriculture*, Indian Society for Dryland Agriculture, Central Research Institute for Dryland Agriculture, Hyderabad, February 18–20, available at http: //www.crida.ernet.in/conf-completed/PublicationsClimate per cent 20change per cent 20/leadpapers.pdf, viewed on October 20, 2010.

Rathore, L. S., Singh, K. K., Saseendran, S. A., and Baxla, A. K. (2001), "Modeling the Impact of Climate Change on Rice Production in India," *Mausam*, 52, 1, pp. 263–74.

Reserve Bank of India (2010), *Handbook of Statistics on Indian Economy*, Reserve Bank of India,availableathttp: / /www.rbi.org.in/scripts/AnnualPublications. aspx?head=Handbook per cent 20of per cent 20Statistics per cent 20on per cent 20Indian per cent 20Economy, viewed on October 25, 2010.

Rhoades, D.F. (1985), Offensive-defensive interactions between herbivores and plants: their relevance in herbivore population dynamics and ecological theory. *American Naturalist* 125: 205-238.

Rosenzweig, C and Parry, M.L. (1994) "Potential impact of climate change on world food supply", Nature, 367 (6450), pp 133-138.

Sachs, E.S., Benedict, J.H., Stelly, D.M., Taylor, J.F., Altman, D.W., Berberich, S.A., and Davis, S.K. (1998), Expression and segregation of genes encoding Cry1A insecticidal proteins in cotton.*CropScience*38: 1-11.

Sahoo, S. K. (1999), "Simulating Growth and Yield of Maize in Different Agro-Climatic Regions," M.Sc. thesis (unpublished), Indian Agricultural Research Institute, New Delhi; ascited in Mall, Singh, Gupta, Srinivasan and Rathore.

Saseendran, A. S. K., Singh, K. K., Rathore, L. S., Singh, S. V., and Sinha, S. K. (2000), "Effects of Climate Change on Rice Production in the Tropical Humid Climate of Karala, India," *Climatic Change*, 44, pp. 495–514.

Schlenker W, Lobell DB (2010) Robust negative impacts of climate change on African agriculture. Environ Res Lett 5(1): 014010.

Selvaraju, R. (2003), Impact of El Nino- Southern Oscillation on Indian food grain production. *International Journal of Climatology*23: 187-206.

Sharma, H.C. and Ortiz, R. (2000), Transgenics, pest management, and the environment.*Current Science* 79: 421-437.

Sinha, S. K., and Swaminathan, M. S. (1991), "Deforestation Climate Change and Sustainable Nutrients Security," *Climatic Change*, 16, pp. 33–45.

Sivakumar, M.V.K., Das, H.P. and Brunini, O. (2005), Impacts of present and future climate variability and change on agriculture and forestry in the arid and semi-arid tropics.*Climatic Change* 70: 31–72.

SunandaMoka, M. G. Jadhav, P. R. Jaybhaye and B. V. Asewar (2015), Effect of Weather Parameters on Incidence of Leaf Miner in Soybean *J. Agric. Res. Technol., 40* (2) *: 314-317*

Thomas, C.D., Cameron, A., Green, R. E., Bakkenes, M., Beaumont, L.J., Collingham, Y.C., Erasmus, BFN, de Siqueira, MF, Grainger, A., Hannah, L, Hughes, L, Huntley, B., van Jaarsveld, A.S., Midgley, G.F., Miles, L., Ortega-Huerta, M.A., Peterson, A.T., Phillips, O.L., and Williams, S.E. (2004), Extinction risk from climate change.*Nature*427: 145-148.

Verchot, L. V., Noordwijk, M. V., Kandji, S., Tomich, T. and Ong, C. (2007), Climate change: linking adaptation and mitigation through agroforestry. *Mitigation and Adaptation Strategies for*

Veteli, T.O., Kuokkanen, K., Julkunen-Tiitto, R., Roininen, H., and Tahvanainen, J. (2002), "Effects of Elevated CO_2 and Temperature on Plant Growth and Herbivore Defensive Chemistry," *Global Change Biology*, 8, 12, pp. 1240–252.

Watanabe Tsugihiro and Takashi Kume. (2009). A general adaptation strategy for climate change impacts on paddy cultivation: special reference to the Japanese context. Paddy Water Environ (2009)7: 313–320

Yao Z., Z. Zhou, X. Zheng, B. Xie, B. Mei, R. Wang, K. Butterbach-Bahl, and J. Zhu (2010). Effects of organic matter incorporation on nitrous oxide emissionsfrom rice-wheat rotation ecosystems in China. Plant and Soil 327, 315 – 330.doi: 10.1007/s11104 - 009 - 0056 - 4, ISSN: 0032-079X, 1573 – 5036.

Zickfeld, K *et al.* (2005),"Is the Indian summer monsoon stable against global change?" Geophysical Research Letters Vol 32, L 15707.

Chapter 22

Weather Forecasting and Climate Smart Agriculture

Sentiyanger Jamir[1]*****, *Shivani Kumari*[2], *Mapellu. S. Likhitha Reddy*[3] *and Wajid Hasan*[4]

[1]*Department of Agronomy, SHUATS, Prayagraj, Uttar Pradesh*
[2]*Department of Agronomy, SASRD, Nagaland University, Nagaland*
[3]*Department of Agronomy, UBKV, West Bengal*
[4]*Krishi Vigyan Kendra, Jehanabad, BAU, Sabour, Bihar*

Agriculture is one prime sector that impacted heavily due to global climate change. India being a massive country with different climate, seasons, crops and farming systems all over the country. Factors such as rainfall and radiation are having direct impact on agriculture systems and practices. Wet and dry spells also as extreme events causes enormous impacts on the agriculture sector. It finds that even at just 1°c of warming a negative impact for major crops like wheat, rice and maize could be seen. Despite change in climate and it's adverse impact on crops/animals, income to farmers shouldn't decrease. It is essential to provide opportunities to farmers to have multiple source of income from agriculture, animal, husbandry, fisheries and poultry. So, all doesn't fail together. All this can be achieved through Weather forecasting and climate smart agriculture (CSA).

Keywords: *Climate smart agriculture, Weather forecasting, Sustainable, Warming.*

Introduction

Weather plays an important role in agricultural production it has a profound influence on the growth development and yields of a crop incidence of pest and diseases water needs and fertilizer requirements in terms of difference in nutrient mobilization as a result of water stresses, as well as the timeliness and effectiveness of preventive measures and cultural operations on crops. Weather aberations may cause number of damages to crops and soil erosion. The quality of crop producing during movement from field storage and transport to plug depends on weather. That inclemency may effect the standard of produce during transport, and viability and vigour of seeds and planting material during storage. Access to near real-time agronomic information and cropping system performance could potentially lead to increased profitability while simultaneously improving environmental performance (Archontoulis *et al.*, 2015 and 2016)

Usefulness of Weather Forecasts

Occurrences of erratic weather are beyond human control. There is a possibility to adapt to or mitigate the effects of adverse weather if a forecast of the expected weather can be had in advance. Agronomic strategies to deal with changing weather are available. For example delay in start of crop season is often countered by using short duration varieties or crops and thicker sowings. However, once the crop season starts the resources and technology get committed and therefore the only option then left is to adopt crop-cultural practices to attenuate the consequences of mid-seasonal hazardous weather phenomena on the idea of advanced intimation of their occurrences. For example, lots of pesticides being wash away by rain can be prevented if timely rainfall is predicted on time. Thus, the usefulness of medium range weather forecasts with a validity period that permits farmers to arrange and perform appropriate cultural operations to deal with or cash in of the forecasted weather is warranted. The advancement in Information Technology is getting penetrated into the rural areas, forecasting of timely and precise weather forecasts for farmers is on the rise and needs. Improved methods of forecasting crop production can also be beneficial in making marketing decisions that could improve farm profitability (Anderson, 1973; Jones *et al.*, 2000; Brandes *et al.*, 2016; Johnson *et al.*, 2016).

Essential Requirements of Weather Forecasts for Agriculture

Forecasting of late start of the crop season initiate agronomic changes from the normal at the field level. Organization and execution of such a technique comes under the category of high cost decisions and can take quite some time. So, pre-seasonal forecasting needs to have a validity period of at least 10 days and not less than a week. Measures to counter the effects of forecasted hazardous weather such as wind direction, speed, pests, diseases *etc.* take some time and hence mid-seasonal forecasts must preferably be communicated 5 days and not at less than 3 days in advance if possible. Dissemination of weather forecasting after their formulation to farmers and agricultural users needs to be fast with minimum possible temporal lag. Some of the measures like pre-seasonal agronomic corrections, control operations against pests and diseases, supplementary irrigation and pre-planning of crop harvests are getting to be high cost decisions. Therefore, the weather forecasts must not only be timely but must even be very accurate. Weather forecasts must ideally be issued for little areas. In the case of well-organized weather systems the specified areal delineation of forecasts are often realized. In other cases the area(s) to which the weather forecasts are getting to be applicable must be unambiguously stated.

Elements of Agricultural Weather Forecasts

An agricultural weather outlook should ask all weather elements, which immediately affect farm planning or operations. The elements will vary from place to put and from season to season. Usually a weather forecasting has the following parameters, they are:

- ☆ Amount and sort of coverage of sky by clouds
- ☆ Rainfall and snow
- ☆ Maximum, minimum and dew point temperatures
- ☆ Relative humidity
- ☆ Wind Speed and Direction
- ☆ Extreme events like heat and cold waves fog, frost, hail, thunderstorms, wind squalls and gales, areas with low pressure, different levels of intensities of depressions, tornados *etc.*

An agricultural weather outlook should contain the subsequent information also:

- ☆ Bright hours of sunshine
- ☆ Solar radiation
- ☆ Dew
- ☆ Leaf wetness
- ☆ Pan evaporation
- ☆ Soil moisture stress conditions and supplementary irrigation for rainfed crops
- ☆ Information on irrigation timing and amount in terms of pan evaporation
- ☆ Specific information about the evolution of meteorological variables into the cover layer in some specific cases
- ☆ Micro-climate inside crops in specific cases.

Format of Forecast

Formats of forecasts for agriculture are highly variable in several agricultural contexts in function of the strong variability of users, crops, agro-techniques, *etc.* Specialised forecasts are often mentioned crops, farming, forestry, fisheries and horticulture. Problems with forecasts can't be barren of a technical slant but the forecast has got to be frames in as simple a dialogue as possible to enable the farmer to readily grasp its content. Therefore, use of "intermediaries" (employed by the National Meteorological Services and/or the extension wing of agricultural services) as an important link between the forecasters (and their products) and therefore the farmers should be educate on the utilization of forecasts as agro-meteorological services for field operations must be provided for. There are yield forecasting approaches that rely on crop models, which are driven by a combination of current and historical weather data (Cantelaube *et al.*, 2005; Chipanshi *et al.*, 2015; Ferrise *et al.*, 2015) A forecast produced for educational purpose and released weekly by University of Milano. This product consists of three main parts:

1. A general evolution;
2. Forecasting for seven days (precipitation, cloud coverage, wind, air temperature and also phenomena, like fog, frost, then forth);
3. Forecasting of water net primary, net balance.

Forecasts for Agricultural Purposes

Specific description of the processing of weather forecasts of single weather variables for agricultural purposes is presented below.

Sky Coverage

Forecast of sky coverage are often defined by adopting some standard classes, like sky clear (0–2 octas), partly cloudy (3–5 octas), mostly cloudy (6–7 octas) and overcast (8/8). It's also important to supply information about the character of prevailing clouds. For instance, high clouds produce a depletion of worldwide radiation quite different from that produced by mid- or low clouds. it's important to offer a thought of the expected variability of sky coverage in space and time also. A probabilistic approach can also be adopted so as to extend the usefulness of this type of data.

Bright Sunshine

Sun shining through clouds won't affect crop performance, because during this case the reduction are going to be in diffuse radiation from the sunlit sky and therefore the latter is merely a fraction of total global radiation. So in cloudiness forecasts the fraction of cloudiness in the sun should even be laid call at addition to the whole cloudiness.

Solar Radiation

The main parameters, extraterrestrial radiation (Ra) and possible daylight hours (N), required to derive radiation (Rs) from bright hours of sunshine (n) are readily available on a weekly basis for any location and period (Venkataraman, 2002). The connection between the ratio of Rs/Ra and n/N t may be a straight-line type. The worth of the constants, however, varies with seasons and locations but can readily be determined.

Precipitation

Snow and rainfall are probably two of the foremost difficult forecast variables. Quantitative forecasting of rainfall, especially of heavy downpours, is extremely difficult and realizable only in rare instances and using highly sophisticated Doppler radars. For crop operations, however, the quantitative forecasting of rain isn't half as important because the forecasting of non-occurrence of rains (dry spells) and therefore the sort of rain spell which will be expected. Forecasts of rain are often defined by adopting some standard classes supported the climate and therefore the agricultural context of the chosen area. A probabilistic approach is sort of important so as to maximise the usefulness of this forecast. It's possible to supply daily information like this:

1. Mostly cloudy or might be overcast with some rainfall (Class 3, highly probability);
2. Partly cloudy with a bit of rainfall unlikely (Class 2, very low probability);
3. Sky clear with absence of precipitation.

Fog can contribute significantly to crop water needs and should be measured by covering the funnel of a rain gauge with a gaggle of fine wires. Dew is a crucial parameter influencing leaf wetness duration and it therefore plays a task in facilitating entrance of disease spores into crop tissues. Dew is useful in contributing to water needs of crops in winter and in helping the survival of crops during times of soil moisture stress, because the quantum of dew collected per unit area of crop surface is repeatedly quite that recorded with dew gauges. Dew is additionally desirable for using pesticides and fungicides in sort of dust. The environmental condition required for dew formation are an equivalent as those for fog formation, except that there must be an absence of air turbulence within the air layers on the brink of the bottom and therefore the crop canopy temperature must be less than the screen temperatures. Thus, nomograms employed by forecasters for predicting fog are often used to predict dew, within the absence of low-level air turbulence, by factoring into the temperature criteria the expected depression of the crops minimum temperatures under the screen minimum.

Temperature

Forecasting of air temperature is significant for several agrometeorological applications. Forecasts of the temperature of soil, water, crop canopies or specific plant organs also are important in some specific cases. Crop species exhibit the phenomenon of thermoperiodicity, which is that the differential response of crop species to daytime, nocturnal and mean air temperatures (for example, Solanaceae to night time temperatures, Papilionaceae to day time temperatures and Gramineae to mean air temperatures). It's possible to derive mean day and night-time temperatures from maximum and minimum temperature data. Forecasts of temperature are generally expressed as a variety of expected values (for example, 32°C –36°C for max and 22°C –24°C for minimum). If the forecast is directed at mountainous territories, temperature ranges might be defined for various altitudinal belts, taking under consideration also the results of aspect. Special care could be reserved for temperature forecasts at particular times of the agricultural cycle, taking into account the values of cardinal and important temperatures for reference crops. Other thermal variables with a selected physiological meaning (for example, accumulation of thermal units or chill units) are often the topic of specific forecasts. The bottom temperature above which the accumulations will apply, however, varies with crop types (for example, wheat, maize and rice: 4.5°C, 10°C and 8°C, respectively). Forecasting dates of attainment of specific phenological stages of plants, statistic data showing actually realized heat or chill accumulations by various crops up to the time that forecasts are issued need to be maintained. A approach can then be adopted to forecast the probable dates that specific crops will reach particular phenological stages.

Humidity

Forecasting of ratio is often important in some specific cases. Probability of critical values (very high or very low) also can be important.

Wind Speed and Direction

Forecasting of wind speed is vital for several different agricultural activities. Wind direction are often defined also. It's important to offer a thought of the expected variability in speed and direction of wind. The monthly windrose at a station may be a climatological presentation that indicates the frequency of occurring the wind from the above each of the eight accepted points of the compass and frequencies of occurring of the given wind speed ranges in each of the above eight directions. Wherever possible, the windroses must be checked out before forecasts are issued. For agricultural purposes, wind speed and direction are required at a height of two m. But weather forecasts of wind ask heights greater than 2 m. Change in wind direction between 2 m and therefore the forecast height won't occur. Wind speed at 2 m are going to be considerably less than at the forecast height, however. Ready tables to convert wind speeds at any height to the speed at 2 m are available and should be wont to forecast wind at a height of two m. Speeds must be forecast in km/h for a height of two m.

Leaf Wetness

Leaf wetness is the result of rainfall, dew or fog. Duration of this phenomenon are often important so as to plan different activities, like the appliance of pesticides and harvesting of crops. Leaf wetness may be a parameter that's scarcely recorded. Variety of empirical methods cited by (Matra *et al.,* 2005) are wont to derive leaf wetness durations from meteorological parameters. One can obtain the hourly march of temperatures from maximum and minimum temperatures (Venkataraman, 2002). The temperatures during night hours need to be decreased by a worth adequate to Depression of the radiation minimum below the screen minimum. As mentioned earlier, dewpoint temperature may be a conservative parameter. Thus, the amount of hours during which dewpoint temperature is above the adjusted air temperature will give leaf wetness duration. The time taken for the moisture deposited on the crop leaves to evaporate also has got to be included within the leaf wetness duration. The quantity of moisture deposited on the crop could also be repeatedly quite that indicated by instruments. Therefore the estimated moisture deposition has got to be multiplied by a crop factor and therefore the product divided by the evaporative power of the morning air. As a rule of thumb, two hours after sunrise could also be added to the estimated duration of leaf wetness.

Evapotranspiration

Forecast of evapotranspiration are often important to enhance knowledge of the water status of crops. This type of forecast is founded on the right forecast of radiation, temperature, ratio and wind speed. For real-time use, forecast of evapotranspiration has got to be founded on a forecast of pan evaporation, as discussed. The evaporative power of air (EPA) determines the height water needs of vegetative crops and is that the datum to which all measurements of evapotranspiration (ET) should relate. The Food and Agriculture Organization of the United Nations (FAO, 1998) has advocated the utilization of reference evapotranspiration ET0 as a typical measure of EPA.

Water Balance

Quantitative forecast of probability of water excess or stress for rainfed crops, and of the timing and amount of irrigation for irrigated crops, is extremely useful. This type of forecast for rainfed crops is predicated on correct forecasting of precipitation and evapotranspiration. The water balance approach to reach soil moisture excess or deficiency would require daily forecasts of rain within the first month of crop growth and on a short-period basis thereafter. The influence of physiological control over crop water uptake during maturity (Hattendorf *et al.*, 1988; Venkataraman, 1995) is additionally important. Since irrigation water is applied before crop water consumption for forecasts of irrigation scheduling, forecasts of evapotranspiration and certain rainfall amounts on a short-period basis will do.

Extreme Events

The low level of predictability of utmost events performing at meso or microscale (frost, thunderstorms, hail, tornadoes, then forth) may be a crucial limitation to the usefulness of forecasts for agriculture. So as to offer correct information to farmers, the adoption of a probabilistic approach might be important.

Agriculture Advisories

Agricultural advisories are an act of advice to crop growers/livestock producers based on possible future weather and climate conditions, regarding "what to do" or "what not to do" to maximize advantages and minimize losses in production. Weather and climate forecasts have little importance unless they're operationally used. Good samples of climate forecasts as agrometeorological services, together with other information, are often found in (Abdalla *et al.*, 2002; Harrison, 2005) and Meinke and Stone (2005) for instance. So that maximum advantage are often taken of weather forecasts, agrometeorological advisories are issued in consultation with experts of other concerned disciplines and take into consideration the past, present and predicted weather and its spatial-temporal behaviour. An appropriate forecast on the condition of weather has tremendous outcomes in terms of advance management of the negative impacts of vagaries of weather. This is because the value of weather-related risk reduction before the very fact is far smaller than the post-facto management of the losses (Rathore *et al.*, 2006). These advisories recommend implementation of certain practices or the utilization of special materials to assist effectively prevent or minimize possible weather-related crop damage or loss, for example, spraying advice based upon past and forecast weather to combat crop diseases and insects; sowing advice for better germination and plant stand; and harvesting advice to get optimum crop maturity, quality, and the like. Also recommend initiation of cultural practices which are weather-sensitive.

Agrometeorological advisories are being issued by many developed and developing countries on various spatial and temporal scales. Geographically large countries like China, India, the Russia and the United States now have national bodies or organizations that issue advisories on a county/state/agro-climatic region basis. Some developed countries (European countries, Russian Federation, United States) that have advanced computing and communication systems may consider catering to the tiny temporal scale for agriculturally related advice and frequent updating of advisories (in terms of hours), while developing countries like India issue advisories covering a span of 3–10 days, which enable the farmers to take ameliorative measures. For the agricultural sector, location-specific weather forecasts within the medium range are, therefore, vital. These services may contain advice on all the farming operations or some specific operations, such as pest management irrigation scheduling, and livestock management. On the idea of local agrometeorological and farming information and therefore the weather forecasts, the specialists discuss the choices and consequent effects then choose the recommendation to be given to farmers which are within the scope and knowledge of their expertise and skills. All of these constitute the agricultural advisory (Singh *et al.*, 1999).

Climate Smart Agriculture

The agriculture sector is facing serious challenges. Farmers are compel to produce more food on less land for a rapid growing populations, under a changing climate and with the need to reduce greenhouse gas emissions to meet the target of keeping temperature increase below 2°C. Reducing greenhouse emissions from the agriculture sector can be challenging in South Asia where the sector is an important contributor to the economy. As part of the UNFCCC Paris Climate Agreement's objective of limiting global warming to 2°C, annual emissions reduction from agriculture must reach 1 gigatonne of carbon dioxide equivalents (GtCO2e/Yr) by 2030. Achieving this target means we need to reach the world's half a billion farmers by 2030 with new knowledge,skills, finance, and capacity building to transform livelihoods. Already known capabilities can only deliver 21–40 per cent of this target. A transformation is required within subsequent decade within the way agricultural systems and broader food systems function. Such a change can only be achieved by international cooperation on advanced technologies, climate smart agriculture (CSA), capacity enhancement, and learning. Climate smart agriculture (CSA) as explained by Food and Agricultural Organizations (FAO) during The Hague Conference on Agriculture, Food Security and global climate change in 2010, takes part within the achievement of sustainable progress. By combining three ways of sustainable progress (financial, community based and ecological) and finding solution to food safety and climate issues. That is why, it had been developed on the three major pillars: i)Sustainably increasing agricultural productivity and incomes; ii) adapting and building resilience to climate change; iii) reducing and/ or removing greenhouse gases emissions, where possible. CSA is an approach to improve the scientific rule and investment setting to attain sustainable agricultural progress to ensure the food availability under climate change (FAO, 2012). CSA is a way for transforming and reorienting agricultural development under the realities of climate change (Lipper *et al.*, 2014). Smart agriculture is a way to get sustainable developments as well as green economy goals. It is oriented towards food security and takes steps to conserve natural assets. It is closely related to perception of improved growth, as FAO develops it for crop yields (FAO, 2011). Consideration of the impacts of world climate variations on sustainable green growth may help to predict and follow better farming techniques to enhance agricultural yields (Fraser, 2008). Sustainability is an important principle for international environmental policy and decision making in the twenty first century (Braimoh *et al.*, 2010).

Strategies for Climate Smart Agriculture

Climate smart agriculture are often adapted through variety of techniques. as an example, Methane (CH4) emissions as a results of rice farming are often altered by water management efficiently, including the promotion of mid-season aeration through short-term drainage; recuperating management of organic wastes by promoting aerobic decomposition via the method of composting or incorporate soil for the amount of the off-term drainage; use of rice varieties with a couple of infertile tillers, high oxidative roots and better harvest index; and use of fermented fertilizer as biogas slurry compost rather than unfermented one (Pathak and Wassmann, 2007). Land use management practices like reduced tillage, manure and wastes enhance the mixing of soil biodiversity, assembly and partial coverage can play an important role within the accumulation of carbon within the soil (Pathak and Wassmann, 2007). Problem of global climate change are often solved by climate smart agriculture like developing new varieties which are tolerant to heat, salinity and immune to floods and drought, improving water management, following new agricultural technologies like resource conservation and effective management techniques, crop diversification, improved pest management, crop modelling and better meteorology.

Efficient Resource Management

Resource management is extremely significant feature of CSA and future climate. Food wastage are found in all stages of the food production till food utilization and comsumption. Almost one third a part of food produced is wasted (Gustavsson *et al.*, 2011). The energy consumed in annually world food losses are almost 38 you look after the ultimate energy utilized by the entire organic phenomenon. Things such as food chains, from agricultural, transport, conservation, processing, cooking and consumption are likely areas for improving energy use efficiency (FAO, 2011). Within the continent of Africa, 90 per cent of the removed wood is employed for manufacturing household articles and also, for cooking. Also efficient energy saving in cooking stoves can help to decrease deforestation.

Availability of Technical Knowledge of Farmers

In South Asia farmers mostly belong to poor families and are limited to their resources, in order that they are experimenting with climate variability for hundreds of years. They need abundance of data techniques with the amount of steps which will be utilized in the event of technologies for controlling climate vulnerability. This type of data is required to satisfy the fashionable needs. Conventional environmental knowledge of individuals improved and developed within the test of your time could give different ideas and feasible options for adaptation procedures. Studies have concluded that the observation of traditional farmers was quite evolved within the affairs of earthquakes, landslides and drought and that they have developed better techniques to alleviate the effect of natural and climatic variations.

Role of Institutions for CSA Improvement

Those institutions are essential which create and transfer useful information and guide people to interpret the new technologies into understanding and work there on. The institutions like farmer farm schools that guide and facilitate farmers for the implementation of latest techniques; Shows of farm radio that share agricultural information which is definitely accessible and weather-related ideas and skills to local rural people; agricultural plots exhibitions to the community and exchange of ideas between farmers. The profits achieved by using those management techniques which are sustainable for the land usually take time to seem. During this point, the farmers need to tolerate the entire expenditure that has expenditure of labour, land and cash (McCarthy *et al.*, 2011). Since poor farmers haven't any resources to access credit and markets, and that they are unable to adapt to those new techniques for the success of CSA the strong institutions need to maintain agricultural markets and financing mechanisms which are vital. The foremost efficient technique allows researchers, private sector investors, community members and policy makers to collectively describe troubles that are planned to resolve (Kristjanson *et al.*, 2009).

Forecasting

Weather forecast also as early warning systems are going to be helpful to stop threats of climate losses. Information and technology (ICT) can efficiently help administrators and researchers in planning the contingency programs. Within the field of forecasting a replacement technique Multi Criteria Analysis (MCA) tool has implemented to gauge global climate change policy alert on mitigation and adaptation options (de Bruin *et al.*, 2009). The foremost appropriate thoughts of the choice makers can implement in their deciding, which may be a multi level MCA implemented to label and find possible adaptation techniques.

Crop Modeling

Crop modelling may be a creative and newly developed tool for dealing risks in Agriculture. Computer aided simulation models plays a very important role in approaching the techniques that build crop management responses and forecast crop yield. Simulation models significantly assist the work of higher crop development and recommend valid crop management. Models can guide to seek out the possible risk of climate variation on future crop yields, climate smart agriculture development and mitigation procedures. Like new management options and suitable genotypes which are important factors for better yield, crop simulation models are essential tools in field experiments to make new crop management approaches. During this point of view, two crop management systems are APSIM (Agricultural Production System Simulator) and DSSAT (Decision network for Agro Technology Transfer) mostly utilized in the entire world (Ahmad *et al.*, 2014).

Integration of Modelling and Forecasting

Although there remain some uncertainties due to lack of data, crop simulation models can appreciably estimate and quantify the impact of specific water stress conditions (possible thanks to climate change) on crop productivity if these are well calibrated and validated in field experiments (Wolf *et al.*, 1996; Grossman-Clarke *et al.*, 2001;). Crop models allow variation of environmental factors like the water regime, temperature and simulate the crop response via many calculated growth parameters like crop yield. Thanks to the complexity of the matter, research continues and enhancements are constantly being made to models, for instance, drought impact assessment.

GIS Mapping

GIS (Geographical Information System) is employed in analysis and mapping which helped within the estimation and computation of the storm course and flooding related to hot cyclones. In the study it includes population allocation, infrastructure and other under threat resources. The images and pictures also are utilized in the investigation of seashore thanks to hot cyclones and rising sea levels. Risk and hazard maps are often created at different possible scales to point out the threat allocation across different countries. These regions are often site specific; include provincial or municipal administrative areas and other small national landscapes, like river basins, coastlines and lakes.

Conclusion

FAO estimated that if the present production and consumption rates continue, agricultural production should increase 60 per cent by 2050 to meet the needs of food of world's population. To meet such target of food security and agricultural development goals with adaptation to climate change will be required to lower emission intensities per output. Improving food protection by moderate climate change, sustainably use the natural resource, use all products more competently, have less inconsistency and greater constancy in their outputs. More fruitful and more flexible agriculture requires a most important change in the way of use of land, water, soil nutrients and genetic resources management thought weather forecasting in agriculture and climate smart agriculture techniques.

REFERENCES

Abdalla, A. T., C.J. Stigter, N. I. Bakheit, M. C. Gough, H. A. Mohamed, A. E. Mohammed and M. A. Ahmed. (2002). Traditional underground grain storage in clay soils in Sudan improved by recent innovations. *Tropicultura*, 20: 170–175.

Ahmad, M. I., Ali, A., Ali, M. A., Khan, S. R., Hassan, S. W., Javed, M. M. (2014). Use of Crop Growth Models in Agriculture: A Review. *Science International*, pp: 26.

Anderson, L. G (1973). The Economics of Extended-Term Weather Forecasting. *Monthly Weather Review*, 101, 115-125

Archontoulis, S., Dietzel, R., Castellano, M., VanLoocke, A., Moore, K., Puntel, L., Cordova, C., Togliatti, K., Huber, I., Licht, M. (2015) Forecasting yields and in season crop water nitrogen needs using simulation models. In Proceedings of the 27th ICM conference, Dec 2-3 2015, Ames, IA

Archontoulis, S., Licht, M., Castellano, M., Dietzel, R., VanLoocke, A., Ordonez, R., Iqbal, J., Puntel, L., Cordova, C., Togliatti, K., Martinez-Feria, R., Huber, I., Helmers, M. (2016a). Understanding the 2016 yields and interactions between soils, crops, climate, and management. **In** Proceedings of the 28th ICM conference, Nov 30- Dec 1 2016, Ames, IA

Braimoh, A. K and Osaki, M. (2010). Land use change and environmental sustainability. *Sustainability Science* 5: 5-8.

Brandes, E., McNunn, G. S., Schulte, L. A., Bonner, I. J., Muth, D. J., Babcock, B. A., Sharma, B., Heaton, E. A. (2016). Subfield profitability analysis reveals an economic case for cropland diversification. *Environmental Research Letters*, 11, doi: 10.1088/1748-9326/11/1/014009

Cantelaube, P. and Terres, J. M. (2005). Seasonal weather forecasts for crop yield modelling in Europe. Tellus, Series A: *Dynamic Meteorology and Oceanography*, 57A, 476-487, doi: 10.1111/j.1600-0870.2005.00125.x

Chipanshi, A., Zhang, Y., Kouadio, L., Newlands, N., Davidson, A., Hill, H., Warren, R., Qian, B., Daneshfar, B., Bedard, F., Reichert, G. (2015). Evaluation of the Integrated Canadian Crop Yield Forecaster (ICCYF) model for in-season prediction of crop yield across Canadian agricultural landscape. *Agricultural and Forest Meteorology*, 206, 137-150, doi: 10.1016/j.agrformet.2015.03.007

De Bruin, K., Dellink, R. B., Ruijs, A., Bolwidt, L., Van Buuren, A., Graveland, J., De Groot, R. S., Kuikman, P. J, Reinhard, S., Roetter, R. P., Tassone, V. C., Verhagen, A., Van Ierland, E. C. (2009). Adapting to climate change in The Netherlands: An inventory of climate adaptation options and ranking of alternatives. *Climatic Change*, 95: 23-45

Ferrise, R., Toscano, P., Pasqui, M., Moriondo, M., Primicerio, J., Semenov, M., Bindi, M. (2015). Monthly-to-seasonal predictions of durum wheat yield over the Mediterranean Basin. *Climate Research*, 65, 7-21, doi: 10.3354/cr01325.

Food and Agriculture Organization of the United Nations (FAO). (1998). Crop Evapotranspiration: Guidelines for Computing Crop Water Requirements (R.R. Allen, L.S. Pereira, D. Raes and M. Smith). Irrigation and Drainage Paper No. 56, Rome, FAO.

FAO. (2011). Save and grow: a policymaker's guide to the sustainable intensification of small holder crop production. Rome.

FAO. (2012). The Sahel crisis: executive brief. March. Rome.

Fraser, D. (2008). Understanding animal welfare. *Acta Veterinaria Scandinavica*, 50(Suppl 1), S1.

Gustavsson, J., Cederberg, C., Sonesson, U., Van Otterdijk, R., Meybeck, A. (2011). Global food losses and food waste: extent, causes and prevention. Rome, FAO.

Harrison, M. (2005). The development of seasonal and inter-annual climate forecasting. *Climatic Change*, 70: 201–220.

Hattendorf, M. J., Redelfs, M. S., Amos, B., Stone, L. R., Gwin, R. F. (1988). Comparative water use characteristics of six row crops. *Agronomy journal*, 80: 80–85.

Johnson, M. D., Hsieh, W. W., Cannon, A. J., Davidson, A. (2016). Crop yield forecasting on the Canadian Praries by remotely sensed vegetation indices and machine learning methods. *Agricultural Forecast Meteorology*, 218-219, 74-84, doi: 10.1016/j.agformet.2015.11.003

Jones, J. W., Hansen, J. W, Royce, F. S., Messina, C. D. (2000) Potential benefits of climate forecasting to agriculture. *Agriculture, Ecosystems, and Environment*, 82, 169-184, doi: 10.1016/S0167-8809(00)00225-5

Matra, A.D., R.D. Magarey and S. Orlandini, 2005: Modelling leaf wetness duration and downy mildew simulation on a grapevine in Italy. Agric. For. Meteorol., 132: 84–95.

McCarthy, N., Lipper, L., Branca, G. (2011). Climate smart agriculture: smallholder adoption and implications for climate change adaptation and mitigation. FAO, MICCA Series No. 4. Rome, FAO.

Meinke, H. and Stone, R.C. (2005). Seasonal and inter-annual climate forecasting: the new tool for increasing preparedness to climate variability and change in agricultural planning and operations. *Climate Change*, 70: 221–253.

Rathore, L.S., Parvinder, M. and Kaushik, S. (2006). Impact Assessment of the Agro-meteorological Advisory Service of the National Centre for Medium Range weatherForecasting(NCMRWF).http: //www.agrometeorology.org/filesfolder/repository/ncmrwf.pdf.

Singh, S.V., Rathore L.S. and Trivedi, H.K.N. (1999). A Guide for Agrometeorological Advisory Services. New Delhi, National Centre for Medium Range Weather Forecasting.

Venkataraman, S. (2002). Tabular aids for computation of derived agrometeorological parameters on a weekly basis. *J. Agrometeorol.*, 4: 1-8.

Wassmann, R., Pathak, H. (2007). Introducing greenhouse gas mitigation as a development objective in rice-based agriculture: II. Cost benefit assessment for different technologies, regions and scales. *Agril. Syst.*, 94: 826-840.

Chapter 23

Climate Resilient Smart Agriculture for Sustainable Livelihood

Dewali Roy[1], Aditya Pratap Singh[2], Yengkhom Linthoingambi Devi[3] and Til Kumari Limboo[4]*

[1]*Department of Soil Science and Agricultural Chemistry, ICAR-Indian Agricultural Research Institute, New Delhi*
[2]*Department of Genetics and Plant Breeding, Bidhan Chandra Krishi Viswavidyalaya, Mohanpur – 741252, Nadia, West Bengal*
[3]*Department of Genetics and Plant Breeding, College of Agriculture, Indira Gandhi Krishi Vishwavidyalaya, IGKV, Raipur, Chhattisgarh*
[4]*Krishi Vigyan Kendra, Gyalshing – 737111, West Sikkim, Sikkim*

Climate change is a global phenomenon and its threat is now well recognized in the scientific communities. With ever increasing world population, achieving sustainability in food and nutritional security is of prime importance. In the present chapter, we are discussing different prospects of climate resilient agriculture in view of climate change and how climate resilient smart agriculture can help in attaining sustainable livelihood.

Keywords: *Climate change, Climate resilient smart agriculture, Sustainable livelihood.*

Climate Change: A Global Threat

The definition of climate change provided by UNFCCC is 'a change which is attributed directly or indirectly to human activity that alters the composition of the global atmosphere and in addition to natural climate variability as observed over comparable time periods'. Although the Earth's climate is relatively stable for the past 10,000 years or so and the changes are occurring at natural rate but since 1900s there are more rapid changes which have taken place and mainly thought to be man-made (McCarthy *et al.*, 2001). Best estimates predict that global annual mean temperature will increase in the range of 1.8-4°C during the year 2100, which may result to variability in rainfall and gradually enhance frequency of extreme weather events such as cold waves, heat waves, floods and droughts. Global mean surface temperature will rise between 1.8°C to 4.0°C by 2100 (Maheshwari *et al.*, 2014). The pre industrial level of CO280 ppm will jump to 700 ppm by the end of the century (IPCC, 2014). Decrease in ice mass in Antarctica- 127 Giga tonnes/yr and increase in sea level-3.3 mm/yr worldwide are also few symptoms of climate change (Field, *et al.*, 2014). Among different gases, carbon dioxide is responsible for 77 per cent of global warming over a 100 year period and hence the most important GHG (Mendelsohn, *et al.*, 2000).

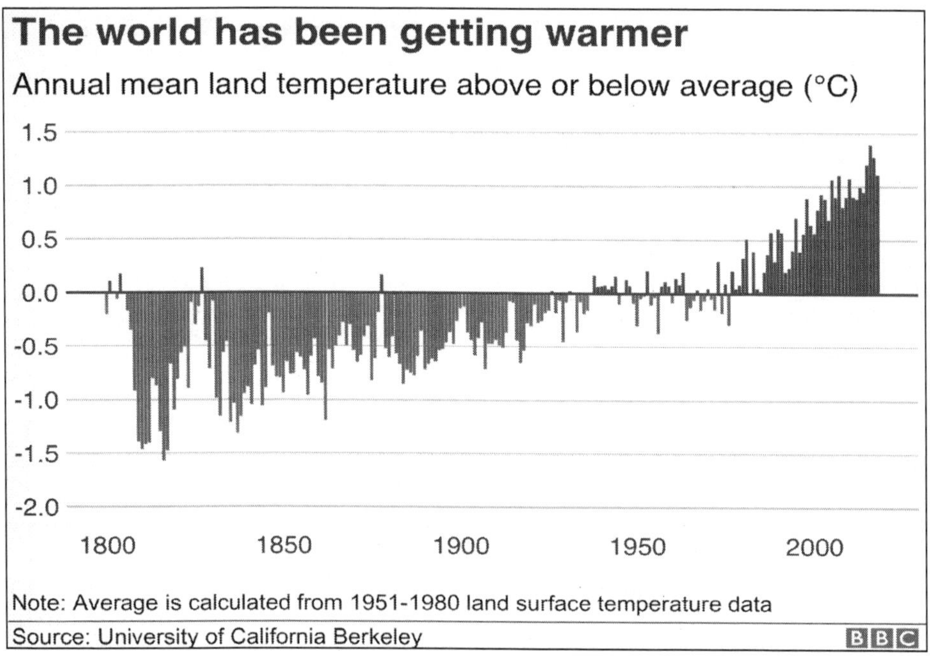

Figure 23.1: Average Increase of Mean Temperature.

In the coming decades, climate change may be expected to further exacerbate the risks of more natural disasters - intense storms, floods and long-lasting droughts which can erode existing community coping capacities and hamper them to prepare for, respond to and recovering from successive hazard events (Ray *et al.*, 2007).

Negative impacts of climate change can be both slow onset and sudden onset such as rise in sea levels, increasing temperature, ocean acidification and salinization, land and forest degradation, loss of biodiversity and desertification (slow onset) as well as extreme weather events, such as heat waves, tropical cyclones, droughts and floods (sudden onset). Another alarming threat is that gradually these all event related to climate change will reduce economic development and public health, degrade ecosystems and increase migration vulnerability among communities especially in areas that are disaster prone and fragile.

Impact of Climate Change on Agriculture and Food Security

The relationship and interaction between climate change and agriculture is complex both ecologically and politically. The hazard events of climate change have a direct impact on people's agricultural production (crop farming, livestock keeping) and livelihood security and inherently linked to the productive capacity of agricultural production systems worldwide. These slow onset events adversely affect agriculture by not only decreasing in overall productivity but also leads to loss of space (land, water, forest, *etc.*) for agricultural activities which can drive out people who rely on it. The yield reduction is likely brought out by several governing factors including pests, weeds and diseases, loss of biodiversity, rise in sea level, saline water intrusion in coastal belts, poor quality of irrigation water, decline in soil fertility, and irregularities in onset of monsoon, heat wave, cold wave, drought, flood and cyclone (Rao *et al.*, 2013).

Agriculture, particularly in India has nearly 60 per cent rainfed area and which is highly risky venture with vagaries of monsoon. Uncertain and irregular rainfall increases riskiness of agriculture with more pronounced effect on rainfed agriculture is practiced as they are fragile, degraded and sloppy lands which are thirsty as well as hungry and prone to erosion (Prasad *et al.*, 2014).

The inherent relationship between climate change and agriculture has another dimension that highlights the facts that the global climate is also being influenced by greenhouse gas (GHG) emissions from conventional industrial agriculture, which is responsible for between 11 to 15 percent of total global emissions. Simultaneously, this sector is also responsible for nearly 60 percent of nitrous oxide (N_2O) emissions, mainly from chemical fertilizers and for about 50 per cent of methane emissions (CH_4), much of which comes from livestock. Such impacts is global but much of the damage will be in developing countries where 11 per cent of arable land could be affected by climate change, including a reduction of about 16 per cent of agricultural GDP and deforestation driven by agricultural expansion for food, fiber and fuel (Venkateswarlu *et al.*, 2011).

Poverty is one of the biggest barriers to human development and economic growth. Almost 2.1 billion people still live in poverty and 767 million are extreme poor (Friel *et al.*, 2009). Evidence shows that climate change is highly impacting the rural poor by damaging their infrastructure, depressing crop yields, jeopardizing fish stocks and eroding natural resources and endangering species which resulted significant damage to agriculture, water resources, ecosystems and human health

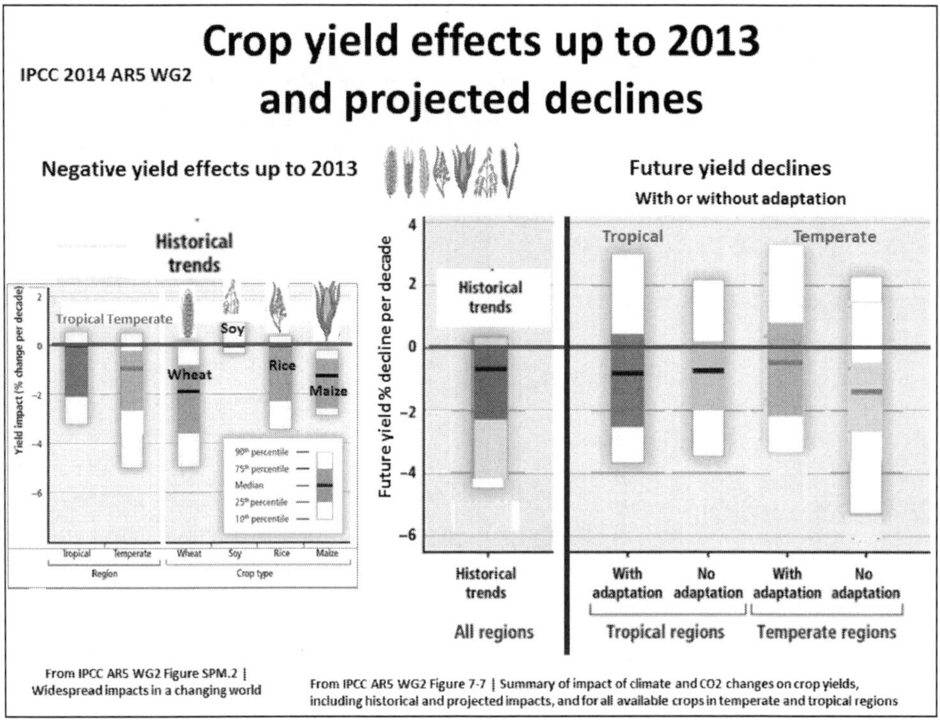

Figure 23.2: Effect of Climate Change Strategies on Crop Yield.

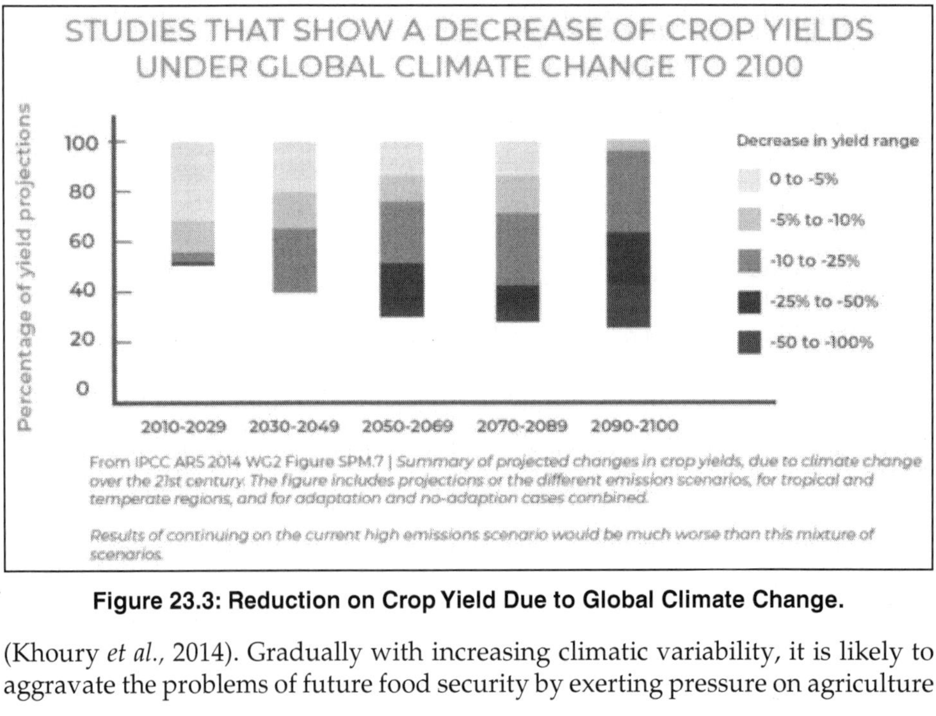

Figure 23.3: Reduction on Crop Yield Due to Global Climate Change.

(Khoury *et al.*, 2014). Gradually with increasing climatic variability, it is likely to aggravate the problems of future food security by exerting pressure on agriculture

(Singh *et al.*, 2017). In India, significant negative impacts have been observed with medium-term (2010-2039) climate change that is predicted to reduce yields by 4.5 to 9 per cent depending on the magnitude and distribution of warming. Since agriculture contributes roughly 16 percent of India's GDP, a 4.5 to 9 per cent negative impact on production signifies a cost of climate change to be roughly up to 1.5 per cent of GDP per year (Maheswari *et al.*, 2014). Such impact of climate change on agriculture will lead to problems with food security and may threaten the livelihood activities upon which much of the population depends.

Climate Resilient Smart Agriculture (CRSA): The Way Forward

Climate Resilient smart Agriculture (CRSA) can be defined as 'agriculture that reduces poverty and hunger in the face of climate change, improving the resources it depends on for future generations.' (Bulkeley *et al.*, 2013). It is an approach that guides actions needed to transform and reorient agricultural systems to effectively support development and ensure food security in a changing climate (Pathak *et al.*, 2012). It aims to tackle three main objectives: sustainably increase of agricultural productivity and incomes; adaptation and build resilience to climate change; remove greenhouse gas emissions (Haddad *et al.*, 2015). As stated in the FAO report, without urgent action, millions more people will be at risk of hunger and poverty.

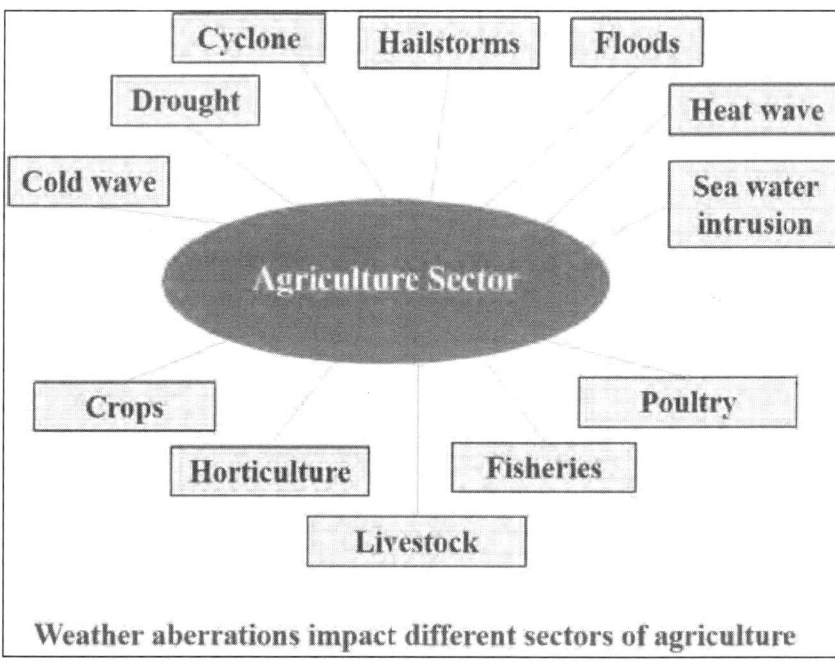

Figure 23.4: Effect of Weather Abnormalities on Climate Change.

It is evident after observing the limitations of contemporary climate change policy, that there must be an alternative way towards that are more socially just, economically viable, and environmentally sound that improves the livelihoods and builds the resilience of smallholder farmers (Singh *et al.*, 2017). CRSA contributes to global climate change mitigation not solely by capturing carbon in soil organic matter and above-ground biomass, however additionally by avoiding emission of carbonic acid gas and different GHGs from farms by reducing direct and indirect energy use (Rao *et al.*, 2013). Additional considerably, as proof shows, agro ecological farming strategies will cushion the negative impacts of utmost weather-related events as a results of global climate change. Though several ancient farming communities are able to deal with environmental condition extremes throughout history in truth, several farmers cope and inure global climate change by minimizing failure through numerous techniques like adapt drought tolerant native varieties, harvest fresh water, conservation, mulching, and active mixed cropping, crop rotation, and agro forestry. Still with none planned adaptation, the implications of semi permanent global climate change may be severe on the sustenance security of the poor. Therefore, it's of utmost importance to reinforce the resilience of Indian agriculture to global climate change (Beddington *et al.*, 2011).

Quality Criteria for CRA

During a initial spherical of consultation, professionals from NGOs, analysis and donors highlighted that effective CRA approaches ought to contribute to 1. Productivity and ecologic resilience, 2. Adaptational capability of stakeholders (including SDC), 3. Mitigation of greenhouse gases, 4. National commitments/ policies in regard to UNFCCC method (National Adaptation Plans and across the country Determined Commitments), 5. Contributions of CRA and Food Systems to different SDGs (McCarthy *et al.*, 2001)

Key Parts of Climate Resilient Agriculture

1. Capability (vulnerability assessment) and sanctionative policy and institutions; a pair of 2. Robust farmer organisations and networking; 3. Climate informed advisories and early warning; 4. Digital agriculture; 5. Climate Resilient and low-emission practices and technologies (incl. sensible water management interventions and trendy irrigation technologies to deal with scarceness and enhance water productivity by adopting a multi-stakeholder approach to boost property water resources management at basin level (water stewardship), 6. Prioritization and pathways of modification (vulnerability integration); 7. Credit and insurance; 8. Expanded personal sector activity and Public personal Partnerships. However consultants highlight additionally the importance of a landscape approach to climate and disaster resilience (Venkateswarlu *et al.*, 2011).

Bringing Resiliency in Climate Smart Agriculture system for Sustainable Livelihood

Resilience is the ability of a system to soak up shocks and recover as quickly as doable to traditional conditions once external atmosphere improves. Planned adaptation is important to extend the resilience of agricultural production to global climate change. Management practices that increase agricultural production beneath adverse atmospheric condition tend to support global climate change adaptation as a result of they increase resilience and scale back yield variability beneath variable climate and extreme events.

Table 23.1: Effect of Climate Change on Livelihood (Glantz *et al.,* 2009)

Community's Perception of Change	*Impacts on Livelihood System*	*Coping and Adaptation*	*Potential Future Risks*
Decrease in rainfall and change in timing of rainfall	Reduction in agricultural productivity and agro biodiversity	Improvising with cash crops, delayed sowing, more dependency on market	Increased food and livelihood insecuruty
Longer dry spells	Drying up of springs and less flow in springs and streams	Traditional water storing system and delayed sowing	Scarcity of drinking water and water for agriculture, workload on women and children. crop failure
Rise in temperature and decreased water availability	Lack of fodder and less productive land	Shift to smaller livestock, less land area under cultivation, dependence on market for food	Food insecurity, increased drudgery, dependence on cash income
Warmers winters and significantly less snowfalls, more precipitation in the form of rain instead of snow	Increased incidence of crop infestation with disease, pest and weeds	Increased use of insecticides and pesticides	Increased food and livelihood insecurity
Temperature and rainfall timing	Phonological changes in bee population	No coping mechanism	Degradation of orchards, Impact on horticulture
Better survival of animals due to less harsh winter	Increasing attacks onf animals in agricultural field	Replacing crops, no coping mechanism	Crop loss, less agricultural productivity
Reduced forest cover and biodiversity	Reduced snowfall and increase in temperature	Afforestation, No coping mechanism	Impact on livelihood dependent on non wood forest products.

Although, smallholder farmers often have other livelihood activities, but agriculture is usually the most important in terms of income and food security. Climate risks are only one of the various challenges facing individuals UN agencies livelihoods rely upon agriculture and who area unit resource-constrained. Global climate models must be downscaled to ensure regional needs and seasonal weather forecasting requires improvement (Prasad *et al.,* 2014).

Resilient livelihoods are livelihoods have the capability to resist or absorb the impact of climate changes calamities, adapt to dynamic climate conditions and reply to disasters consequently. Social protection programmes, disaster risk reduction and addressing gender-differentiated vulnerabilities are few ways that of accelerating the resilience of rural livelihoods and guaranteeing a longer-term positive impact on food access (Pathak *et al.,* 2012). According to the FAO's high level panel of experts, a food security oriented agricultural development strategy ought to stress on smallholder farming, formulate context-dependent channels for a productivity revolution and shift to higher-value agriculture by diversifying rural-based non-farm economic activity, connecting higher to markets, and establishing an energetic rural economy (Ayers *et al.*, 2014).

Targeted investments and insurance programmes are also needed and one of the major factor to increase the resilience of farming livelihood as an effective early warning systems. The financial needs of smallholders farmers, ranging from working capital expenditures for fertilizers and seeds, to medium- and long-term investments, can be made at the policy and institutional level. For example, policies related to weather-based index insurance schemes more readily available to smallholder farmers. In recent years, weather index insurance has gained vital international attention (Prasad *et al.*, 2014). It will facilitate to secure the financial gain of smallholders UN agency are notably prone to climate variability. The weather-based index insurance will act as a contingency insurance therein a selected event which may trigger associate insurance payment. As an example, the Normalized Distinction Vegetation Index (NDVI) derived from data collected by satellites, provides a sign of vegetation health and therefore potential crop yields and wont to give index-based drought insurance. Another option of financial assistance can be by Providing Decent Rural Employment (DRE) opportunities in rural areas prone to the negative effects of climate change which can increase resilience of the rural community, thereby reducing unwanted migration

Potential Strategies to Boost CRSA in Small Holder Farming and Secure Food Security

Enhancing resilience in agriculture against climate risk is of preponderant importance for safeguarding livelihoods of marginal farmers. Over the years, an array of practices and technologies has been developed by experts towards bringing stability in agricultural production against the onslaught of climate variations. Adoption of climate smart practices associated technologies by farmers seems to be additional of a necessity than an choice. Such quite approaches ensures adaptation gains and offers advantages to farmers along side doable reduction in GHG emissions and warming potential of agriculture (Rao *et al.*, 2013, Maheswari *et al.*, 2019). Such as:

1. Rejuvenation of farming in cyclone and flood prone coastal areas
2. Community paddy nursery as a contingency measure for delayed planting
3. Direct seeded rice for promoting water use efficiency
4. Drum seeding of rice for water saving and timeliness in planting
5. Drought tolerant paddy cultivars to tackle deficit precipitation
6. Short period crop varieties appropriate for late sowings
7. Crop diversification for keep security and resilience to climate variability agro-ecosystems through land shaping
8. Flood tolerant varieties impart resilience to farmers in flood-prone areas
9. Rising the resilience of poor farmers reclaiming arable wastelands.

10. Community tanks/ponds as a way of augmentation and management of village level water resources.
11. Enhancing resilience through improvement in conveyance potency.
12. Captive rearing of fish a livelihood opportunity in flood-prone areas.
13. Shelter management for small ruminants to tackle heat stress and rain storm.
14. Improved planting ways for enhancing water use efficiency and crop productivity.
15. Developing cultivars tolerant to heat and salinity stress and tolerant to flood and drought. (Maheswari *et al.*, 2013).

CRSA works on the right to food, environmental conservation and long-run community that facilitates food security at the native level and effectively contribute national and international climate change policies that helps independency and sustainability in agricultural systems worldwide (Smith *et al.*, 2007). In order to scale-up CRSA with the existing resources, agricultural policies and practices must be transformed and revised. During this context, our government with support from donors and international establishments is aiming on the subsequent initiatives:

These strategies should outline (Pathak *et al.*, 2012):

- Role of governments to prioritize their support towards smallholder farmers specifically women to promote sustainable agriculture and the kind of support smallholder farmers will receive from the government.
- How farmers will be involved in policy design and implementation.
- The kind of investments that is needed to promote those strategies and to bring the transition further away from conventional farming.
- How farmers' own know-how and creativity will be incorporated in the process of building sustainable alternatives.

Design and Implementation of Climate Change Risk Reduction Strategies Based on

1. Identifying the main impacts of climate change on smallholder agriculture, as well as the risks that may be faced by smallholder farmers in the future. (Venkateswarlu *et al.*, 2011).
2. The implementation of early warning systems to extend the state of farmers to sudden rains, floods, and dry periods.

Phasing out Input Subsidy Schemes for Agro-chemicals (Chemical fertilisers and pesticides) to Promote Sustainable Agriculture

Such subsidy programmes should include to:

- Embrace gender equity and women's rights as a matter of principle.
- Promote soil conservation.
- Promote sustainable water management.

- ✰ Conserve Agrobiodiversity preservation;
- ✰ Support livelihood diversification.
- ✰ Strengthen decentralized processing and improve smallholders' access to markets.
- ✰ Support farmers organisations and collective action

Re-orient Agricultural Research and Extension Services and Build 'Knowledge Hubs' to Support Smallholder Farmers in Promoting Sustainable Agriculture within the Context of Global Climate Change

National governments must focus on:

1. Increasing extension services, improve training and reach greater number of farmers than present situation.
2. Providing Training to female extension agents to support women farmers to make them aware about climate change and provide services that are tailored to their routines and needs.
3. Ensuring that the new extension services are farmer-interest and build bridges between local and scientific knowledge. Beside, encouraging local innovation to reduce the dependency on external and exported inputs.
4. Developing such seed varieties that are resistant to droughts, floods, and pests.
5. Stepping up support for improved water and soil management and incentivize practices such as rainwater harvesting, on-farm water preservation, and low-cost irrigation systems. (Maheswari *et al.*, 2019).

Besides, major attempts should be taken to promote community banks of grain, seeds, biomass, fodder, and storage facilities at the local level, to increase food security and food sovereignty, preserve local varieties, facilitate farmers access of to quality seeds, and prevent local genetic materials from becoming lost because of climate change (Field *et al.*, 2014). School feeding programmes and public food distribution systems could procure food from smallholder farmers practicing sustainable agriculture (Lobell *et al.*, 2008).

Conclusion

With the adoption of the 2030 Agenda on Sustainable Development (SDG), countries have shifted the focus towards their commitment to fight against poverty, hunger and malnutrition, acknowledging that tackling climate change is key to moving people out of poverty. One of the major targets of SDG pays special attention to building resilient livelihoods and helping the rural poor reduce their exposure and vulnerability to climate change and natural disasters. As part of this, FAO is supporting countries to achieve effective pro-poor development in the face of climate change, by promoting multi-sectoral, pro-poor and climate informed development policies, sustainable and inclusive production, diversification and decent rural employment, as well as risk informed and shock responsive social protection systems, with a view to leaving no one behind.

REFERENCES

Ayers, J., Huq, S., Wright, H., Faisal, A. M., and Hussain, S. T. (2014). Mainstreaming climate change adaptation into development in Bangladesh. *Climate and Development*, *6*(4), 293-305.

Beddington, S. J. (2011). The future of food and farming. *International Journal of Agricultural Management*, *1*(2), 2-6.

Bulkeley, H., and Betsill, M. M. (2013). *Cities and climate change* (Vol. 382). London: Routledge.

FAO, (2009). Coping with a changing climate: considerations for adaptation and mitigation in agriculture.

FAO, (2014). World Food Situation. Food & Agriculture Organization,. http://www.fao.org/worldfoodsituation/ foodpricesindex/en/.

Field, C. B., and Barros, V. R. (Eds.). (2014). *Climate change 2014–Impacts, adaptation and vulnerability: Regional aspects*. Cambridge University Press.

Friel, S., and Baker, P. I. (2009). Equity, food security and health equity in the Asia Pacific region. *Asia Pacific journal of clinical nutrition*, *18*(4), 620-632.

Glantz, M. H., Gommes, R., and Ramasamy, S. (2009). Coping with a changing climate: considerations for adaptation and mitigation in agriculture. *Environment and Natural Resources Management Series, Monitoring and Assessment-Food and Agriculture Organization of the United Nations*, (15).

Haddad, L. J., Hawkes, C., Achadi, E., Ahuja, A., Ag Bendech, M., Bhatia, K., and Flores-Ayala, R. (2015). *Global Nutrition Report 2015: Actions and accountability to advance nutrition and sustainable development*. Intl Food Policy Res Inst.

Intergovernmental Panel on Climate Change (IPCC) (2014). Climate Change.

Khoury, C. K., Bjorkman, A. D., Dempewolf, H., Ramirez-Villegas, J., Guarino, L., Jarvis, A., and Struik, P. C. (2014). Increasing homogeneity in global food supplies and the implications for food security. *Proceedings of the National Academy of Sciences*, *111*(11), 4001-4006.

Lobell, D. B., Burke, M. B., Tebaldi, C., Mastrandrea, M. D., Falcon, W. P., and Naylor, R. L. (2008). Prioritizing climate change adaptation needs for food security in 2030. *Science*, *319*(5863), 607-610.

Maheswari, M., Rao, M. S., Rao, V. U. M., Rao, C. S., Reddy, K. S., Ramana, D. B. V., and Sikka, A. K. (2013). National Initiative on Climate Resilient Agriculture (NICRA) Research Highlights (2013-14).

Maheswari, M., Sarkar, B., Vanaja, M., Srinivasa Rao, M., Prasad, J. V. N. S., Prabhakar, M., and Alagusundaram, K. (2019). Climate resilient crop varieties for sustainable food production under aberrant weather conditions. *ICAR-Central Research Institute for Dryland Agriculture, Hyderabad. 64p.*

McCarthy, J. J., Canziani, O. F., Leary, N. A., Dokken, D. J., and White, K. S. (Eds.). (2001). *Climate change 2001: impacts, adaptation, and vulnerability: contribution of*

Working Group II to the third assessment report of the Intergovernmental Panel on Climate Change (Vol. 2). Cambridge University Press.

Mendelsohn, R., and Tiwari, D. (2000). *Two essays on climate change and agriculture: A developing country perspective* (No. 145). Food and Agriculture Org.

Pathak, H., Aggarwal, P. K., and Singh, S. D. (2012). Climate change impact, adaptation and mitigation in agriculture: methodology for assessment and applications. *Indian Agricultural Research Institute, New Delhi*, 1-302.

Prasad, Y. G., Maheswari, M., Dixit, S., Srinivasarao, C., Sikka, A. K., Venkateswarlu, B., and Mishra, A. (2014). Smart practices and technologies for climate resilient agriculture.

Rao, V. U. M., and Rao, B. B. (2013). Climate change impact on Indian agriculture-adaptation and mitigation strategies.

Ray, A. Impacts, Adaptation, and Vulnerability: A Summary of the Report from IPCC Working Group 2 and Implications for the Western US.2007.

Singh, M., Sharma, N., Tiwari, N. K., and Kumar, N. (2017). Impact of climate change on global food security-challenges, adaptation measures and mitigating strategies. *International Journal of Pure and Applied Bioscience*, *5*(4), 381-394.

Smith, P., Martino, D., Cai, Z., Gwary, D., Janzen, H., Kumar, P. and Towprayoon, S. (2007). Policy and technological constraints to implementation of greenhouse gas mitigation options in agriculture. *Agriculture, Ecosystems and Environment*, *118*(1-4), 6-28.

Venkateswarlu, B., Kumar, S., Rao, S. D. C. S., Kokate, K. D., and Singh, A. K. (2011). Demonstration of climate resilient technologies on farmers' fields. Action Plan for 100 Vulnerable District.

World Bank. (2013). Policy Brief: Opportunities and Challenges for Climate Smart Agriculture in Africa.

Chapter 24

Climate Smart Agriculture (CSA): Needs and Pathways for Food Security

Shivani Kumari[1], Mallepu. S. Likhitha Reddy[2], Sentiyanger Jamir[3], M.R. Meshram[3] and L.K. Sanodiya[3]*

[1]Department of Agronomy, SASRD, Nagaland University, Medziphema, Nagaland
[2]Department of Agronomy, UBKV, West Bengal
[3]Department of Agronomy, NAI, SHUATS, Prayagraj, Utttar Pradesh

Climate Smart Agriculture 'CSA' strategy is primarily the adaptation and reorientation of agricultural systems in order to facilitate food security in the light of the current realities of climate change. It aims at improving the capacity of soils and agricultural systems to mitigate greenhouse gas emissions and to absorb carbon. CSA has huge potential to help smallholder farmers who are most vulnerable in achieving food security while adapting to changing conditions and militating against further climate change. Climate related event tends to jeopardize food security in terms of availability and access. It can increase food insecurity, in terms of both availability and access. The primary goal of Climate smart agriculture which can be achieved through enhanced productivity, adaptation, and mitigation. Several agricultural technologies are delineated such as minimum tillage, intercropping, crop-rotation, Precision farming, Permaculture, crop establishment methods i.e. System of Rice Intensification (SRI) in rice, water saving techniques like sprinklers and drip irrigation should be practiced at the field level which can help farmers to mitigate impact of climate change and variability in agriculture. In general, the CSA options integrate innovative and traditional technologies, practices and services which are suitable for particular region and reduce the effect of climate change and provide the opportunities to stand such changing scenario.

Keywords: *Adaptation, Climate change, Food security, Mitigation, Technology.*

Introduction

The world's population is anticipated to increase by one-third by 2050 according to projections of FAO. Among these majority of population are residing in the developing countries. Simultaneously, a lot of individuals are living in cities. If current financial gain and consumption growth trends continue, UN agency estimates that agricultural production have to be compelled to increase by 60 per cent by 2050 to satisfy the expected demands for food and feed. Agriculture should thus remodel itself if it's to feed a growing worldwide population and supply the premise for economic development and alleviation of poverty. Climate change has made this task tougher under a business-as-usual scenario, due to undesirable effect on agriculture, requiring spiralling alteration and related costs. In 1961, global agriculture was accountable to feed 3.5 billion individuals by tilling 1.37 billion ha of land. Since then population increased more than 50 per cent globally whereas crop cultivable area increased solely 12 per cent and almost no more accessibility if to take care of natural habitat for sustaining environmental changes. Water has been emerging as the most precious commodity on earth as agriculture utilizes approximately 70 per cent of the world's fresh water and will need 17 per cent more by 2025 to produce adequate food supplies. Currently, world has only 17 per cent of arable irrigated land contributing nearly 40 per cent of total agricultural output (FAO, 2009).

Food sufficiency/abundance is outlined as 'a scenario that prevails when all individuals, at all times, have physical, social and economic access to ample, healthy and nourishing food that meets their daily dietary needs and food choices for an energetic and healthy life' (FAO, 2001). Hence, food security needs that (a) food must be available in sufficient quantities, taking into account domestic production, imports and national stocks; (b) household livelihoods must be adequate to provide people with access to food supplies and (c) the supplies available must satisfy the specific dietary and health needs of people (FAO, 2001). Inadequate, old fashioned and traditional methods of food handling causing more than 30 percent food loss before reaching to the fork of end user and leaving many to sleep hungry every night. Additionally, global climate change is compelling to transfer the agricultural outgrowth agenda over the world. Changing temperature and precipitation and the rising frequency of extreme climate events are projected to significantly reduce global food production during the current century (Barkhordarian *et al.*, 2013).

The Food and Agriculture Organization (FAO) characterizes CSA as "agriculture frameworks that reasonably expands profitability, versatility (variation), eliminates ozone depleting substances (relief) while upgrading the accomplishment of public food security and improvement objectives" (FAO, 2010). Food and Agriculture Organization (FAO) assesses that there is need to build absolute farming creation by 60 per cent to take care of the total populace. With a considerable lot of the assets required for practical food security previously extended, the food security challenges are enormous. All the while, environmental change has affected horticultural creation around the world and locally in negative manner. Environment dangers to editing, domesticated animals and fisheries are relied upon to increment in coming many years, especially in low-pay nations where supportive limit is more fragile (Salamin *et al.*, 2010).

Adapting agriculture to climate change is indispensable to achieve food security and dawdling climate change. Climate change has an adverse impact on agricultural food production system leading to food insecurity. By accepting climate hazards scientist and social activist focused on solution for changing scenario. Climate smart agriculture (CSA) is a methodology for changing and reorienting agricultural frameworks to help food security under the new real factors of environmental change (Nwanze, 2011). Environment shrewd rural practices could likely counter this pattern. CSA is a methodology for changing and reorienting rural frameworks to help food security under the new real factors of environmental change. Boundless changes in precipitation and temperature designs undermine agrarian creation and increment the weakness of individuals reliant on agribusiness for their livelihoods, which incorporates a large portion of the world's poor (Campbell *et al.*, 2014; Santos *et al.*, 2002).

Environmental change disturbs food markets, presenting populace wide danger to food supply. Ahead threats can be reduced by training and extension of technology to farmers for increasing the adaptive capacity of them as well as increasing resilience and resource use efficiency in agricultural production system under climate change. In order to derive regionally appropriate solutions, CSA includes consideration of all three targets, from the local to the world scale and over short and very long time horizons. CSA pathways studied and observed by researcher result in higher resilience and lowers risks to food insecurity; also it helps in promoting sustainable agricultural development and mitigating the negative impacts of climatic changes on agricultural productivity in India (Kumari *et al.*, 2019). The two major principles defining CSA practices include increasing resource use efficiency in agricultural systems and enhancing the resilience of agricultural systems and the people who depend upon them. Efficiency in resource use is a crucial element approach for sustainable agricultural intensification. More effective utilization of assets, for example, nitrogen compost, feed for domesticated animals, land and water helps in expanding the net re-visitation of ranchers and in this way the wages while tension on scant assets and emanations per unit delivered are diminished. Expanding versatility includes diminishing weakness just as improving versatile limit. CSA is unflinchingly worried about how food is delivered, not who approaches land, water, work and different contributions to create it, nor how it is consequently dispersed and burned-through (Taylor, 2018). CSA techniques necessitate that versatility and asset use productivity are sought after together, albeit explicit innovations and institutional plans may influence just either. Or maybe, effectiveness and versatility should be considered in a general frameworks point of view that thinks about various spatial and fleeting scales.

Commencement of Climate Smart Agriculture

Food security we consider only when all human on earth are able to access enough safe and nutritious food to meet better healthy life while utilizing limited resources in sustainable manner, in ways the planet can sustain into the future. Food security, however, faces a series of obstacles in both production and distribution that research will be vital to solve. The impacts of climate hazards such as extremely heavy rainfall or no rain at all, extreme variation in temperature, frequent droughts, floods and hailstorms *etc.* are felt severely in underdeveloped and developing nations. To enhance food production under changing climatic conditions like aberrant weather, rising CO_2 level, rising temperature and rising sea level, we require the reorientation of agriculture from current practices to more sustainable and environmental friendly practices. Climate resilience agriculture helps to improve food security under changing climatic situation while also reducing food waste globally (CCAFS, 2013) and minimizing danger on natural resources. The first enunciation of the conceptualization of the CSA was implemented in the 2009 FAO study 'Food Security and Agricultural Mitigation in Developing Countries: Synergy Capture Options' launched during Barcelona Workshop on Climate Change took place in November of that year. In 2010, at the Hague Conference on Agriculture, Food Security and Climate Change; the FAO paper entitled "Climate Smart" Agriculture, Policies, Practices and Financing for Food Security, Adaptation and Mitigation" was launched as a background paper (FAO, 2010). These first expressions of the idea of climate smart agriculture argue that the agricultural industry, not just because of its high vulnerability to the impact of climate change, is essential to responding to climate change but also because it is a main contributor accounting between 19 and 29 per cent of total greenhouse gas emissions (Vermeulen *et al.*, 2012; Newell *et al.*, 2019). It also argued that the technological transition of the agricultural sector is sustainable. The sector is essential for achieving food security, and it is therefore important to formulate responses to climate change within this target. However, CSA developments have not only been at international level, with CSA programs undertaken at national and regional level, typically in collaboration with international organizations such as the FAO, the World Bank, local and international NGOs and the CGIAR Climate Change and Food Security Programme.

Climate Change Influences Every One of the Four Elements of Food Security

Food Production and Availability

The environment has a direct effect on food production through changes in agro-ecological conditions and indirectly on growth and income distribution, and thus on demand for agricultural products. Changes in land appropriateness, expected yields (*e.g.*, CO_2 preparation) and productivity of current cultivars are likely. Movements in land suitableness are probably going to prompt expansions in appropriate cropland in higher scopes and decrease of potential cropland in lower scopes.

Stability of Food Supplies

It is predicted that weather patterns will become more unpredictable than at present, with increase in frequency and intensity of catastrophic events. The stability of food supply and food security can be adversely affected by greater volatility in crop yields and local food supplies. Climatic variances will be more noticeable in semi-dry and sub-humid areas and are probably going to decrease crop yields and domesticated animals numbers and efficiency. As these zones are generally in sub-Saharan Africa and South Asia, the most impoverished areas with the most significant levels of constant undernourishment will be presented to the highest level of uncertainty.

Access to Food

Access to food refers to the ability of individuals, communities, and countries to purchase food insufficient quantity and quality. Falling real prices for food and rising real incomes over the last 30 years have led to substantial improvements in access to food in many developing countries. Possible food cost increment and declining pace of income growth coming about as a result of natural change may turn around this pattern.

Food Utilization

Climate change can trigger a vicious cycle in which infectious diseases, including water-borne diseases, induce or increase starvation, which in turn renders the affected community more susceptible to these diseases. Results may include reduce in labor potency and rise in impoverishment, morbidity and transience.

How Climate Change is Impacting Food Security in India

Low productivity is the principal issue of Indian agriculture. It has been noted that meeting demand for food grains for the future via domestic production will be very difficult with current production patterns (Kumar *et al.*, 2009). There is an urgent need to increase productivity in all sectors of agriculture to meet India's rising food requirement. But given the vulnerability of Indian agriculture to climate change, farm practices need to be reoriented to provide better climate resilience (Chakrabarty, 2016). Agricultural policy should focus on improving crop productivity through development and dissemination of crop varieties which are nutrient and water efficient and are more tolerant of temperature precipitation fluctuations.

Need of Climate Smart Agriculture for Future Food Security

FAO report 2016 stated that "Climate change is already affecting agricultural productivity, production stability, income and food security. Without urgent action, millions more people will be at risk of hunger and poverty." Agriculture is key economic sector and major source of employment for the people residing in the rural regions (Rural Poverty Report, 2011). FAO projected that to meet the expected demand of food and feed, agricultural production needs to be increased by 60 percent by 2050 (Alexandratos and Bruninsma, 2012). Agriculture faces a set of biophysical and socioeconomic stressors, including climate change (Lipper *et al.*, 2014). Climate change and natural resource degradation posses threat to agricultural production and so access of food to urban and rural population. Scientists projected that before the finish of 2050 there may be yearly harvest decrease up to minus 25 per cent on a normal, a few ranchers might even lose their total reap in 1 of 10 years on average because of outrageous and surprising occasions like dry spells, higher precipitation, warmth or cold waves. Therefore, agriculture must change according to present situation to meet the demand of food security and resist the vagaries of climate change. Climate smart agriculture (CSA) is an inescapable task to ensure food supply for an anticipated nine billion people by 2050 (World Bank 2011a; FAO 2013a). Food security and development has been identified as primary goals of Climate smart agriculture (FAO, 2013a), which can be achieved through increased productivity, adaptation, and mitigation (Kumari, 2020). To achieve food security and agriculture development goals, agriculture sectors need to overcome three intertwined challenges:

1. Increasing agricultural productivity in sustainable way to meet global food demand.
2. Adaptation and building resilience to the impacts of climate change from farm to national levels.
3. Lowering the accumulation and emission intensities of greenhouse gases in the atmosphere. (Beddington *et al.*, 2012a and Beddington *et al.*, 2012b).

To tackle the above challenges, there has been growing support at international and national levels for this approach. CSA in the nutshell is an approach to developing the technical, policy and investment conditions to achieve sustainable agricultural development for food security under climate change. A variety of sustainable approaches to agriculture, such as sustainable agriculture, sustainable intensification and conservation agriculture are the cornerstone and technical basis of CSA on which it is built upon. Ideally it is not possible to achieve the three pillars of CSA in one go. Trade- offs must be necessitated at the time of CSA execution. This requires us to identify synergies and weigh the costs and benefits of different options based on stakeholder objectives identified through participatory approaches (Figure 24.1).

Climate Smart Approaches to Crop Production

For each crop framework, there are endless environmental change transformation and alleviation alternatives that can reasonably improve yields and limit the unsafe ecological effects of production. They will vary for each cultivating family workers, contingent upon its adapting and versatile components. Climate change adaptation and mitigation management methods and technologies include practices that are specifically focused on adaptation and wider-range activities to reduce production risks and minimize emission of pollutants. Particular climate smart crop production approaches include:

- ☆ Increasing diversity and complexity within the agricultural ecosystem, which can be done in many ways (*e.g.* extending the variegation of crops or crop species), at numerous spatial scales.
- ☆ Improving sustainable soil and land management (*e.g.* carefully channeling the expansion of crop and grazing land to mitigate the loss of carbon storage that results from land-use change).
- ☆ Increasing energy use efficiency.

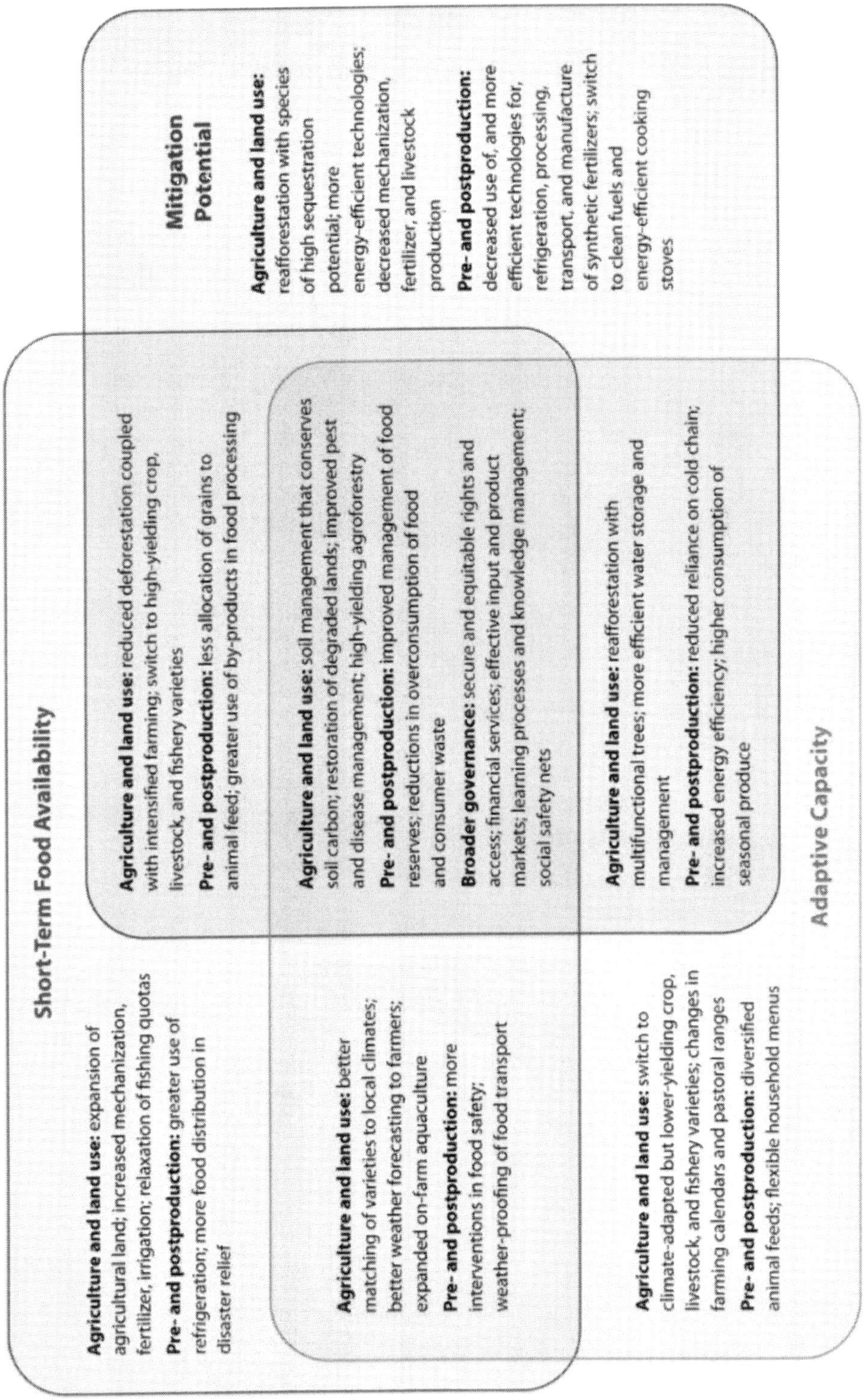

Figure 24.1: Synergies and Trade-offs for Adaptation, Mitigation and Food Security (*Source*: Vermeulen *et al.*, 2012).

- ☆ Promoting sustainable mechanization (*e.g.* continuing to increase the availability of appropriate machinery to minimize greenhouse gas emissions from different farming and manufacturing activities in conjunction with proper agronomic management).).
- ☆ Developing simple and robust scientific tools to guide the decision-making of farmers on a seasonal and long-term basis.
- ☆ To cushion against the impacts of climate change nutrient smart, water smart, weather smart, carbon smart, energy smart and knowledge smart technologies are being promoted through Climate Resilient Agriculture Programme.

CSA practices and technologies make agriculture more sustainable and climate resilient and a viable source of livelihood and food security for millions of farmers in the country. Zero budget natural farming (ZBNF) as a climate resilient farming system can enhance food and nutritional security, enabling farmers to improve soil fertility and yields through lower costs (Kumari *et al.*, 2020), risk and irrigation requirements, thus protecting the ecosystem by improving soil organic matter, water retention and biodiversity and reducing air and water pollution as well as greenhouse gas emissions.

Ethics for Sustainable Food and Agriculture (FAO, 2014a)

1. **Improving efficiency in the use of resources** - This includes the use of natural resources, energy external inputs, and labor. Altering current practices can do a lot to improve the efficiency of numerous food and agrarian production frameworks.
2. **Direct action to conserve, protect and enhance natural resources** - Food and agricultural production depends on natural resources and therefore the sustainability of production depends on the sustainability of the resources themselves. Much should be possible to lessen negative effects and improve the status of natural assets.
3. **Protect rural livelihoods and improve equity and social well-**being - Ensuring that producers have adequate access to and control of productive resources, and addressing the gender gap, can contribute essentially to lessening destitution and food uncertainty in rustic regions.
4. **Enhance the resilience of people, communities and ecosystems, especially to climate change and market volatility** - Extreme weather events, market volatility and civil strife impair the stability of agriculture. Sustainability can also be improved by policies, technology and practices that create producers' resilience to threats.
5. **Responsible and effective governance is essential for the sustainability of both the natural and human systems** - It requires policies and techniques that are Sector-based continuity, alignment of legal structures and investments, and capacity building of organizations and related stakeholders at all levels. It depends on shareholders conversation, alliances and the use of processes aimed at creating agreements on the goals of sustainable development.

Climate Smart Strategies

Climate resilient agriculture will essentially involve judicious and improved management of natural resources, namely land, water, soil and genetic resources through adoption of best practices, appropriately integrated and timely management of farming systems and farm mechanization. Conservation and advancement of land, water, biodiversity and environment assets are pre-imperative for feasible agriculture with regards to changing environmental situation. Some promising climate smart technologies which would be significant tool for adaption in agriculture under future climate change scenario includes (Khatri-chhetri et. al., 2017; Nagargade *et al.,* 2017). Adaptation programmes under the Strategy focus on: (i) crop and income loss risk management for improved farmers' livelihood; (ii) disaster risk management to build resilience of farmers; (iii) adapting crop varieties and corresponding farming practices to enhance farmers' adaptive capacities. Mitigation targets include reduced CH_4 emissions from rice fields and reduced land degradation and soil erosion through conservation agriculture.

- Precise nutrient management by using green seeker, leaf colour chart and chlorophyll meter.
- Site-specific nutrient management
- Use of improved crop varieties for obtaining higher yield.
- Retention of residue on soil surface.
- Practicing minimum tillage by using zero till seed drill or happy seeder
- Drought tolerant varieties.
- Direct-seeded rice.
- *In situ* moisture conservation.
- Water harvesting and its storage.
- Laser land levelling.
- Use of using tensiometer to manage irrigation water.
- Accessing weather information through SMS.
- Integration of legume in cropping system.
- Crop diversification.
- Use of solar pump.
- Use of a mobile phone app helps to calculate how much fertilizer to apply throughout the growing season.

Effect of CSA Technologies on Global Food Security and Greenhouse Gas Emissions

A study was conducted to analyse the effect of adopting CSA technologies and practices in three extensively crops *i.e.* wheat (*Triticum aestivum*), maize (*Zea mays*) and rice (*Oryza sativa)* in long run on global food security and GHG emissions. These crop accounts for 41 per cent of the global harvested area and approximately 64 per cent of the estimated 2-3Gt CO_2e per year emitted by crop production globally

(Carlson *et al.*, 2017). Here, the CSA framework encourages the use of four practices for wheat, rice and maize. Technologies adopted for wheat and maize includes no-tillage and integrated soil fertility management while for rice alternate wetting and drying and nitrogen use efficiency were considered. The positive effects of adoption of CSA practices may play on a global scale in terms of yields, resource use efficiency (*e.g.*, water and nutrients) and their overall effects on GHG emissions are presented below:

Crop	*CSA Technology Adopted*	*Description*	*Potential Effects on Yields and GHG Emissions*	*References*
Wheat, Maize	No tillage	Minimum or no soil disturbance, often in combination with residue	☆ Positive or neutral effects on yields ☆ Uncertain effect on GHG emissions	Pittelkow *et al.*, 2015; Powlson *et al.*, 2014
Wheat, Maize	Integrated soil fertility management	Combination of chemical fertilizers, crop residues, and manure/Compost	☆ Positive effects on yields ☆ Variable effects on GHG emissions	Gentile *et al.*, 2008
Rice	Alternate wetting and drying	Prolonged flooding disruptions across the season, As the upper soil layer dries out until subsequent re-flooding, the water decreases	☆ Lower to no significant changes in yields. ☆ High confidence in lower GHG emissions due to reduction of methane emissions	Huda *et al.*, 2016; Tyagi *et al.*, 2010
Rice	Nitrogen use Efficiency	Practices and technologies to enhance the uptake of inorganic soilnitrogen by crops	☆ Positive results on yields ☆ Reduction of GHG emissions	Huda *et al.*, 2016; Bandaogo *et al.*, 2015; Gaihre *et al.*, 2015

Source: De Pinto *et al.*, 2020.

Agricultural Practices which Fit into the CSA Profile Based on FAO Recommendation Involves

CSA Practice	*Definition*
Conservation agriculture	Reducing the degree of destruction of the soil andenabling crop residue on the field to stabilize the soil after harvesting
The use of organic manure	Boost soil fertility by means of manures and other organic matter
Crop diversification	Planting varieties of crops of the same or different species
Crop rotation	Sowing various crops on the same agricultural land area inconsecutive planting seasons
Use of cover crops	Planting cover crops
Agroforestry	Trees or shrubs are planted on or near farmland or grazing land
Mulching	Covering the soil between plants with layer/s of material

CSA Practice	Definition
Use of drought and heat tolerant crops	Planting drought and heat tolerant crops
Efficient manure management	Efficient management of manure on livestock production
Use of wetland	Cultivating on wetlands where even the soil is water-filled with high fertility
Soil conservation techniques	Minimizing pollution by acidification, salinization or other chemicals to prevent soil loss through erosion or ameliorating soil fertility loss
Integrated crop-livestock management	Integrating crop and livestock production
Improved grazing	Effective pasture development (improved varieties are sown for pasture and rotational grazing)
Diet improvement for animals	Achieving more protein output with reduced feed consumption and lower emissions

Climate Smart Agriculture in India

To meet the challenges posed by climate change on the agricultural system, India has accorded high priority in understanding the impacts of climate change and developing adaptation and mitigation strategies. The Prime Minister's National Action Plan on Climate Change (NAPCC) established during the year 2008 and has identified agriculture as the priority area of eight National Missions in India. The NAFCC, National Adaptation Fund for Climate Change was founded in August 2015 to meet the cost of climate change adaptation for the Indian State and Union Territories, which are especially susceptible to climate change's adverse effects. Climate change and its variability are emerging as the major challenges influencing the performance of agriculture and its sustainability (De Salvo, 2013).

Long-term changes in moving climate designs bring about evolving environment, which compromise agricultural efficiency through high-and low-temperature systems, expanded precipitation inconstancy and rising ocean levels that conceivably disintegrate seaside freshwater reserve and increment in the danger of flooding. The challenge gets articulated on account of surprising or outrageous changes in environment, exemplified by dry spells, floods, heat waves, *etc.* frequencies of which are anticipated to increase later on (World Bank, 2013). Developing nations, similar to India, are more susceptible against such stuns due to their hefty reliance on agribusiness and absence of specialized and monetary assets to adapt up to them (Lal *et al.*, 2012). Environmental change (and a worldwide temperature alteration) impacts human existence in all all areas. In this specific circumstance, ICAR began NICRA undertaking to conquer the environmental change weakness.

National Initiative for Climate Resilient Agriculture ICAR (NICRA) (Vanaja *et al.*, 2019)

ICAR initiated this programme during February 2011 with the objectives mentioned below:

a. Increase the resilience of Indian agriculture to climate change and climatic variability by development and implementation of production and risk management strategies.

b. Demonstration of sitespecific technologies on farmer's field.

c. Enhancing the capacity of scientists and other stakeholders for climateresilient agriculture research and its application.

The Following are the Smart Agricultural Practices in India Promoted by ICAR

1. Rejuvenation of farming in cyclone and floodprone coastal agroecosystems through land shaping
2. Staggered paddy nursery as a contingency measure for drought
3. Water efficient directseeded rice cultivation technology
4. Drumseeded rice for improving water use efficiency
5. Shortduration rice cultivar for drought tolerance
6. Droughttolerant shortduration finger millet varieties for late season drought in south interior Karnataka
7. Shortduration crop varieties suitable for late sowings
8. Crop diversification for livelihood and food security, sustainability
9. Floodtolerant varieties impart resilience to farmers in floodprone areas
10. Improving the resilience of poor farmers reclaiming cultivable wastelands
11. Community tanks/ponds as a means of augmentation and management of village level water resource
12. Individual farm ponds for improving livelihoods of small farmers
13. Jalkund- low cost rainwaterharvesting structures
14. Check damstoring excess runoff in streams
15. Rainwater harvesting and recycling through temporary check dam
16. Enhancing resilience through improvement in conveyance efficiency
17. Recharge of wells to improve shallow aquifers
18. Integrated Farming System modules
19. Captive rearing of fish seed—a livelihood opportunity in floodprone areas
20. Management practices to tackle cold stress in backyard poultry
21. Shelter management for small ruminants to tackle heat stress and rain storm
22. Smallfarm mechanization through Custom Hiring Centres for farm machinery
23. Improved planting methods for increasing water use efficiency and crop productivity
24. Zerotill drill wheat to escape problem of terminal heat stress

25. In situ incorporation of biomass and crop residues for enhancing soil microbial population and soil health
26. Village level seed banks for reducing seed shortages
27. Improved fodder cultivars to solve the problem of fodder scarcity

Way Forward Adopting CSA and Path of Food Security

a. Farmers will need to be supported both technically and financially as they adapt CSA practices, relevant to their geography.
b. Multilateral lending institutions, such as the World Bank and Asian Development Bank (ADB) must play an important role in directing funding.
c. The private sector too must be made a partner in implementation schemes.
d. Zero Budget Natural Farming helps in developing resilience and mitigating the influence of climate change.
e. Global partnerships and knowledge sharing are critical pillars of the mitigation strategy.
f. The Global Alliance for Climate Smart Agriculture (GA-CSA) will be a key player in propagating policies and action plans for CSA adaptation.

Conclusion

A decline in crop and livestock productivity could lead to an increase in food prices and threaten the food security of low-income families. Furthermore, climate change and recent disasters are new drivers of migration to cities and neighboring countries. CSA technologies and practices has been initiated in several places as adaptation strategies to cope with the risks of climate change to ensure stable and sustainable agricultural production. It has been found that the farmers who are using CSA strategies were found to achieve higher output, yield and return compared to the one who did not. There are exciting chances for scaling out and tremendous possibilities of these techniques for improving crop yields and farm income and diminishing ozone harming substance emanations.

REFERENCES

Alexandratos, N. and Bruninsma, J. (2012). World agriculture towards 2030/2050: The 2012 revision. ESA Working Paper No. 1203. Rome, Italy: Food and Agriculture Organisation of the United Nations.

Bandaogo, A., Bidjokazo, F., Youl, S., Safo, E., Abaidoo, R. and Andrews, O. (2015). Effect of fertilizer deep placement with urea supergranule on nitrogen use efficiency of irrigated rice in Sourou Valley (Burkina Faso), Nutrient Cycling in Agroecosystems, 102(1): 79-89.

Barkhordarian, A., von Storch, H. and Bhend, J. (2013). The expectation of future precipitation change over the Mediterranean region is different from what we observe, Climate Dynamics, 40: 225-244.

Beddington, J., Asaduzzaman, M., Clark, M., Fernandez, A., Guillou, M., Jahn, M., Erda, L., Mamo, T., Van Bo, N., Nobre, C.A., Scholes, R., Sharmam R. and Wakhungu, J. (2012b). Achieving food security in the face of climate change: final report from the Commission on Sustainable Agriculture and Climate Change. Copenhagen, Denmark, CGIAR Research Program on Climate Change, Agriculture and Food Security (CCAFS).

Beddington, J.R., Asaduzzaman, M., Clark, M., Fernandez, A., Guillou, M.D., Howlett, D.J.B., Jahn, M.M., Lin, E., Mamo, T., Negra, C., Nobre, C.A., Scholes, R.J., Van Bo, N. and Wakhungu, J. (2012a). What next for agriculture after Durban? Science, 335.

Campbell, B.M., Thornton, P., Zougmore, R., van Asten, P. and Lipper, L. (2014). Sustainable intensification: What is its role in climate smart agriculture? Current Opinion in Environmental Sustainability, 8: 39-43

Carlson, K. M., Gerber, J. S., Mueller, N. D., Herrero, M., MacDonald, G. K., Brauman, K. A., Havlik, P., O'Connell, C. S., Johnson, J. A., Saatchi, S. and West, P.C. (2017). Greenhouse gas emissions intensity of global croplands, Nature Climate Change, 7(1): 63-68.

CCAFS. Big Facts on Climate Change, Agriculture and Food Security. Copenhagen,Denmark: CGIAR Research Program on Climate Change, Agriculture and Food Security (CCAFS); 2013.

Chakrabarty, M. (2016). Climate change and food security in India. Observer Research Foundation (ORF): New Delhi, India.

De Pinto, A., Cenacchi, N., Kwon, H-Y., Koo, J. and Dunston, S. (2020). Climate smart agriculture and global food-crop production. PLoS ONE 15(4): e0231764.

De Salvo, Raffael, R. and Moser, R. (2003). The impact of climate change on permanent crops in an alpine region: A Ricardian analysis, Agricultural System, 118: 23-32.

FAO (2001). The state of food insecurity in the World 2001. Rome: Food and Agriculture Organisation (FAO) of the United Nations.

FAO. (2009). Joint guidelines for crop and food security assessment missions (CFSAMs). (available at http://ftp.fao.org/docrep/fao/011/i0515e/i0515e.pdf)

FAO. (2010). "Climate smart" agriculture: Policies, practices, and financing for food security, adaptation, and Mitigation. Rome, FAO.

FAO. (2013a). Climate Smart Agriculture: Sourcebook. Rome, Italy: Food and Agriculture Organization of United Nations.

FAO. (2014a). Building a common vision for sustainable food and agriculture – principles and approaches. Rome.

Gaihre, Y. K., Singh, U., Islam, S. M., Huda, A., Islam, M. R., Satter, M. A., Sanabria, J., Islam, M.R. and Shah, A.L. (2015). Impacts of urea deep placement on nitrous oxide and nitric oxide emissions from rice fields in Bangladesh, Geoderma, 259: 370-379.

Gentile, R., Vanlauwe, B., Chivenge, P. and Six, J. (2008). Interactive effects from combining fertilizer and organic residue inputs on nitrogen transformations, Soil Biology and Biochemistry, 40(9): 2375-2384.

Huda, A., Gaihre, Y. K., Islam, M. R., Singh, U., Islam, M. R., Sanabria, J., Satter, M. A., Afroz, H., Halder, A. and Jahiruddin, M. (2016). Floodwater ammonium, nitrogen use efficiency and rice yields with fertilizer deep placement and alternate wetting and drying under triple rice cropping systems, Nutrient Cycling in Agroecosystems, 104(1): 53-66.

Khatri-Chhetri, A., Aggarwal, P.K., Joshi, P.K. and Vyas, S. (2017). Farmers' prioritization of climate smart agriculture (CSA) technologies, Agricultural systems, 151: 184-191.

Kumar, P., Joshi, P. K. and Birthal, P. S. (2009). Demand projections for food grains in India, Agricultural Economics Research Review, 22(347-2016-16842), 237-244.

Kumari, S. (2020). Climate Resilient Agricultural Practices for increasing Productivity and Food Security. In: Proceeding: National Web-Conference on Sustainable Soil and Water Management for Bio-diversity Conservation, Food Security and Climate Resilience, Soil Conservation Society of India, 29-30 December 2020. New Delhi, India.

Kumari, S., Tzudir, L. and Meshram, M. R. (2020). Carbon Farming: Need of Future, Biotica Research Today, 2(12): 1280-1282.

Lal, P.N., Mitchell, T., Aldunce, P., Auld, H., Mechler, R., Miyan, A., Romano, L. E. and Zakaria, S. (2012). National systems for managing the risks from climate extremes and disasters. In: Field CB, Barros V, Stocker TF, Qin D, Dokken DJ, Ebi KL, Mastrandrea MD, Mach KJ, Plattner GK, Allen SK, Tignor M, Midgley PM, editors. Managing the Risks of Extreme Events and Disasters to Advance Climate Change Adaptation. A Special Report of Working Groups I and II of the Intergovernmental Panel on Climate Change. Cambridge and New York: Cambridge University Press, 339-392.

Lipper, L. and Zilberman, D. (2018). A Short History of the Evolution of the Climate Smart Agriculture Approach and Its Links to Climate Change and Sustainable Agriculture Debates. In Lipper, L., McCarthy, N., Zilberman, D., Asfaw, S., and Branca, G., editors, Climate Smart Agriculture : Building Resilience to Climate Change, pp 13-30. Springer International Publishing, Cham.

Lipper, L., Thornton, P., Campbell, B.M., Baedeker, T., Braimoh, A., Bwalya, M., Caron, P., Cattaneo, A., Garrity, D., Henry, K. and Hottle, R. (2014). Climate smart agriculture for food security, Nature climate change, 4(12): 1068-1072.

Nagargade, M., Tyagi, V. and Kumar, M. (2017). "Climate smart agriculture: an option for changing climatic situation." Plant Engineering by Snjezana Juric, IntechOpen. 143-165.

Newell, P., Taylor, O., Naess, L. O., Thompson, J., Mahmoud, H., Ndaki, P., Rurangwa R and Teshome, A. (2019). Climate Smart Agriculture? Governing the Sustainable Development Goals in Sub-Saharan Africa, Front. Sustain. Food Syst, 3: 55.

Nwanze, K. F. Viewpoint, Smallholders can feed the world, International Fund for Agricultural Development. (Rome, Italy, 2011).

Pittelkow, C. M., Liang, X., Linquist, B. A., Van Groenigen, K. J., Lee, J., Lundy, M. E., Van Gestel, N., Six, J., Venterea, R. T. and Van Kessel, C. (2015). Productivity limits and potentials of the principles of conservation agriculture, Nature, 517(7534): 365-368.

Powlson, D. S., Stirling, C. M., Jat, M. L., Gerard, B. G., Palm, C. A., Sanchez, P. A. and Cassman, K.G. (2014). Limited potential of no-till agriculture for climate change mitigation, Nature Climate Change, 4(8): 678-683.

Rural Poverty Report 2011 (International Fund for Agricultural Development, 2011).

Salamin, N., Wuest, R. O., Lavergne, S., Thuiller, W. and Pearman, P. B. (2010). Assessing rapid evolution in a changing environment. Trends in Ecology and Evolution, 25: 692–698.

Santos, F. D., Forbes, K. and Moita, R. (2002). Climate change in Portugal: scenarios, impacts and adaptation measures. SIAM project report, Gradiva, Lisbon, Portugal, p. 456.

Kumari, S., Singh, T.P. and Prasad, S. (2019). Climate Smart Agriculture and Climate Change, International Journal of Current Microbiology and Applied Sciences, 8(3): 1112-1137.

Taylor, M. (2018). Climate smart agriculture: what is it good for? The Journal of Peasant Studies, 45 (1): 89-107.

Tyagi, L., Kumari, B. and Singh, S. N. (2010). Water management- A tool for methane mitigation from irrigated paddy fields, Science of the Total Environment, 408(5): 1085-1090.

Vanaja, M., Bal, S. K., Nagasree, K., Narsimlu, B., Raju, B. M. K., Reddy, K. S., Prasad, J. V. N. S., Ravindra Chary, G., Tiwari, G., Shirur, M. and Krishna Rao B. (2019). Climate Smart Agriculture, Feed The Future - India Triangular Training (FTF-ITT), 20th August-3rd September, 2019, ICAR- Central Research Institute for Dryland Agriculture, Hyderabad, India.

Vermeulen, S.J., Campbell, B. M. and Ingram, S. G. I. (2012). Climate Change and Food Systems, Annual Review of Environment and Resources, 37:195-222.

World Bank. (2011a). Increased productivity and food security, enhanced resilience and reduced carbon emissions for sustainable development opportunities and challenges for a converging agenda: country examples. Agricultural and Rural Development Department.

World Bank. (2013). Turn Down the Heat: Climate Extremes, Regional Impacts, and the Case for Resilience. A Report for the World Bank by the Potsdam Institute for Climate Impact Research and Climate Analytics. Washington, DC: World Bank.

Index

G

H

I

L

M

N

O

P

R

S